FARADAY DISCUSSIONS OF THE CHEMICAL SOCIETY
NO. 60   1975

Electron Spectroscopy of Solids and Surfaces

THE FARADAY DIVISION
CHEMICAL SOCIETY
LONDON

A GENERAL DISCUSSION

ON

Electron Spectroscopy of Solids and Surfaces

15th, 16th and 17th July, 1975

A GENERAL DISCUSSION on the Electron Spectroscopy of Solids and Surfaces was held at The University of British Columbia, Vancouver, B.C., on the 15th, 16th and 17th July, 1975 with the generous support of the Chemical Institute of Canada, Canadian Government and Industry. The President of the Faraday Division, Professor T. M. Sugden, C.B.E., F.R.S., was in the chair at the first session: about 165 Fellows of the Faraday Division and others attended the meeting. Visitors to Canada included:

Dr. S. Andersson, *Sweden*
Dr. P. Bagus, *U.S.A.*
Dr. P. E. Bierstedt, *U.S.A.*
Prof. M. Boudart, *U.S.A.*
Dr. D. Briggs, *U.K.*
Dr. G. Brodèn, *U.S.A.*
Dr. C. R. Brundle, *U.K.*
Dr. T. A. Carlson, *U.S.A.*
Dr. G. Celsai, U.S.A.
Mr. M. D. Chinn, *U.S.A.*
Dr. T. J. Chuang, *U.S.A.*
Dr. D. T. Clark, *U.K.*
Dr. J. P. Coad, *U.K.*
Dr. J. W. Coburn, *U.S.A.*
Prof. J. E. Collin, *Belgium*
Mr. D. M. Collins, *U.S.A.*
Dr. A. Couper, *U.K.*
Mr. J. G. Cunningham, *U.K.*
Dr. N. S. Dalal, *U.S.A.*
Dr. J. M. Dale, *U.S.A.*
Dr. S. Evans, *U.K.*
Prof. C. S. Fadley, *U.S.A.*
Prof. S. Fain, *U.S.A.*
Prof. R. D. Feltham, *U.S.A.*
Prof. U. Garbatski, *U.S.A.*
Dr. R. P. H. Gasser, *U.K.*
Prof. C. R. Ginnard, *U.S.A.*
Prof. Y. Harada, *Japan*
Prof. R. G. Hayes, *U.S.A.*
Dr. I. H. Hillier, *U.K.*
Dr. L. D. Hulett, *U.S.A.*
Prof. S. Ideka, *Japan*
Dr. P. M. James, *U.S.A.*
Dr. J. S. Johannessen, *U.S.A.*
Prof. F. Jona, *U.S.A.*
Dr. R. W. Joyner, *U.K.*
Dr. K. S. Kim, *U.S.A.*
Dr. S. P. Kowalczyk, *U.S.A.*
Dr. U. Landman, *U.S.A.*
Mr. B. Law, *U.K.*
Dr. I. J. Lawrenson, *U.K.*
Dr. S.-T. Lee, *U.S.A.*
Prof. J. W. Linnett, *U.K.*
Dr. R. L. Martin, *U.S.A.*

Dr. M. G. Mason, *U.S.A.*
Prof. R. Mason, *U.K.*
Prof. G. D. Mateescu, *U.S.A.*
Mr. J. M. McDavid, *U.S.A.*
Dr. G. E. McGuire, *U.S.A.*
Miss B. E. Mills, *U.S.A.*
Dr. P. R. Norris, *U.K.*
Dr. I. B. Ortenburger-Miller, *U.S.A.*
Mr. S. H. Overbury, *U.S.A.*
Dr. D. E. Parry, *U.K.*
Dr. J. B. Peri, *U.S.A.*
Prof. W. T. Peria, *U.S.A.*
Prof. P. G. Perkins, *U.K.*
Dr. D. L. Perry, *U.K.*
Mr. P. Pianetta, *U.S.A.*
Dr. A. Pidcock, *U.K.*
Dr. C. M. Quinn, *U.K.*
Prof. J. J. Rehr, *U.S.A.*
Dr. D. W. Rice, U.S.A.
Mr. N. V. Richardson, *U.K.*
Prof. M. W. Roberts, *U.K.*
Dr. D. E. Sayers, *U.S.A.*
Prof. M. E. Schwartz, *U.S.A.*
Mr. C. Shaw, *U.S.A.*
Dr. K. Shimokoshi, *Japan*
Dr. D. A. Shirley, *U.S.A.*
Dr. A. J. Signorelli, *U.S.A.*
Prof. G. A. Somorjai, *U.S.A.*
Prof. R. Srinivasan, *India*
Prof. E. A. Stern, *U.S.A.*
Prof. F. S. Stone, *U.K.*
Mr. R. W. Streater, *U.S.A.*
Prof. T. M. Sugden, *U.K.*
Prof. D. T. Thomas, *U.S.A.*
Prof. F. C. Tompkins, *U.K.*
Dr. M. J. Tricker, *U.K.*
Dr. M. J. Van Der Wiel, *The Netherlands*
Dr. M. A. Van Hove, *U.S.A.*
Dr. C. D. Wagner, *U.S.A.*
Dr. R. F. Willis, *The Netherlands*
Prof. N. Winograd, *U.S.A.*
Dr. D. P. Woodruff, *U.K.*
Dr. D. A. Young, *U.K.*
Mr. K. K. Yu, *U.S.A.*

ISBN: 0 85186 888 6
ISSN: 0301-7249

© The Chemical Society and Contributors 1976

Printed in Great Britain at the University Press, Aberdeen

CONTENTS

page 7 *Relaxation and Final-State Structure in XPS of Atoms, Molecules and Metals*
By S. P. Kowalczyk, L. Ley, R. L. Martin, F. R. McFeely and D. A. Shirley

18 *Surface Analysis, Peak Intensities and Angular Distributions in XPS*
by C. S. Fadley

30 *Satellite Structure in the Photoelectron Spectra of Transition Metal Compounds Ionized in the K Shell of the Metal Ion*
by T. A. Carlson

37 *Multiplet Structure in X-Ray Photoelectron Spectra of Rare Earth Elements and their Surface Oxides*
by W. C. Lang, B. D. Padalia, L. M. Watson, D. J. Fabian and P. R. Norris

44 GENERAL DISCUSSION—Mr. K. Y. Yu, Dr. C. R. Brundle, Dr. I. H. Hillier, Dr. T. A. Carlson, Dr. R. L. Martin, Prof. D. A. Shirley, Prof. C. S. Fadley, Dr. D. P. Woodruff, Dr. R. W. Joyner.

51 *XPS and UPS Studies of the Adsorption of Small Molecules on Polycrystalline Ni Films*
by C. R. Brundle and A. F. Carley

71 *Photoelectron Spectroscopic Study of the Interaction of Nickel and Oxygen*
by P. R. Norton and R. L. Tapping

81 *Electron Spectroscopic Studies of the Initial Oxidation of Zinc*
by D. Briggs

89 *Interaction of Oxygen with Cu(100) Studied by Low Energy Electron Diffraction (LEED) and X-Ray Photoelectron Spectroscopy (XPS)*
by M. J. Braithwaite, R. W. Joyner and M. W. Roberts

102 *Valence Band Studies of the Copper + Oxygen System*
by S. Evans, D. E. Parry and J. M. Thomas

112 *Chemisorption of Oxygen on Iridium*
by G. Brodén and T. N. Rhodin

119 *Chemisorption of Organic Molecules on Single Crystal Surfaces of Transition Metals*
by T. A. Clarke, I. D. Gay, B. Law and R. Mason

127 *The Adsorption of Methanol, Formaldehyde and Ammonia on W(100) Studied by Ultra-Violet Photoelectron Spectroscopy*
by W. F. Egelhoff, J. W. Linnett and D. L. Perry

137 GENERAL DISCUSSION—Mr. K. Y. Yu, Dr. C. R. Brundle, Dr. C. M. Quinn, Prof. R. Mason, Prof. C. S. Fadley, Prof. M. W. Roberts, Prof.

G. A. Somorjai, Dr. U. Landman, Dr. P. R. Norton, Dr. R. L. Tapping, Dr. S. Evans, Dr. D. Briggs, Dr. P. Pianetta, Dr. R. W. Joyner, Prof. F. S. Stone, Dr. M. J. Tricker, Prof. J. M. Thomas, Mr. N. V. Richardson, Mr. D. M. Collins, Dr. G. Brodén, Dr. D. L. Perry.

page 173   *Electronic Structure of Some Periodic Polymers Studied by ESCA and Band Structure Calculation*
M. H. Wood, R. Lavery, I. H. Hillier and M. Barber

183   *Application of ESCA to Studies of Structure and Bonding in Polymers*
D. T. Clark, A. Dilks, J. Peeling and H. R. Thomas

196   GENERAL DISCUSSION—Dr. T. A. Carlson, Dr. D. T. Clark, Dr. I. H. Hillier, Dr. S. Evans.

201   *Roles of Lateral Interactions between Adatoms and Local Surface Geometry in Determining the Binding Energies of Core Electrons in Adsorbed Species*
by C. M. Quinn and N. V. Richardson

210   *Structure of Clean Crystalline Surfaces and Chemisorbed Overlayers*
F. Jona

218   *Data Averaging and Pseudo-kinematical Approaches to the LEED Surface Structure Problem*
D. P. Woodruff

230   *Fourier Transforms in Surface Structure Determination from LEED*
U. Landman

239   GENERAL DISCUSSION—Dr. U. Landman, Dr. D. P. Woodruff, Prof. E. A. Stern.

245   *Inelastic Low-Energy Electron Scattering from Solid Surfaces and Adsorbed Species*
by R. F. Willis

255   *Electronic Excitations in Adsorbed Alkali Metal Layers*
S. Andersson and U. Jostell

269   *Beam Effects in AES Revealed by XPS*
J. P. Coad, M. Gettings and J. C. Rivière

279   *Auger Electron Spectroscopy of Alloy Surfaces*
G. A. Somorjai and S. H. Overbury

291   *Chemical Shifts of Auger Lines, and the Auger Parameter*
C. D. Wagner

301   GENERAL DISCUSSION—Dr. C. M. Quinn, Dr. W. F. Egelhoff, Prof. J. W. Linnett, Dr. D. L. Perry, Dr. D. P. Woodruff, Dr. R. A. Armstrong, Dr. U. Landman, Prof. E. A. Stern, Dr. R. W. Joyner, Dr. C. D. Wagner, J. S. Johannessen, Dr. P. R. Norton, Prof. M. W. Roberts, Dr. J. P. Coad, Prof. S. C. Fain, Prof. G. A. Somorjai, Dr. T. A. Carlson, Mr. K. Y. Yu, Dr. C. R. Brundle, Prof. C. S. Fadley, Prof. R. N. O'Brien.

319   AUTHOR INDEX

Relaxation and Final-State Structure in XPS of Atoms, Molecules, and Metals

By S. P. Kowalczyk, L. Ley, R. L. Martin, F. R. McFeely and D. A. Shirley

Department of Chemistry and Lawrence Berkeley Laboratory,
University of California, Berkeley, California 94720

Received 4th April, 1975

Photoemission from a many-electron system is a many-electron process, even though the transition operator may affect only one electron directly. Relaxation and "shake-up" structure are related by a sum rule: when one is present, the other must also be present. Shake-up structure is shown to be accurately predictable in atomic neon and molecular HF if the CI calculations are done carefully. In metals the sum rule also applies but final-state effects usually appear as relaxation energy, which is large even for valence electrons. Finally, in rare-earth metals discrete shake-up structure is observable in the $4p$ region.

1. INTRODUCTION

Photoelectron spectroscopy has progressed well beyond the one-electron approximation. Nevertheless, much of our common parlance on the subject involves expressions such as "relaxation" and "shake-up" derived from one-electron pictures. In this paper we discuss photoemission spectra from a more general viewpoint, emphasizing the relation between relaxation and shake-up and the fundamental similarity in this regard of atoms, molecules, and solids. Section 2 deals with the accurate calculation of shake-up energies and intensities in atoms and molecules. Metals are discussed in Section 3, and data showing shake-up structure in rare-earth $4p$ spectra are presented in Section 4.

2. ATOMS AND MOLECULES

The photoelectron spectrum of an atomic or molecular species yields information about its various ionic states. Although, in principle, there is an infinite manifold of such states, the cross section for photoionization discriminates against all but a few. Generally speaking, a comparison of the electronic structure of the ionic state to that of the ground state permits the identification of " primary " ionic states and " satellites ".

The primary states usually correspond to the most intense peaks in the spectrum and are those directly related to Koopmans' description of photoionization.[1] They thus provide very direct information about the shell structure of the ground state. Koopmans' frozen-orbital ionic state is not, however, a true many-electron eigenstate of the Hamiltonian. The electron density in the eigenstate has actually rearranged. The stabilization afforded by this relaxation reduces the binding energy from that predicted by Koopmans' Theorem.[2] Even though relaxation is an artificial concept which arises when we compare the actual final state to an approximation to it, the fact that Koopmans' assumption represents a well-defined first approximation (and a rather good one) makes it useful. Relaxation can then be envisioned as a secondary process whereby the other electrons react to the influence of the departure of the photoelectron.[3]

Satellite peaks appear on the high-binding-energy side of each core-level primary peak. The most intense of these correspond to states which can be imagined as being reached by a simultaneous core orbital ionization and (monopole) valence electron excitation. The reason for this monopole " selection rule " is that to a first approximation the transition moment to such a state is dictated by the overlap integral between the valence orbital in the initial-state and the final-state orbital which the electron is " shaken up to ". Once again, it is the total wavefunction of the state which is meaningful and this orbital picture is just a helpful approximation. If the total wavefunction of the final state has a component which connects it with the initial state via the transition operator, it will be observed. It may be that this component is not adequately described in a one-electron picture, and in fact satellites have been observed which appear to involve a dipole excitation of a valence electron,[4, 5] two-electron excitations relative to the primary state,[6, 7] etc.

Recent work on the relative intensities of F 1s satellites in the hydrogen fluoride molecule [8, 9] has shown the importance of configuration interaction effects in determining quantitative cross-section ratios. The HF molecule in its ground state is described by the Hartree-Fock configuration $1\sigma^2 2\sigma^2 3\sigma^2 1\pi^4$, with the F 1s hole state represented by the single configuration $1\sigma^1 2\sigma^2 3\sigma^2 1\pi^4$. The particular model used to obtain shake-up states employs a configuration expansion technique. One chooses a set of configurations which are expected to approximate closely the structure of the excited state. In the HF case these would be

$$1\sigma^1 2\sigma^2 3\sigma^1 4\sigma^1 1\pi^4; \quad 1\sigma^1 2\sigma^2 3\sigma^2 1\pi^3 2\pi$$
$$1\sigma^1 2\sigma^2 3\sigma^1 5\sigma^1 1\pi^4; \quad 1\sigma^1 2\sigma^2 3\sigma^2 1\pi^3 3\pi$$
$$\vdots \qquad \qquad \vdots$$

For each of these configurations two linearly-independent configuration state functions of $^2\Sigma^+$ symmetry can be formed. The Hamiltonian matrix in this configuration basis —using, for example, the orbitals optimized for the F 1s hole state—is formed and diagonalized. The lowest root of this CI matrix will be predominantly the F 1s hole state configuration, while the higher roots should be fairly good approximations to the shake-up states. These higher roots are usually dominated by two or three configuration state functions and can be interpreted as being reached by either a particular excitation or, possibly, a small number of one-electron excitations. The relative intensities are then computed in the overlap approximation.[8, 10] The effective intensity of the final-state is given simply by the overlap integral between the final-state CI wavefunction and an initial-state in which the 1σ electron has been annihilated.

The results of this approach are compared with experiment in the column labelled method A in table 1. Although the general appearance of the spectrum is reproduced, nearly all the intensities are predicted to be a factor of two weaker than the experimental result. The reason for this lies in the fact that we have used a single determinantal initial-state; components in the true many-electron initial-state which contribute substantially to the satellite intensities are not described by the Hartree-Fock function. The results obtained using a correlated initial-state wavefunction (method B in table 1) are, essentially, in quantitative agreement with experiment.

We are at present using this model to predict the satellite intensities in the Ne 1s hole-state satellite spectrum. Preliminary results using the analogue of method A are shown in table 2. An examination of this table shows that the overall appearance of the spectrum is predicted quite nicely. The intensities relative to the primary hole-state are once again too small.[12a] We are currently examining possible causes for the low theoretical intensities. Because the satellites are very weak, additional small configuration-interaction effects in both the final- and initial-states could have important consequences for the computed intensities.

TABLE 1.—HF SATELLITE PEAK INTENSITIES IN THE OVERLAP APPROXIMATION [a]

state[c]	description[b]	method A I_n(theor)[d]	method B I_n(theor)[d]	I_n(expt)[e]	E(theor)[f]	E(expt)[f]
0	$1\sigma^1 2\sigma^2 3\sigma^2 1\pi^4$	(100.0)	(100.0)	(100.0)	693.5	694.0(5)
1	$(3\sigma \to 4\sigma)_{lower}$	0.0	0.1	—	23.89	—
2	$(3\sigma \to 4\sigma)_{upper}$	1.2	2.0	1.9(3)	25.90	22.4(2)
3	$(1\pi \to 2\pi)_{lower}$	1.5	3.0	3.0(4)	29.57	26.50(9)
4	$(3\sigma \to 5\sigma)_{lower}$	0.0	0.0	—	30.89	—
5	$(1\pi \to 2\pi)_{upper}$	3.6	6.2	5.7(5)	32.35	29.90(7)
6	$(1\pi \to 3\pi)_{lower}$	0.0	0.1	—	32.72	—
7	$(3\sigma \to 5\sigma)_{upper}$	0.7	1.2	1.0	33.31	30.87
8	$(3\sigma \to 6\sigma)_{lower}$	0.0	0.0	—	33.74	—
9	$(1\pi \to 4\pi)_{lower}$	2.8	4.1	3.8(5)	34.84	32.7(3)
10	$(3\sigma \to 7\sigma)_{lower}$	0.5	0.7	0.7	35.43	33.3
11	$(1\pi \to 3\pi)_{upper}$	0.0	0.0	—	35.72	—

[a] From ref. (9).
[b] In order of increasing energy.
[c] The descriptions are somewhat oversimplified. In many of the states configurations with a large overlap with the initial state have very small coefficients but supply relatively large contributions to the computed intensity.
  The 3σ orbital in HF is the bonding combination of $F(2p_\sigma)$ and $H(1s)$, while the 1π orbital is $F(2p_\pi)$. The virtual orbitals are roughly described as follows: 4σ-antibonding combination of $F(2p_\sigma)$ and $H(1s)$; 5σ–$F(3s)$; 6σ–$F(3p_\sigma)$; 7σ–$F(3ds)$; 2π–$F(3p_\pi)$; $3\pi \to F(3d_\sigma)$; 4π–$F(4p_\pi)$.
[d] All intensities are normalized to peak 0. Absolute values of $(S_0^{11})^2$ are 0.781 (method A) and 0.720 (method B).
[e] Error in last place given parenthetically.
[f] The first entry is the absolute binding energy of the F $1s$ hole state in eV; the others are incremental energies relative to this.

The intensities of the satellite peaks associated with a given primary state are related to the relaxation energy in the primary state through an approximate sum rule derived by Manne and Åberg [13]:

$$E_R = \frac{\sum\limits_{i=1}^{\infty} (I_i/I_0)\Delta_i}{\sum\limits_{i=0}^{\infty} (I_i/I_0)}. \tag{2.1}$$

Here E_R is the relaxation energy, (I_i/I_0) is the intensity of the satellite peak relative to the primary peak, and Δ_i is the energy separation between the satellite and the primary peak. The summations are taken over all the discrete states; they convert to an integration over any continua. The denominator simply reflects a normalization condition.

This expression should be rather precise for the core-level peaks observed in conventional XPS experiments using soft X-rays ($h\nu \approx 1.5$ keV). In principle, everything on the right hand side of eqn. (2.1) is experimentally observable, and the sum rule provides the apparatus for determining the relaxation energy without specific recourse to a Hartree-Fock calculation. In practice, the relationship is not really useful in this respect because the intensity distribution of the " shake-off " continuum is not readily deduced from the spectrum. Nevertheless, the sum rule can provide a great deal of qualitative information about the photoemission process. Eqn (2.1) expresses a "lever-arm" relationship. A given value of E_R can be manifested as intense satellites near the primary peak, weak satellites distant from the primary peak, a broad continuum of satellites on the high binding energy side of the primary peak, etc.

TABLE 2.—PRELIMINARY RESULTS FOR THE Ne 1s SATELLITE INTENSITIES

state	description	$I_n(\text{theor})^a$	$I_n(\text{expt})^b$	$E(\text{theor})^c$	$E(\text{expt})^d$
0	$1s^1 2s^2 2p^6$	(100.0)	(100.0)	(868.6)	(870.4)
1	$(2p \to 3p)_{\text{lower}}$	1.32	3.15(8)	36.8	37.35(2)
2	$(2p \to 3p)_{\text{upper}}$	1.43	3.13(10)	40.0	40.76(3)
3	$(2p \to 4p)_{\text{lower}}$	1.13	2.02(10)	41.6	42.34(4)
4	$(2p \to 5p)_{\text{lower}}$	0.34	0.42(6)	43.3	44.08(5)
5	$(2p \to 6p)_{\text{lower}}$	0.10	$\sim .2^*$	44.2	45.10(7)
6	$1s^1 2s^2 2p^5 \, (^3P)$	—	—	45.2	47.4(5)
7	$(2p \to 4p)_{\text{upper}}$	0.49	0.96(11)	45.5	46.44(5)
8	$(2p \to 5p)_{\text{upper}}$	0.13	0.17(5)	47.4	48.47(7)
9	$(2p \to 6p)_{\text{upper}}$	0.04	—	48.4	—
10	$1s^1 2s^2 2p^5 \, (^1P)$	—	—	49.5	51.7(5)
11	$(2s \to 3s)_{\text{lower}}$	0.09	0.57(5)	61.3	59.8(1)
12	$(2s \to 4s)_{\text{lower}}$	0.00	—	68.2	—
13	$(2s \to 3s)_{\text{upper}}$	0.16	0.49(6)	68.4	65.9(1)
14	$2s \to 5s)_{\text{lower}}$	0.00	—	70.8	—
15	$(2s \to 4s)_{\text{upper}}$	0.02	—	75.7	—

a Based on method A, see text.
b From ref. (7). The starred entry, (state 5), is an approximation we have made from visual inspection of the spectrum in ref. (7). Gelius reports an intensity of 0.50(15), which we feel must be a misprint.
c The first entry is the computed Ne 1s binding energy in eV; all others are incremental relative to this.
d From ref. (7).

3. METALS

Turning from small molecules to metals, we should first emphasize that there is no fundamental difference between an isolated metal crystal and a molecule in the gas phase. Therefore, all the avove conclusions reached about molecular photoemission hold equally true for the metal crystal. The language of molecular theory is, however, inappropriate for discussing metals; instead, solid-state theory which deals with a hypothetical substance, the infinite crystal, must be employed.

While one can argue formal equivalence of the small molecule and the crystal, the high symmetry and macroscopic extent of the crystal in combination produce some qualitatively different effects. The most notable of these is the occurrence of plasmons, elementary collective excitations of the valence electron gas, without analogue in small molecules. In addition, the valence-electron canonical orbitals extend over the entire crystal, and their energy spacing is nearly continuous, yielding an immense number of nearly degenerate "configurations". We shall treat photoemission from core-states and band-states separately.

3.1 CORE LEVELS

Fig. 1 shows a typical spectrum of a core level. We note the following features:

(1) an asymmetric primary peak at considerably lower binding energy (with respect to the vacuum level) than in the free atom,

(2) a flat, constant tail to higher binding energy, and
(3) one or more surface and bulk plasmon peaks.

This loss structure, in contrast to that observed in the gas phase, all results from the total process of electron emission from the crystal. While all of these peaks may, in principle, be treated on an equal footing, we shall, following common practice, treat photoemission by a semi-classical three-stage model:

(1) optical excitation
(2) transport to the surface, and
(3) escape into the vacuum.

This model, as Mahan [14] and Fiebelman and Eastman [15] have shown, may be derived from a Golden Rule expression for photoemission. It has the desirable property of separating the one-electron band-structure effects from the many-body and surface effects. We shall deal primarily with step (1) of the process.

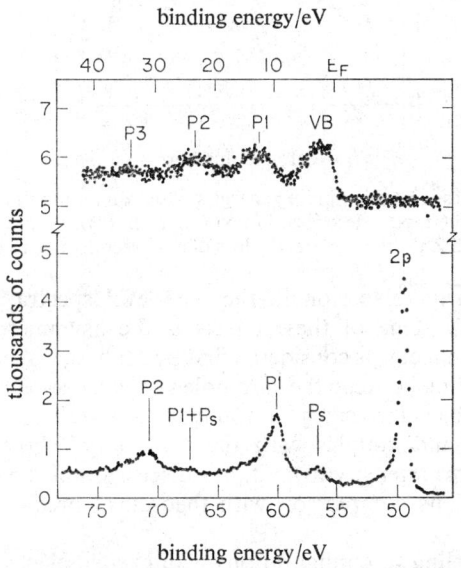

FIG. 1.—Typical photoemission spectrum of a metal valence band (top) and of a metal core level (bottom). (After ref. (22).)

The difference in binding energies between free-atom and metallic-core levels implies the existence of an extra-atomic relaxation energy. Hedin and Johansson [16] have shown that the binding energy for an atomic electron in orbital $|i\rangle$ is given by

$$E_B^{(i)}(\text{atom}) \cong -\varepsilon_i - \tfrac{1}{2}\langle i|V_p^a|i\rangle \qquad (3.1)$$

where ε_i is the orbital energy and V_p^a is the " polarization potential " given by

$$V_p^{(a)} = V_a^* - V_a, \qquad (3.2)$$

the difference in the total Hartree-Fock potentials for the ion and atom. Similarly, for a metal we can express the binding energy as

$$E_B^i(\text{metal}) = -\varepsilon_i - \tfrac{1}{2}\langle i|V_p^a|i\rangle - \tfrac{1}{2}\langle i|V_p^{ea}|i\rangle. \qquad (3.3)$$

The extra-atomic polarization potential, V_p^{ea}, is appreciable because a semi-localized state has formed around the ion from the bottom of the valence bands, self-consistently screening the core hole from the lattice (analogous to Friedel alloy theory).[17] We approximate V_p^{ea} by the Hartree-Fock potential of the atomic orbital out of which the lowest conduction band state is formed. Since hole-state calculations on the final-state ions [necessary for the calculation of (V^*-V)] are not available, we approximate the coulomb and exchange integrals necessary to evaluate $\langle i | V_p^{ea} | i \rangle$ by an equivalent-cores approximation. The result for the $3d$ transition series is shown in fig. 2. Note that the model represents an overestimate, since the screening charge is not localized as an atomic state but rather semi-localized.

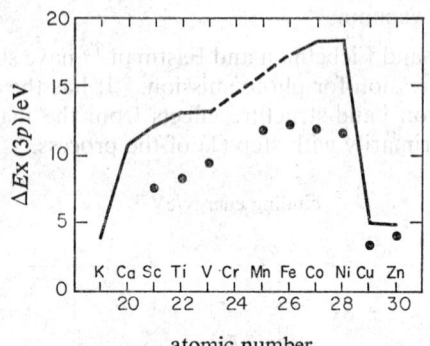

FIG. 2.—Experimental excess metallic binding energies based on an atomic orbital approximation to the screening state, using the model described in text (after ref. (3)). Note the drop in extra-atomic relaxation energy in Cu and Zn which reflects the loss of $3d$-screening due to the filling of the $3d$-band.

Since there is clearly relaxation in the core-level spectra, one expects to find "shake-up structure". One of these effects is the asymmetry exhibited by core level lines.[18,19] This effect was considered first by Mahan,[20] Anderson[21] and others. It arises from the coupling between the core-hole and valence-band electron-hole pairs. In the language of molecular theory, a configuration with an excited electron-hole pair in the valence band couples with the "primary" hole-state without these excitations, and acquires intensity from it. A discussion of the quantitative measure of the asymmetry and its comparison with theoretical predictions is given by Ley et al.[22]

In addition to coupling to configurations with excited electron-hole pairs, several workers[23,24] have described the coupling between the primary hole-state and plasmons in the valence electron gas. This results in the appearance of satellite peaks at higher binding energy at integral multiples of the plasmon energy. However, all of the plasmon peaks observed in the XPS spectra[25] do not represent the intensity predicted by the Lundqvist and Langreth theories and the intensity sum rule. In fact, the sum rule is not very useful for solids, a point which may be appreciated by considering the total states of the N-particle system. The photoemission process may be written as

$$\psi_{GROUND\ STATE}^N + \text{photon} \rightarrow \psi_{EXCITED}^N,$$

where the excited N-particle state includes the continuum state. We can then conceptually separate the excited state°

$$\psi_{EXCITED}^N = \psi_{EXCITED}^{N-1} + e^-(T).$$

The sum rule can then be derived from a consideration of the manifold of these excited "final" state ions. In the molecular case we infer the energies of these various

" final " ionic states by measuring the kinetic energy of the emitted electron. The point is that this" final " state ψ^N_{EXCITED} is not final at all; it shows time-dependent decay. This is no problem for molecules as these decay modes involve $\psi^{N-1}_{\text{EXCITED}}$ only, and thus the kinetic energy of the photoelectron still reflect the energies of the quasi-final states. This is not true in solids. An excitation is created in the crystal, yielding a ψ^N_{EXCITED}; however, the system may be decaying to the ground state significantly long before the electron leaves the crystal and thus the entire ψ^N_{EXCITED} participates in this decay. In a one-electron spirit, one would say that the initial core-hole excitation decayed (partially) into plasmons, phonons, particle-hole excitations, etc., and thus the measured kinetic energy no longer reflects that of the " quasi-final " state. It is this process of the decay of the " quasi-final state " that is described by steps (2) and (3) of the semi-classical three-step model of photoemission.

3.2 VALENCE BANDS

The primary difference between valence band-states and core-states is that the valence band levels are describable in terms of Bloch states which extend over the entire lattice. One might, therefore, be tempted to conclude that thee is no relaxation in the valence bands. A careful consideration of this point by Ley et al.[22] has shown this contention to be false. For a mono-valent free electron metal, (e.g., Na), the average binding energy of a valence electron relative to the vacuum level is given by

$$E^V_B = \phi + \tfrac{2}{5}(E_0 - E_F), \qquad (3.4)$$

where ϕ is the work function, E_F the Fermi level, and E_0 the bottom of the band. It is found experimentally that $E^V_B < E^A_B$ (the atomic binding energy). In fig. 3, we note that we can write an expression for ϕ:

$$\phi = E_C + E_A(V) - E_R - (\bar{E}_{VB} - E_F). \qquad (3.5)$$

The cohesive energy E_C appears because removal of one electron breaks one bond.

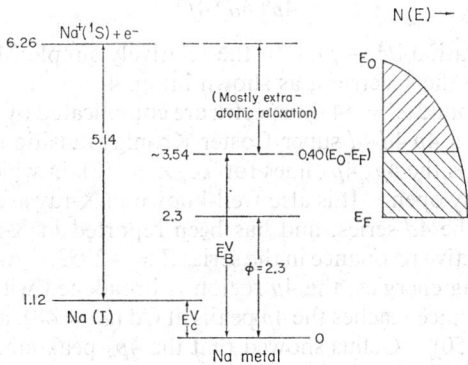

FIG. 3.—Energy level diagram relating the binding energy of a $3s$ electron in atomic Na to that of a $3s$ electron in the metal valence band (after ref. (22)).

Approximating E_R by the localized hole picture above, one obtains very good agreement with the experimental ϕ. Comparing eqn (3.5) with the expression obtained by Wigner and Bardeen[26] for ϕ we find

$$E_R = 0.6\, e^2/r_s - 0.458\, e^2/3r_s = 3.05 \text{ eV}. \qquad (3.6)$$

This Wigner-Bardeen result is a de-localized electron result that reflects the coulomb and exchange energy, respectively, of an *itinerant* hole propagating through the lattice. The localized-hole model yields

$$E_R = \tfrac{1}{2} \langle 3s | V_P | 3s \rangle = \tfrac{1}{2} F^\circ (3s, 3s)_{\text{ATOMIC Na}} = 2.93 \text{ eV}. \quad (3.7)$$

The near equality of these two results suggests that the relaxation energy is insensitive to the degree of localization of the initial-state.

4. CORRELATION STATES IN RARE-EARTH $4p$ SPECTRA

To tie several of the ideas of the previous two sections together, we present below a series of spectra for the $4p$ region in rare-earth metals. In these spectra the $4p_{\frac{1}{2}}$ hole state is obliterated as a single peak and appears instead as a number of well-defined individual peaks that arise through final-state correlations. With these spectra we extend to elements $Z \sim 45$–75 the interesting collective-resonance plus correlation effects reported by Gelius [7] for elements $Z = 52$–56.

First we note that intra-shell correlation effects have shown up earlier in photo-emission spectra, particularly in splitting up the lower-spin member of a multiplet such as the $3p^5 3d^5$; ^5P [27] or $3s^1 3p^6 3d^5$; ^5S [6] states in Mn^{3+}. In the latter case the formal analogy to the effect reported by Gelius is striking. The Mn^{3+} final-state is

$$3s^1 3p^6 3d^5; {}^5S.$$

The two-electron excitation $p^2 \to sd$ yields

$$3s^2 3p^4 3d^6; {}^5S$$

in two ways.[6] To relate this to Gelius' spectra for Xe $4p_{\frac{1}{2}}$, the quantum numbers n and l need only be raised by one, obtaining the main configuration

$$4p^5 4d^{10} 4f^0$$

and correlation states built on the configuration

$$4p^6 4d^8 4f^1$$

obtained by the excitation $d^2 \to pf$. In the relatively simple Mn^{3+} case, a sum rule is clearly operative in the spectrum, as shown in fig. 4.

For the $4p$ shell around $Z = 54$ the spectra are complicated by a collective resonance which arises through a $4p\, 4d\, 4d$ super Coster-Kronig transition. Gelius found that this resonance overlaps the $4p_{\frac{1}{2}}\, 4p_{\frac{3}{2}}$ lines for Te ($Z = 52$), in which the Coster-Kronig channel is energetically open. It is also well-known in X-ray spectroscopy for lighter elements down into the $4d$ series, and has been reported in X-ray photoemission.[28] Fig. 5 shows the collective resonance in the series $Z = 42$–52. As the resonance moves up from lower binding energies, the $4p$ region is broadened with nearly total loss of structure. The resonance reaches the $4p$ peaks at Cd ($Z = 48$), and gives the broadest structure at Sn ($Z = 50$). Gelius showed that the $4p_{\frac{3}{2}}$ peak emerged below the resonance (i.e., at lower binding energy) in ($Z = 53$), and the $4p_{\frac{1}{2}}$ structure at Xe ($Z = 54$). This structure, which was distributed among several peaks, was also observed by Gelius in salts of Cs and Ba ($Z = 55$ and 56), in which the collective resonance moved progressively to higher binding energies.

Fig. 6 shows the $4p$ region for several rare-earth metals. Final state splitting is still clearly present in La and Ce, in fact the $4p_{\frac{1}{2}}$ structure closely resembles that in Cs and Ba. The structure is broadened in Ce, probably through multiplet splitting by the $4f$ electron. In the other open-shell rare-earths multiplet splitting is dominant, but in Yb and Lu, at the end of the series, the $4p_{\frac{1}{2}}$–$4p_{\frac{3}{2}}$ structure has reformed, and by

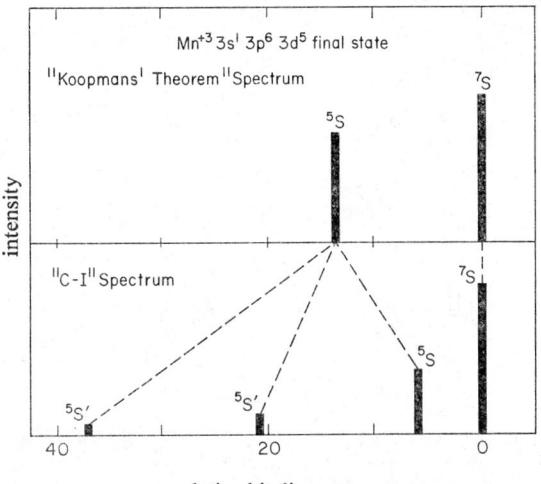

FIG. 4.—(a) Schematic representation of the one-electron (Koopmans' Theorem) spectrum for the Mn^{3+} $3s^1$ $3p^6$ $3d^5$ final-state. (b) Schematic representation of the observed spectrum of the Mn $3s$ region in MnF_2 (after ref. (6)). This illustrates the partitioning of energy and intensity throught the sum rule. The satellites labelled $^5S'$ are correlation states arising from strong interaction of the $3s^2$ $3p^4$ $3d^6$ configuration with the $3s^1$ $3p^6$ $3d^5$ final-state.

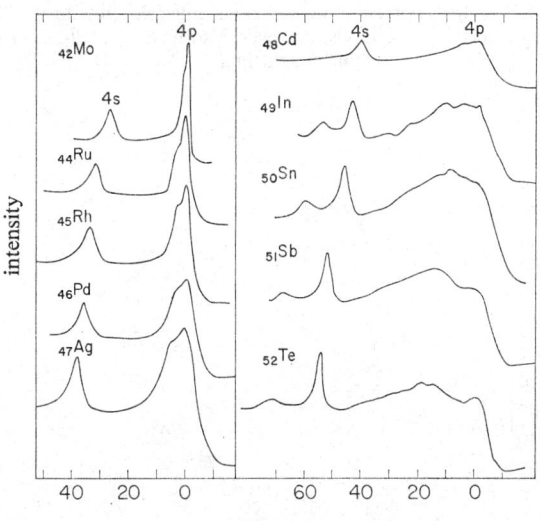

FIG. 5.—Left panel shows the XPS spectra of the $4s$ $4p$ region of the metals Mo ($Z = 42$) to Ag ($Z = 47$). Right panel shows XPS spectra of the $4s$ $4p$ region of the eleemnts Cd ($Z = 48$) to Te ($Z = 52$) which exhibit the strong collective resonance discussed in the text.

Ta ($Z = 73$) the lines once again exhibit a simple spin-orbit doublet. This systematic behaviour from $Z = 42$ to 73 can be understood in terms of the successive opening and closing of the $4p$ $4d$ $4d$ collective resonance channel and the $4d^2 \rightarrow 4p$ $4f$ correlation channel. A rough idea of the origins of this behaviour can be obtained from the orbital energies [29] sketched in fig. 7. Lundqvist and Wendin [30] have given a more sophisticated theoretical discussion of effects of this type.

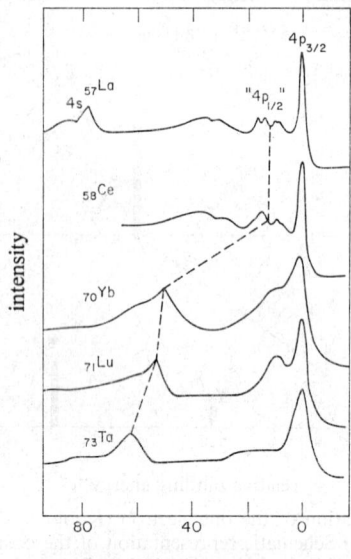

FIG. 6.—The XPS spectra of the 4p region of La ($Z = 57$), Ce ($Z = 58$), Yb ($Z = 70$), Lu ($Z = 71$) and Ta ($Z = 73$) metals. La and Ce exhibit correlation states of the type discussed in Section 4, Yb and Lu, where the $4d^2 \to 4p\,4f$ correlation channel becomes closed with the filling of the $4f$ shell, shows both $4p_{\frac{1}{2}}$ and $4p_{\frac{3}{2}}$ levels. These levels, however, also have a satellite. Ta displays just spin orbit splitting.

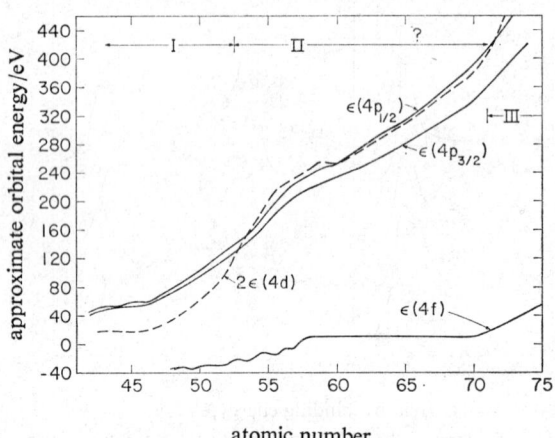

FIG. 7.—Orbital energies from ref. (29), illustrating final-state effects in the 4p shell. Energy for I ($Z = 53$) is adjusted to fit data of Gelius. In Region I the collective resonance coincides with 4p states, because $\varepsilon(4p)+\varepsilon(4f) \sim 2\varepsilon(4d)$. In Region II the same condition is met, but the $d^2 \to pf$ excitation yields a bound state, hence discrete shake-up lines. This structure is obscured for $Z > 58$ by multiplet splitting. The $4f$ channel is closed for $Z > 70$ and the simple $4p_{\frac{1}{2}}$–$4p_{\frac{3}{2}}$ structure is restored by $Z = 73$.

Work done under the auspices of the U.S. Energy Research and Development Administration.

[1] T. Koopmans, *Physica*, 1933, **1**, 104.
[2] We have neglected any differences in the correlation for the two states and have disregarded possible multiplet structure. They are usually small and may be of either sign; see ref. (3).
[3] Relaxation in the ionic states of atoms, molecules, and solids has recently been reviewed. R. L. Martin and D. A. Shirley, *Electron Spectroscopy: Theory, Techniques and Applications*, ed. A. D. Baker and C. R. Brundle, (Academic Press, to be published).
[4] J. Berkowitz, J. L. Dehmer, Y. K. Kim and J. P. Desclaux, *J. Chem. Phys.*, 1974, **61**, 2556.
[5] S. Süzer and D. A. Shirley, *J. Chem. Phys.*, 1974, **61**, 2481.
[6] S. P. Kowalczyk, L. Ley, R. A. Pollak, F. R. McFeely and D. A. Shirley, *Phys. Rev. B*, 1973, **7**, 4009; see also P. S. Bagus, A. J. Freeman and F. Sasaki, *Int. J. Quant. Chem.*, 1973, **7 S**, 83.
[7] U. Gelius, *J. Electr. Spectr.*, 1974, **5**, 985.
[8] R. L. Martin and D. A. Shirley, *J. Chem. Phys.*, 1975.
[9] R. L. Martin, B. E. Mills and D. A. Shirley, *J. Chem. Phys.*, 1975.
[10] This is identical to a multi-determinantal extension of the sudden approximation result,[11] but comes from an application of the usual dipole approximation cross section to the satellite states.
[11] T. Åberg, *Phys. Rev.*, 1967, **156**, 35.
[12] (a) Even at this preliminary stage, however, certain points are evident. For example, we corroborate the earlier assignment (see ref. (12b)) fo the first four satellites as $(2p \to 3p)_{lower}$, $(2p \to 3p)_{upper}$, $(2p \to 4p)_{lower}$, and $(2p \to 5p)_{lower}$. We furthermore predict the state $(2p \to 6p)_{lower}$ to occur at a relative energy of 44.2 eV, in agreement with Gelius' assignment of state 5. Its partner, $(2p \to 6p)_{upper}$, lies at $\Delta E = 48.4$ eV. This state is calculated to have a very small intensity and was not observed by Gelius. The assignment of the peaks at 59.6 and 65.9 eV as involving $2s \to 3s$ excitations is given support by the calculations. We also find the $(2s \to 4s)_{lower}$ state to fall in this region with an unobservable intensity. Examination of table 2 shows that the $(2p \to np)_{lower}$ states form a very definite Rydberg progression leading to the $1s^1 2s^2 2p^5$ (^3P) shake-off limit 45.2 eV from the Ne 1s primary state. The $(2p \to np)_{upper}$ states converge to the ^1P shake-off limit at 49.5 eV.
[12] (b) The assignment was made with the help of multi-configuration Hartree-Fock calculations by P. S. Bagus and U. Gelius reported in K. Siegbahn, C. Nordling, G. Johannson, J. Hedman, P. F. Hedén, K. Hamrin, U. Gelius, T. Bergmark, L. O. Werme, R. Manne and Y. Baer, *ESCA Applied to Free Molecules*, (North-Holland, Amsterdam, 1969).
[13] R. Manne and T. Åberg, *Chem. Phys. Letters*, 1970, **7**, 282.
[14] G. D. Mahan, *Phys. Rev. B*, 1970, **2**, 4334.
[15] P. J. Fiebelman and D. E. Eastman, *Phys. Rev. B*, 1974, **10**, 4932.
[16] L. Hedin and G. Johannson, *J. Phys. B*, 1969, **2**, 1336.
[17] J. Friedel, *Adv. Phys.*, 1954, **3**, 446.
[18] P. H. Citrin, *Phys. Rev. B*, 1973, **8**, 5545.
[19] S. Hüfner, G. K. Wertheim, D. N. E. Buchanan and K. W. West, *Phys. Letters A*, 1974, **46**, 420.
[20] G. D. Mahan, *Phys. Rev.*, 1967, **163**, 612.
[21] P. W. Anderson, *Phys. Rev. Letters*, 1967, **18**, 1049; *Phys. Rev.*, 1968, **164**, 352.
[22] L. Ley, F. R. McFeely, S. P. Kowalczyk, J. G. Jenkin and D. A. Shirley, *Phys. Rev. B*, 1975, **11**, 600.
[23] L. Hedin, B. I. Lundqvist and S. Lundqvist, *Solid State Comm.*, 1967, **5**, 237; B. I. Lundqvist, *Phys. Kond. Mat.*, 1969, **9**, 236.
[24] D. C. Langreth, *Phys. Rev. B*, 1970, **1**, 471.
[25] R. A. Pollak, L. Ley, F. R. McFeely, S. P. Kowalczyk and D. A. Shirley, *J. Electr. Spectr.*, 1974, **3**, 381.
[26] E. Wigner and J. Bardeen, *Phys. Rev.*, 1935, **48**, 84.
[27] C. S. Fadley, D. A. Shirley, A. J. Freeman, P. S. Bagus and J. V. Mallow, *Phys. Rev. Letters*, 1969, **23**, 1397; S. P. Kowalczyk, L. Ley, F. R. McFeely and D. A. Shirley, *Phys. Rev. B*, 1975, **11**, 1721.
[28] G. B. Fisher, R. Shalovy and P. J. Estrup, *Bull. Amer. Phys. Soc. Ser. II*, 1974, **19**, 233.
[29] C. C. Lu, T. A. Carlson, F. B. Malik, T. C. Tucker and C. W. Nestor, *Atomic Data*, 1971, **3**, 1.
[30] S. Lundqvist and G. Wendin, *J. Electr. Spectr.*, 1974, **5**, 513.

Surface Analysis, Peak Intensities and Angular Distributions in XPS

By Charles S. Fadley
Department of Chemistry, University of Hawaii,
Honolulu, Hawaii 96822

Received 10*th March* 1975

Purely instrumental contributions to the angular dependence of XPS peak intensities are considered theoretically and experimentally, and it is found that peak-ratio angular distributions should be free of all instrumental effects, and thus amenable to more direct analysis in terms of specimen-related properties. This is demonstrated experimentally for gold specimens with thin carbon-containing surface layers. The effects of surface roughness on XPS angular distributions are also investigated. For the triangular-periodic surfaces of aluminum diffraction gratings, pronounced surface profile effects on oxide/metal ratio angular distributions are observed; these effects are in good agreement with model theoretical calculations. For unidirectionally-polished aluminum specimens of more random surface profile, oxide/metal ratio angular distributions are distinclty different for angle variation parallel to and perpendicular to the polishing grooves, with much more surface enhancement being possible at low electron emission angles for parallel variation. This difference is qualitatively consistent with theoretical expectations. For the highly random surface contours of Al_2O_3 powder specimens subjected to solution adsorption of Si and Ca, it is also possible to do qualitative depth profiling of the various atoms present according to their relative enhancements at low angles of electron emission.

Several recent studies have explored various aspects of the application of X-ray-photoelectron angular distribution (AD) measurements to surface analysis.[1-7] If attention is restricted to those features in these angular distributions that are controlled primarily by the macroscopic properties of the specimen, several effects with potential utility for surface characterization have been noted: (1) the relative enhancement of surface-atom photoelectron intensities for low (grazing) angles of electron emission [1,2,4-7] or low angles of X-ray incidence [4,5]; (2) the quantitative analysis of AD data from specimens with nominally flat surfaces to yield electron attenuation lengths,[2,3] surface layer thicknesses,[2,4-6] and surface layer coverages [5,6]; and (3) the analysis of AD data from specimens exhibiting some form of surface roughness to provide information analogous to that in (2), as well as certain surface-profile characteristics.[4,5,7]

However, there are still several aspects of such macroscopic AD studies for which little or no experimental verification has previously been presented, and it is with two of these that we shall be concerned here: (1) the assessment of purely instrumental contributions to peak intensity variations with angle, and the development of procedures to correct for such effects; and (2) the precise experimental characterization of surface-roughness effects on angular distributions for a few selected test cases, and an estimation of the overall utility of AD measurements for the determination of rough-surface properties.

THEORY

Both instrumental- and surface-roughness-effects on angular distributions have been discussed from a theoretical point of view by Fadley et al.[4,5] We briefly review a few basic results, neglecting throughout this discussion any effects due to X-ray refraction and reflection at low incidence angles. Certain pertinent quantities are defined in fig. 1, which shows a schematic spectrometer geometry of a rather general type. X-rays are incident on the surface at an angle ϕ_x; electrons are emitted at

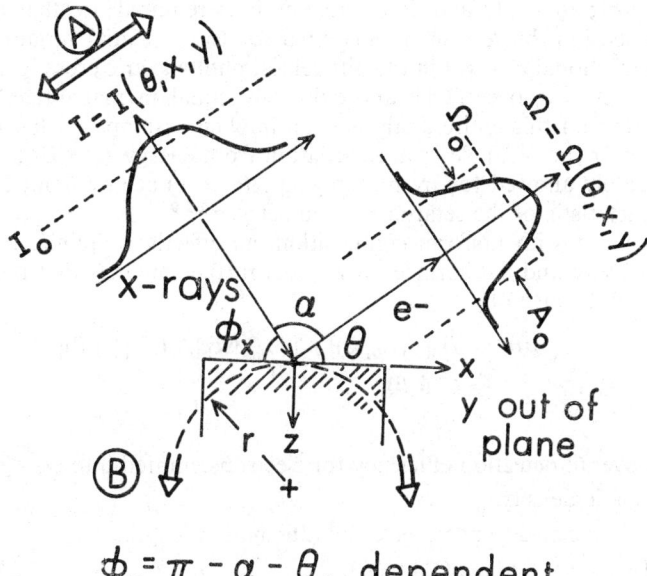

$\phi_x = \pi - \alpha - \theta$, dependent

FIG. 1.—Schematic XPS spectrometer geometry. The specimen is rotated about the y-axis.

an angle θ. The angle α between the mean X-ray-incidence- and mean electron-emission-directions is held fixed. The angles θ and ϕ_x can be varied by rotating the specimen about the y-axis perpendicular to the plane of the figure. Each point on the specimen surface is uniquely specified by the coordinates θ, x, and y. $I(\theta,x,y)$ denotes an arbitrary distribution of X-ray flux over the surface. For emission from each surface point, the spectrometer accepts electrons over a small solid angle given by $\Omega(\theta,x,y)$, again an arbitrary distribution. Idealized approximations for the surface distributions of X-ray flux and solid angle have also been made [2, 3]; these assume that $I = I_0 = $ constant and/or that Ω is non-zero only over the surface region defined by the projection of an effective spectrometer aperture A_0, in which it has a value $\Omega = \Omega_0 = $ constant.

This model predicts that, regardless of specimen type, the angular dependence of any photoelectron peak intensity will contain as a multiplicative factor a purely instrumental response function $R(\theta)$ defined as [4, 5]

$$R(\theta) \equiv \sin\theta \iint_A I(\theta, x, y)\Omega(\theta, x, y) \, dA, \quad (1)$$

in which the integration is over the actual specimen surface area A. That is, for an arbitrary peak k, the intensity N_k can be written as

$$N_k(\theta) = S_k(\theta)R(\theta), \quad (2)$$

where $S_k(\theta)$ is a function directly related to specimen properties. Therefore, in a ratio of any two peak intensities $N_{k'}$ and N_k, the $R(\theta)$ factors should cancel, leaving a measurable quantity primarily linked to the desired specimen characteristics:

$$N_{k'}(\theta)/N_k(\theta) = S_{k'}(\theta)/S_k(\theta). \quad (3)$$

Although the original derivation leading to eqn (1)–(3) assumed a flat surface and a relatively small solid angle Ω of simple shape (e.g., that in a hemispherical electrostatic analyzer), the same basic conclusions regarding $R(\theta)$ can also be derived for an arbitrary rough surface or a larger, more complex, solid angle (e.g., that in a cylindrical mirror analyzer); the only additional assumptions required are that characteristic roughness dimensions be very small in comparison to the spatial variations of I or Ω, and that any fractional changes in the differential photoelectric cross sections $d\sigma_{k'}/d\Omega$ and $d\sigma_k/d\Omega$ occurring over Ω be approximately equal in magnitude.[8] A further minor restriction on this entire analysis is an implicit assumption that any intensity changes caused by possible electron retardation from energy E to E_0 before analysis can be adequately allowed for by multiplying $\Omega(\theta, x, y)$ at any point by a function $F(E_0/E)$ characteristic of the retardation geometry.[4, 5, 9]

For a specimen with uniform composition, an effectively infinite thickness, and an atomically clean and flat surface, this model further predicts that the intensity of the kth peak will be given by [4, 5]

$$N_{k,\infty}(\theta) = D_0[F(E_0/E)] \rho \, [d\sigma_k/d\Omega] \, [\Lambda_e(E)] \, [R(\theta)]$$
$$= C_k R(\theta), \qquad (4)$$

in which

D_0 = an overall detection efficiency for electrons emitted into Ω,
ρ = an atomic density,
$\Lambda_e(E)$ = an energy-dependent electron attenuation length,
$C_k \equiv D_0[F(E_0/E)] \rho \, [d\sigma_k/d\Omega] \, [\Lambda_e(E)]$.

Thus, experimental measurements on such a specimen should yield the form of $R(\theta)$ directly. For an idealized X-ray flux and solid angle, the same intensity will be

$$N_{k,\infty}(\theta) = C_k I_0 \Omega_0 A_0$$
$$= \text{constant with } \theta. \qquad (5)$$

$R(\theta)$ as defined here thus allows for all instrumental deviations from this constancy.

The primary effects by which surface roughness influences peak intensities are [4, 5]: (1) surface shading for electron emission or X-ray incidence, and (2) deviation of the true electron emission angle θ' at a given surface point from the instrumental angle θ measured relative to the planar average of the rough surface (see fig. 2(a)). With neglect of X-ray shading, general expressions for calculating rough-surface AD's have been presented, and these have been applied to certain special cases of the periodic profiles shown in fig. 2(b)–2(e).[4, 5] We shall be concerned here primarily with the particular case of a metal surface with triangular-periodic profile as shown in fig. 2(b) and a uniform covering layer of metal oxide of thickness t(oxide). Further concentrating on a single core level in the metal atom that is chemically shifted between oxide and metal, we arrive at the following expression for the k(oxide) to k(metal) intensity ratio after cancellation of common factors of $d\sigma_k/d\Omega$ and $R(\theta)$ [7]:

$$\frac{N_k(\text{oxide})}{N_k(\text{metal})} = \frac{\rho_M(\text{oxide})\Lambda_e(\text{oxide}) \int_{A^R(\theta)} \{1 - \exp[-t(\text{oxide})/\Lambda_e(\text{oxide}) \sin \theta']\} \sin \theta' \, dA}{\rho_M(\text{metal})\Lambda_e(\text{metal}) \int_{A^R(\theta)} \exp[-t(\text{oxide})/\Lambda_e(\text{oxide}) \sin \theta'] \sin \theta' \, dA}. \qquad (6)$$

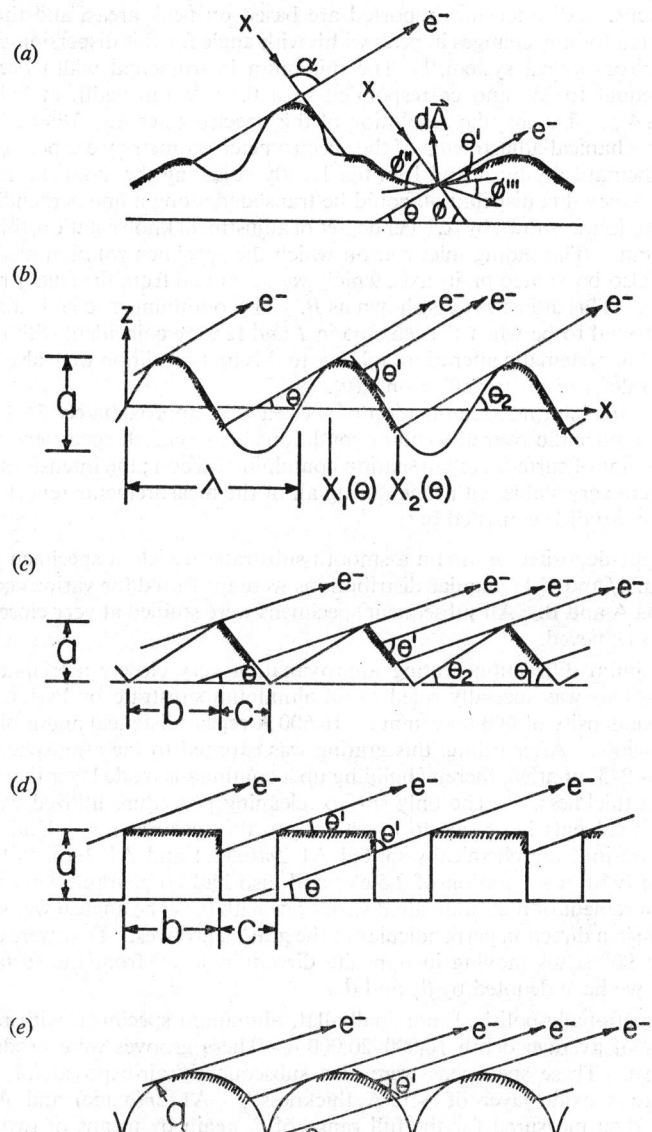

FIG. 2.—Schematic representations of: (a) a general rough surface with X-ray and electron-shading, and (b)-(e) simple-periodic surfaces with electron shading only.

Here, ρ_M is a metal-atom density, Λ_e is an electron attenuation length, $A^R(\theta)$ represents the electron-unshaded portions of the triangular surface for a given instrumental emission angle θ, θ' is the true angle of electron emission at any point on the surface ($0 \leqslant \theta' \leqslant 90°$), and dA is a differential element of area on the surface. The integrals are easily written down in closed form for a triangular profile.

EXPERIMENTAL

A specially-modified [10] Hewlett Packard 5950A spectrometer with monochromatized Al $K\alpha$ X-ray source [11] was used to measure core-peak angular distributions for four different types of specimens. All intensities reported are based on peak areas, and thus are automatically corrected for any changes in peak width with angle for this dispersion-compensated X-ray- and electron-optical system.[10] The minimum instrumental width occurred for θ approximately equal to 25° and corresponded to a 0.80 eV full width at half maximum intensity for Au $4f_{\frac{7}{2}}$. The angular resolution of this spectrometer was $\Delta\theta = 5°$.

Two basic mechanical adjustments of the spectrometer geometry were possible and these are indicated schematically by A and B in fig. 1. By adjusting the monochromator crystal orientation, the X-ray flux distribution could be translated along a line perpendicular to the propagation direction, as shown by A. Per degree of adjustment knob rotation, this translation was 1.25×10^{-3} cm. The sliding inlet rod on which the specimen rotation mechanism was mounted could also be rotated on its axis, which was separated from the true θ rotation axis by $r = 0.714$ cm. This adjustment is shown as B. The optimum mechanical alignment of the system was found to be when the maxima in I and Ω were coincident with the specimen rotation axis,[10] but systematic alterations relative to this best condition were also investigated in an attempt to determine their effect on $R(\theta)$.

The base pressures during accumulation of spectra were approximately 5×10^{-9} Torr for most samples. Under the overall vacuum conditions achieved, all specimens possessed at least a slight amount of surface contamination containing carbon; the intensities due to such contaminants were very stable with time during all of the measurements reported.

The specimens studied consisted of:

(1) Gold vapor-deposited *in situ* on a smooth substrate to yield a specimen of >5000 Å in thickness—Au $4f$ and C $1s$ angular distributions were measured for various combinations of the alignments A and B. All subsequent specimens were studied at very close to the optimum alignments achieved.

(2) An aluminum diffraction grating approximating very closely a triangular-periodic surface profile—This was specially ruled in an aluminum substrate by Perkin Elmer Co., and had a groove density of 600 lines/mm (~16,600 Å repeat distance) and a blaze angle of 10° as defined below. After ruling, this grating was exposed to the atmosphere for an extended period (~2–3 months), thereby building up a continuous oxide layer that should have been 20–24 Å in thickness.[12] The only surface cleaning procedure utilized was ultrasonication in various solvents just prior to insertion into the spectrometer. Under these conditions of preparation, the chemically-shifted Al $2p$(oxide) and Al $2p$(metal) peaks were easily resolvable (with a separation of 2.5 eV) and also had comparable intensities, therby permitting measurement of their individual variations with θ. The angle θ was scanned over its full 180° range in a direction perpendicular to the grating grooves. Data were accumulated in two separate 90° scans moving in opposite directions away from the surface normal; these two scans we have denoted by θ_+ and θ_-.

(3) A unidirectionally-polished, nominally-flat, aluminum specimen, with rather coarse surface grooves of average depth 10,000–20,000 Å—These grooves were produced by 400 grit SiC abrasive. These specimens were also subsequently air-exposed for ~15 min to form a continuous oxide layer of ~15 Å thickness.[12] Al $2p$(oxide) and Al $2p$(metal) intensities were then measured for the full range of θ, again by means of two separate θ_+ and θ_- scans. Full angular distributions were obtained along directions perpendicular to and parallel to the polishing grooves.

(4) Powdered alumina (Al_2O_3) specimens with adsorbed Si and Ca. These specimens were prepared by immersing high purity Al_2O_3 powder of 2–5 μm diam in an aqueous solution containing 75 p.p.m. by weight of Si (as $Si(OH)_4$) and 200 p.p.m. Ca (as $CaCl_2$). Following adsorption, the powder was washed three times with large volumes of ethyl alcohol (or in certain cases, deionized distilled water) to remove all but strongly adsorbed species. (The adsorptions carried out with these specimens represented simulations of certain types of inorganic fertilizer action in soils.[13]) The powder was then dried by heating for 8 h at 50 °C and atmospheric pressure, and pressed into pellets for final XPS characterization. Untreated Al_2O_3 specimens subjected to all preparation steps except solution adsorption were also studied as references. Broad-scan spectra from 0 to 1000 eV binding energy were measured at $\theta = 5°$ and $\theta = 38.5°$. The intensities of all observable core peaks

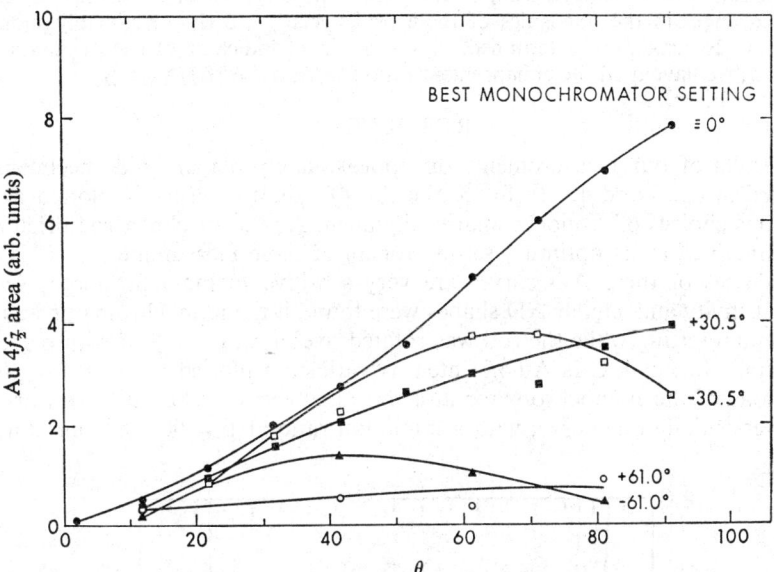

Fig. 3.—Au $4f_{\frac{7}{2}}$ angular distributions for various settings of the X-ray-monochromator adjustment knob (adjustment A in fig. 1).

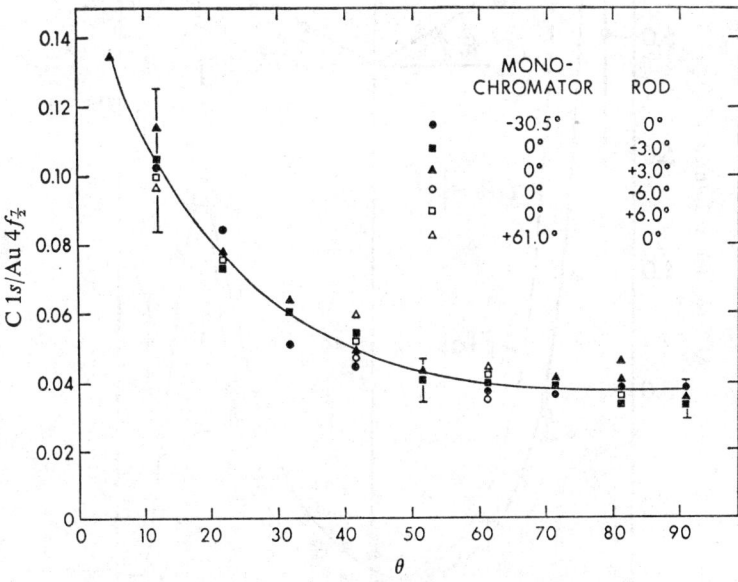

Fig. 4.—C $1s$/Au $4f_{\frac{7}{2}}$ ratio angular distributions for various combinations of the alignments of monochromator (A in fig. 1) and specimen rod (B in fig. 1).

	Monochromator Ⓐ	Rod Ⓑ		Monochromator Ⓐ	Rod Ⓑ
●	−30.5°	0°	○	0°	−6.0°
■	0°	−3.0°	□	0°	+6.0°
▲	0°	+3.0°	△	+61.0°	0°

were measured. These included O 1s, C 1s, Ca 2p, Si 2s, Si 2p, Al 2s, and Al 2p. Angular-dependent changes in the intensities of these peaks relative to the presumably substrate-associated Al 2s peak were determined. As a pertinent indicator of such changes for a given peak k, we have used the enhancement ratio $[N_k/N_{Al\ 2s}]_{5°}/[N_k/N_{Al\ 2s}]_{38.5°}$.

RESULTS

The results of our measurements on approximately planar gold specimens are summarized in fig. 3 and 4. In fig. 3, the Au $4f_{\frac{7}{2}}$ peak intensity is plotted against θ for various choices of monochromator alignment A; the specimen rod orientation B was maintained at its optimum setting for all of these measurements. It is clear that the shapes of these AD curves are very sensitive to monochromator setting. Similar sensitivity and similar AD shapes were found if the monochromator was held at its optimum setting while the rod was rotated over a range of $\pm 6°$ with respect to its optimum. In fig. 4, C 1s/Au $4f_{\frac{7}{2}}$ intensity ratios are plotted versus θ for various combinations of the monochromator and rod alignments. The ratio data are described by essentially one curve, with a minimum ratio at $\theta = 90°$ of approximately 0.038.

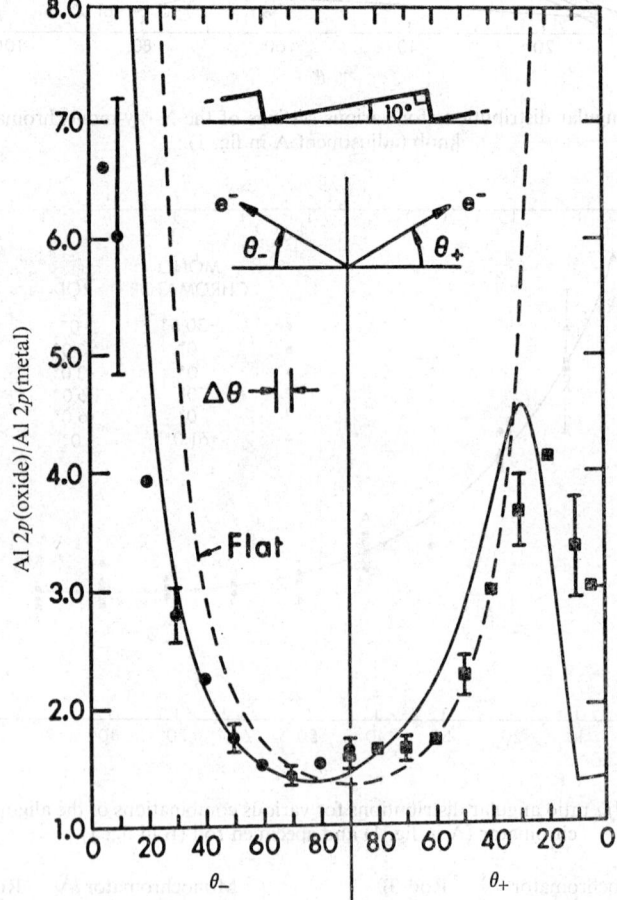

FIG. 5.—Al 2p(oxide)/Al 2p(metal) ratio angular distribution for an aluminum diffraction grating with 10° blaze angle. The curves are theoretical and are based on an empirical value for t(oxide)/Λ_e(oxide) of 1.075.

Fig. 5 presents Al 2p(oxide)/Al 2p(metal) ratio AD data for a diffraction grating with 10° blaze angle, together with flat-surface- and ideal-triangular-surface- theoretical curves. The ideal triangular geometry is shown, together with definitions of the two 90-degree θ scans involved; the latter are denoted by θ_+ and θ_-. The theoretical curves are based on substitution of previously determined values for densities and electron attenuation lengths [14, 15] into eqn (6), together with an empirically-determined value for $t(\text{oxide})/\Lambda_e(\text{oxide})$. This effective oxide layer thickness was derived by requiring the ideal triangular-surface theory to fit the minimum

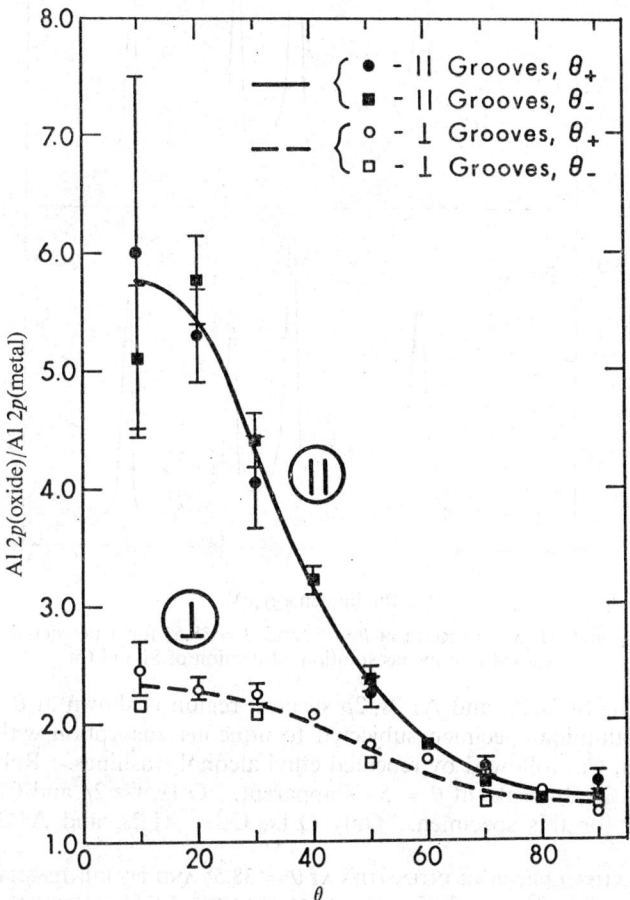

FIG. 6.—Al 2p(oxide)/Al 2p(metal) ratio angular distributions for a unidirectionally-polished aluminum specimen. Separate AD's were measured for θ variation parallel to and prependicular to the polishing grooves.

experimental Al 2p(oxide)/Al 2p(metal) ratio, which occurs for a value of θ_- very near that for which θ' should in theory equal 90° for all unshaded surface ($\theta_- = 80°$). This yielded a value of $t(\text{oxide})/\Lambda_e(\text{oxide}) = 1.075$. The neglect of X-ray shading implicit in the use of eqn (6) is justified by the long penetration depth for Al $K\alpha$ X-rays in aluminum (~90,000 Å) as compared to the groove repeat length of only ~16,600 Å.

In fig. 6, Al 2p(oxide)/Al 2p(metal) ratio AD's for a unidirectionally-polished aluminum specimen are shown. The data obtained by varying θ parallel to the polishing grooves are distinctly different from those for perpendicular variation. θ_+ and θ_- have the same significance as in fig. 5.

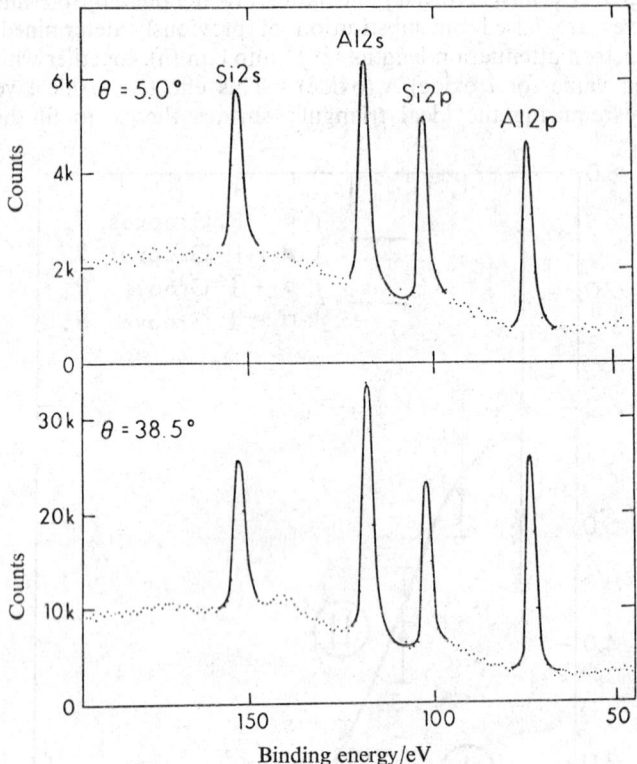

FIG. 7.—Si $2s$, $2p$ and Al $2s$, $2p$ spectra at $\theta = 5°$ and $\theta = 38.5°$ for a powdered Al_2O_3 specimen exposed to aqueous solution adsorption of Si and Ca.

In fig. 7, the Si $2s$, $2p$ and Al $2s$, $2p$ spectral region is shown at $\theta = 5°$ and $\theta = 38.5°$ for an alumina specimen subjected to aqueous adsorption with 75 p.p.m. Si and 200 p.p.m. Ca, followed by repeated ethyl alcohol washings. Relative enhancement of the Si $2s$, $2p$ peaks at $\theta = 5°$ is apparent. O $1s$, Ca $2p$ and C $1s$ peaks were also observed for this specimen. Only O $1s$, C $1s$, Al $2s$, and Al $2p$ peaks were

TABLE 1.—RELATIVE CORE-PEAK INTENSITIES AT $\theta = 38.5°$ AND PEAK-INTENSITY ENHANCEMENT RATIOS BETWEEN $\theta = 5°$ AND $38.5°$ FOR AN UNTREATED Al_2O_3 SPECIMEN AND A SIMILAR SPECIMEN EXPOSED TO AQUEOUS SOLUTION ADSORPTION OF Si AND Ca

	Untreated Al_2O_3		Al_2O_3+Si+Ca	
k	$[N_k/N_{Al\,2s}]_{38.5°}$	$\dfrac{[N_k/N_{Al\,2s}]_{5°}}{[N_k/N_{Al\,2s}]_{38.5°}}$	$[N_k/N_{Al\,2s}]_{38.5°}$	$\dfrac{[N_k/N_{Al\,2s}]_{5°}}{[N_k/N_{Al\,2s}]_{38.5°}}$
C $1s$	0.13	3.0±0.4	0.29	2.6±0.2
Ca $2p$	—	—	0.03	1.8±0.5
Si $2s$	—	—	0.56	1.2±0.1
Si $2p$	—	—	0.62	1.2±0.1
Al $2s$	1	1	1	1
Al $2p$	0.74	1.1±0.1	0.82	0.9±0.1
O $1s$	12.16	0.9±0.1	15.32	0.9±0.1

observed for an untreated alumina specimen. Peak enhancement ratios relative to Al 2s were calculated for all observed peaks in both types of specimens and these are given in table 1. The data shown here were typical of over 15 specimens studied; the results were not changed significantly by using deionized distilled water as a final cleaning agent.

DISCUSSION

The gold specimens represented by fig. 3 and 4 should correspond rather closely to the conditions of uniform composition, infinite thickness, and clean, flat surface described by eqn (4). Although there is a surface contaminant layer present, its thickness as judged by the low C 1s/Au $4f_{\frac{7}{2}}$ ratio at 90° of 0.038 is very small, and prior AD studies on identically-prepared specimens indicate that is it probably also patched.[5] Thus, surface-layer attenuation is probably not a major factor in determining the shapes of the Au $4f_{\frac{7}{2}}$ AD's in fig. 3, and it is thus reasonable to assume that they represent at least rough indicators of the form of the instrumental response function $R(\theta)$ for various spectrometer alignments. It is thus clear from our measurements that $R(\theta)$ can be varied significantly by both the A and B adjustments of fig. 1. The fact that all of the C 1s/Au $4f_{\frac{7}{2}}$ ratio data of fig. 4 are described by essentially one curve in spite of a broad range of variation of $R(\theta)$ thus provides rather convincing evidence of the cancellation of instrumental effects in ratio angular distributions. Ratio data should thus in general be capable of more direct analysis in terms of specimen-related properties.

The Al 2p(oxide)/Al 2p(metal) ratio data for the 10° grating of fig. 5 exhibit marked asymmetry about the central $\theta = 90°$ line, and even show a maximum in the ratio for $\theta_+ \approx 20°$. None of these observations is consistent with flat-surface behaviour, but there is, in general, very good agreement between experiment and theoretical calculations based upon the ideal triangular geometry. There is thus little doubt that the overall shape of the ratio AD is primarily associated with surface-profile effects. A certain amount of disagreement, particularly for low θ_+ and low θ_-, is not surprising in view of the fact that the true grating geometry no doubt deviated in some respects from the ideal. Small deviations were verified qualitatively in scanning electron micrographs. These included slight rounding of the exposed groove edges, which would lead to a discrepancy of the type found at $\theta_+ \approx 0$, and residual small-scale roughness with dimensions $\gtrsim 100$ Å, which would generate lower experimental ratios at $\theta_- \approx 0$. The qualitative shape of both experimental and theoretical curves is easily explained in terms of shading and variation in the average value of θ'. The empirical value of $t(\text{oxide})/\Lambda_e(\text{oxide}) = 1.075$ used in our theoretical calculations is also reasonable, as it yields $t(\text{oxide}) = 18$ Å when combined with a previously-measured value for $\Lambda_e(\text{oxide})$ [14]; this thickness is in good agreement with the 20-24 Å range expected on the basis of our specimen preparation procedure.[12] We have also found similar agreement between the overall shapes of experimental and theoretical ratio AD's for aluminum gratings with 20° and 43° blaze angles.[7] Thus, surface profile effects can produce marked changes in the forms of angular distribution curves, and the theoretical model we have used appears capable of quantitatively predicting such phenomena. With careful analysis, such ratio AD measurements might also prove useful in deriving certain types of surface profile information.

For the much more random surface roughness represented by the unidirectionally-polished aluminum surface of fig. 6, it is nonetheless possible to significantly enhance the oxide signal at low θ values for an angle scan either parallel to or perpendicular to the polishing grooves. The relative enhancement between $\theta = 90°$ and $\theta = 10°$ as defined by $[\text{Al } 2p(\text{oxide})/\text{Al } 2p(\text{metal})]_{10°}/[\text{Al } 2p(\text{oxide})/\text{Al } 2p(\text{metal})]_{90°}$ is 4.0 for parallel orientation and 2.0 for perpendicular orientation. The marked difference in the enhancements possible for the two orientations is easily explained qualitatively by considering the average surface slope perpendicular to the direction of electron emission for very low θ. For emission perpendicular to the groove direction, the unshaded surface at low θ often involves the steeply sloping sides of the grooves, with a resultant larger average θ' value and less oxide enhancement. For emission parallel to the groove direction, on the other hand, the average slope perpendicular to the electron propagation direction will be lower, leading to a lower average θ' and more oxide enhancement. Thus, although surface enhancement at low θ appears possible for surfaces with highly random roughness contours, it can be significantly influenced by directional asymmetry in the roughness distribution. It is also suggested that unidirectional polishing followed by AD measurements parallel to the final polishing direction might represent a method for significantly increasing the amount of surface enhancement possible at low θ.

The data of fig. 7 clearly indicate a slight relative enhancement of the Si $2s$, $2p$ peaks in going from $\theta = 38.5°$ to $\theta = 5°$. This is more quantitatively expressed by the enhancement ratios given in table 1, which are 1.2 for both Si $2s$ and Si $2p$. Enhancement ratios are also given in this table for all of the significant core peaks noted for both this specimen and an untreated reference, and they have been listed in decreasing order as a qualitative indicator of the relative depths from the surface of the atoms involved. That is, an enhancement ratio of unity indicates an atom distributed approximately uniformly throughout the substrate (Al and O). A ratio greater than unity indicates an atom occupying positions nearer the surface. It is thus significant that the Si and Ca peaks both show enhancement ratios greater than 1 (1.2 and 1.8, respectively), thereby verifying that these atoms were to some extent surface-adsorbed. That carbon is predominantly present in an outermost surface layer is also indicated by its extremely large enhancement ratio for both types of specimen (3.0 and 2.6). (Similar effects have also been found in other specimens of this type for which the number and identities of the Si- and Ca-containing solutes, as well as the washing procedure, were varied.[13]) It is worth noting that both values of the O $1s$ enhancement ratio are slightly less than unity, and that this behaviour was found for essentially all specimens of this type studied.[13] As oxygen should, if anything, be somewhat more predominant near the surface, its enhancement ratio might be expected to be greater than 1; the explanation for this slight discrepancy probably lies in the shorter attenuation length for O $1s$ photoelectrons ($E = 950$ eV) as compared to that for Al $2s$ ($E = 1364$ eV). These results thus suggest that, even for the highly irregular surface profiles associated with powdered specimens, low-θ surface enhancement is possible. Furthermore, this enhancement appears sensitive enough to yield qualitative depth profile information of a type demonstrated previously for flat-surface specimens.[4,5]

I express my sincere gratitude to my co-workers in these studies: R. J. Baird, M. Mehta, R. Alvarez, and S. K. Kawamoto. The financial support of the National Science Foundation (Grant GP 3864OX) and the University of Hawaii Research Council is also gratefully acknowledged.

[1] C. S. Fadley and S. Å. L. Bergström, *Phys. Letters A*, 1971, **35**, 375; and in *Electron Spectroscopy*, ed. D. A. Shirley. (North Holland Publishing Co., 1972), p. 223.
[2] W. A. Fraser, J. Florio, W. N. Delgass and W. D. Robertson, *Surface Sci.*, 1973, **36**, 661.
[3] B. L. Henke, *Phys. Rev. A*, 1972, **6**, 94.
[4] C. S. Fadley, R. J. Baird, W. Siekhaus, T. Novakov and S. Å. L. Bergström, *J. Electron Spectr.*, 1974, **4**, 93.
[5] C. S. Fadley, *J. Electron Spectr.*, 1974, **5**, 725.

[6] J. Brunner and H. Zogg, *J. Electron Spectr.*, 1974, **5**, 811.
[7] R. J. Baird, C. S. Fadley, S. K. Kawamoto and M. Mehta, *Chem. Phys. Letters*, 1975, **34**, 49.
[8] C. S. Fadley, *Progr. Solid State Chem.*, 1976.
[9] H. G. Nöller, H. D. Polaschegg and H. Schillalies, *J. Electron Spectr.*, 1974, **5**, 705.
[10] R. J. Baird and C. S. Fadley, manuscript in preparation.
[11] K. Siegbahn, D. Hammond, H. Fellner-Feldegg and E. F. Barnett, *Science*, 1972, **176**, no. 7032, 245.
[12] J. E. Boggio and R. C. Plumb, *J. Chem. Phys.*, 1966, **44**, 1081.
[13] R. Alvarez, *Ph.D thesis*, (University of Hawaii, Agronomy and Soil Science Department, 1975).
[14] F. L. Battye, J. Liesegang, R. C. G. Leckey and J. G. Jenkin, *Phys. Letters A*, 1974, **49**, 155.
[15] H. Kanter, *Phys. Rev. B*, 1970, **1**, 2357.
[16] B. L. Henke and E. S. Ebisu in *Advances in X-ray Analysis*, (Plenum Press, New York, 1973), vol. 17, p. 150.

Satellite Structure in the Photoelectron Spectra of Transition Metal Compounds Ionized in the K Shell of the Metal Ion*

By Thomas A. Carlson

Oak Ridge National Laboratory, Oak Ridge, Tennessee 37830

Received 25th February, 1975

Cu K_α X-rays were used to obtain photoelectron spectra for the K shell of the metal ion in twelve different manganese and iron compounds. Comparison of these spectra with data taken on the $2p$ shell allows one to deduce that the prinicpal source of satellite structure seen in the $2p$ spectra of transition metal compounds of the first row is generally due to electron shake-up rather than multiplet splitting. A discussion is given on various experimental means of distinguishing satellite structure due to multiplet splitting from that due to electron shake-up.

INTRODUCTION

Satellite structure is an important element in the study of X-ray photo-electron spectroscopy. Normally, the ion which has been created by the photo-ejection of an electron from the core shell is left in the ground state. Frequently, however, excited states are also formed, and when this occurs the photoelectron leaves with an energy reduced by the amount of the excitation. The creation of excited states manifests itself in a photoelectron spectrum as satellite structure on the low-energy side of the main or " normal " photoelectron peak.

Two important processes which can give rise to satellite structure are electron shake up and multiplet splitting. In the former case a sudden change in central potential, caused by the removal of a core electron that had been shielding the valence orbital electron from the nuclear charge, gives rise to monopole excitation of the outer-shell electrons. In the latter case, the creation of a core-vacancy leaves a shell with an unpaired spin that can couple with the valence shell, if it is only partially filled. Multiplet splitting is really the consequence of a series of possible spin states rather than the result of specific excitation. Since the most stable state usually has the greatest multiplicity, the ground state will normally be the most intensely populated.

A large number of studies have been carried out on the photoelectron spectra of the $2p$ shells for the first-row transition metal compounds [1-10] in order to observe satellite structure. This is so because the satellite structure associated with the $2p$ shell of the metal ions has been frequently found to be very intense, easy to observe, and strongly dependent on the nature of the chemical species. Both electron shake-up and multiplet splitting have been advanced to explain this structure. It is important to assess the relative roles of these two phenomena. It has been suggested [11,12] that the study of satellite structure in the K shell of the transition metal ions would achieve this purpose since multiplet splitting will be negligible in this situation, while electron shake-up should be very similar to that observed in the photoelectron spectra of the $2p$ subshell. In this paper, data will be presented on six iron compounds and six manganese compounds. In addition to the study of the K-shell photoelectron

* Research sponsored by the Energy Research and Development Administration under contract with the Union Carbide Corporation.

spectra, I shall discuss other ways for clarifying the source of satellite structure and indicate what new experiments ought to be done in the future.

EXPERIMENTAL

Photoelectron spectra were taken with a double-focusing electrostatic spectrometer, which has been previously described.[13] Photoionization in the $2p$ shell was studied with Al K_α rays, while an investigation of the K shell was accomplished with Cu K_α X-rays. In the latter phase of the studies a position sensitive detector [14] was employed to improve the sensitivity.

Data on $FeCl_3$, $FeBr_3$, $K_3Fe(CN)_6$ and $K_4Fe(CN)_6$ have been previously reported. The present paper presents new data on $FeSO_4 \cdot 7H_2O$, $Fe(AcAc)_3$ (AcAc = acetylacetonate), MnF_2, $McCl_2$, M_4Br_2, $K_4Mn(CN)_6$, $MnSO_4 \cdot H_2O$ and $Mn(AcAc)_3$. The compounds were obtained commercially and used without further purification. In some runs the sample holder was water-cooled, but no variations in the spectra were observed. The spectra were also monitored as a function of time to test whether any observable change in the spectra could be detected with radiation damage. None was seen.

Deconvolution of the spectra was done with the aid of a Dupont 310 curve resolver. In the analysis of the $1s$ spectra, data on $K_3Fe(CN)_6$ and $K_4Mn(CN)_6$ were used to provide the shape and f.w.h.m. for the deconvolution of, respectively, the iron and manganese compounds. Both Gaussian and Lorentzian shapes were tried, but the final analysis, generally, was not strongly affected. Special care was taken with MnF_2 because of the low intensity of the satellite structure, and the best fit to the standard peak seemed to be a Gaussian shape with extended tails.

RESULTS AND DISCUSSION

Before turning to the experimental data, a short discussion of results of calculations on multiplet splitting and electron shake-up is in order. First, calculations [12, 15] were made on the multiplet splitting in the free ions of Mn^{2+} and Fe^{3+} as a function of inner-shell vacancy. It was found that the energy splitting was substantial for vacancies created in both the L and M shells, ranging from 3 to 12 eV for a $2p$ hole. However, multiplet splitting for a K vacancy was negligible, being only 0.1 eV. More sophisticated calculations including crystal field effects should not alter the conclusions for a K vacancy, namely, that multiplet splitting will be unobservable in photoelectron spectra taken with Cu K_α X-rays.

Secondly, calculations have been made [12] on electron shake-up plus electron shake-off from the $3d$ shell as a function of the inner-shell vacancy for Fe^{3+}. These calculations were based on the overlap of relativistic Hartree-Fock-Slater wave functions for Fe^{3+} and Fe^{4+} (with a hole in subshell n,l,j). These calculations are not meant to be a realistic appraisal of the shake-up probability for the molecular case. The nature of monopole transitions in molecules involve quite different orbitals than in the atomic case. What the calculations do show is the nature of the dependence of shake-up on the location of the inner-shell vacancy. They indicate that the shake-up probability is essentially constant for photoionization in the $1s$, $2s$, $2p_{\frac{1}{2}}$ and $2p_{\frac{3}{2}}$ shells. This independence of shake-up probability on the core hole has been demonstrated [16] experimentally for the rare gases. The probability for exciting an electron in the $3d$ orbital as the result of photoionization in the $3s$ or $3p$ subshell was calculated to be only about $\frac{1}{3}$ of that for the inner shells. This conclusion is strongly subject to the nature of the molecular excitation and to correlation effects, but it has been generally verified that shake-up satellites in the $3s$ and $3p$ spectra are often considerably less intense than in the $2p$ spectra.

Having demonstrated that the probability for electron shake-up should be similar for photoionization in the $1s$ and $2p_{\frac{1}{2}}$ and $2p_{\frac{3}{2}}$ subshells, it should also be remarked that the energy separation should also be similar. Some changes may arise due to coupling of the inner-shell hole with the monopole excited state. Studies [16] on rare gases, however, suggest that these differences in coupling are small compared to the excitation energy involved in shake-up.

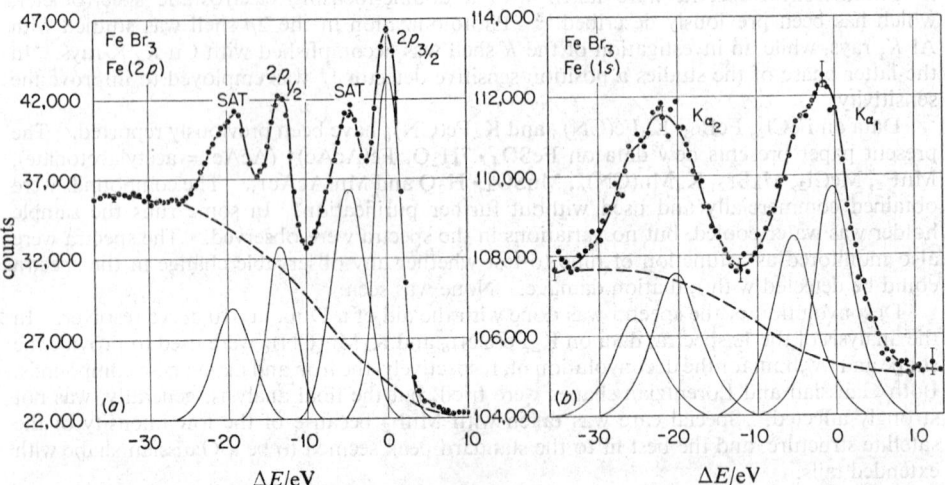

FIG. 1.—Photoelectron spectrum of $FeBr_3$ in the $2p$ shell using Al K_α X-rays, and in the $1s$ shell using Cu K_α X-rays. The background which includes consideration of characteristic energy losses is given by a dotted line. The spectra are deconvoluted using Gaussian shapes. For the $2p$ data the f.w.h.m. are chosen to be 3.6 and 5.1 eV respectively for the main peak and satellite structure. For the $1s$ data the spectrum is deconvoluted with two Gaussians for contributions from both the Cu K_{α_1} and K_{α_2} X-rays. The f.w.h.m. is taken from the experimentally determined value for the Fe($1s$) spectrum for $K_3Fe(CN)_6$. The kinetic energy scale is in eV, arbitrarily set at zero at the highest energy normal peak. Data taken from Carlson et al., ref. (12).

Let us now examine the experimental data. Fig. 1 shows some typical spectra. The resolution for photoelectron spectra in the K shell is poor, the photoelectron peaks being broadened by both the larger natural widths for the Cu K_α X-ray compared with Al K_α X-rays and for the K shell of iron compared with the $2p$ subshell. The counting rates were also poorer. Spectra obtained with $K_3Fe(CN)_6$ and $K_4Mn(CN)_6$ showed little evidence of satellite structure in the $2p$ spectra and were used as standard line width in deconvoluting the spectra taken with the Cu K_α X-rays. Note that the X-ray source contains two K_α X-rays: K_{α_1} and K_{α_2} at 8,048 and 8,028 eV in a ratio of nearly 2 to 1. The photoelectron spectrum in the $2p$ shell has two main lines corresponding to the $2p_{\frac{3}{2}}$ and $2p_{\frac{1}{2}}$ subshells. If electron shake-up is the prominent source of satellite structure, the relative intensity of the satellite peak to the main peak should be the same for both subshells; and, similarly, the excitation energy for creating the satellite structure is about the same.

Table 1 summarizes the data. The intensities are given in terms of the ratio of the peak height of the main satellite structure to its corresponding normal peak. This is somewhat arbitrary since the satellite structure is broader than the main peak and is most likely made up of more than one contribution, but since it is difficult to assay the exact background, the comparison of total intesnities is uncertain. The energy separation is between the normal peak and main peak of the satellite structure. For molecules where strong satellite structure is found in the $2p$ spectrum, similar structure is seen with the $1s$ spectrum. This is a confirmation that electron shake-up is the dominant force in producing satellite structure in the 2p photoelectron spectra of the first-row transition metal compounds.

TABLE 1.—SATELLITE STRUCTURE IN PHOTOELECTRON SPECTRA OF METAL IONS FOR TRANSITION METAL COMPOUNDS[a]

Compound	$2p_{\frac{3}{2}}$		$1s$	
	I	ΔE (eV)	I	ΔE (eV)
$K_3Fe(CN)_6$	—[d]	—	—	—
$K_4Fe(CN)_6$	—[d]	—	—	—
$FeSO_4$	0.72	5.7	0.52	3.9
$Fe(AcAc)_3$	0.66[b]	4.5[b]	0.51	3.6
$FeCl_3$	0.57[b]	4.4[b]	0.57[c]	4.5[c]
$FeBr_3$	0.73[b]	4.7[b]	0.74[c]	4.6[c]
$K_4Mn(CN)_6$	—[d]	—	—	—
$MnSO_4$	0.29	4.8	0.15	5.1
$Mn(AcAc)_3$	0.31[b]	4.1[b]	0.21	4.7
MnF_2	0.08[b]	6.5[b]	0.07	6.7
$MnCl_2$	0.38[b]	5.1[b]	0.35	4.8
$MnBr_2$	0.55[b]	4.8[b]	0.47	5.0

[a] I is the ratio of the peak height of the main satellite structure to that or the "normal" ground-state contribution. ΔE is the separation of those peak heights. Photoelectron data on $2p_{\frac{3}{2}}$ taken with Al K_α X-rays, data on $1s$ taken with Cu K_{α_1} X-rays.
[b] Results taken from Carlson et al., ref. (9).
[c] Results taken from Carlson et al., ref. (12).
[d] Satellite peaks of low intensities and large energy separation are found with cyanides but are believed to be due to characteristic energy losses. No satellite structure is seen in the region around 5 eV.

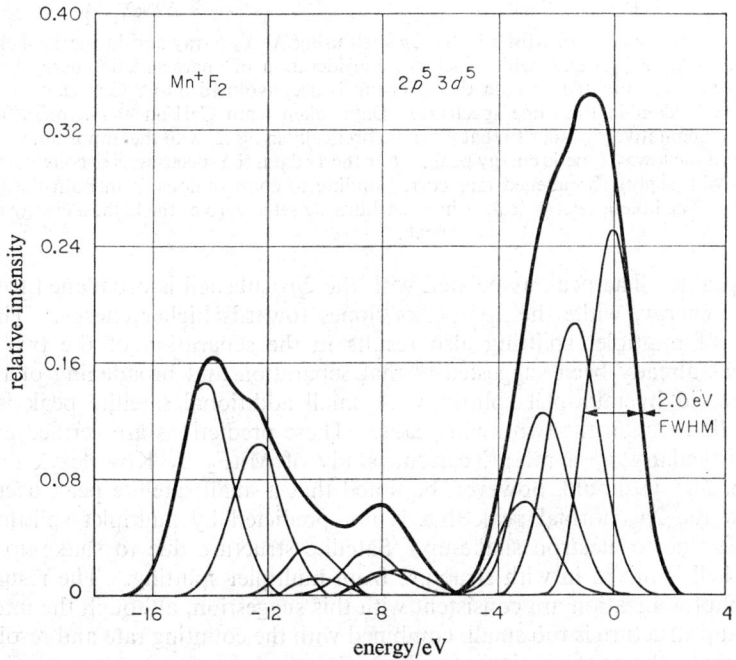

FIG. 2.—Convolution of theoretical calculations (Gupta and Sen, ref. (17)) for the multiplet structure calculated for MnF_2 using a f.h.w.m. of 2.0 eV. The energy scale corresponds to kinetic energy in eV with the highest peak set at zero.

However, the question remains as to what role multiplet splitting plays in the $2p$ spectrum. This question has been largely answered by calculations of Gupta and Sen [17] on MnF_2. They have included both crystal field effects and spin-orbital splitting in their calculation (see fig. 2 for a convolution of their results). They predict a large number of final states, but these states, rather than forming two well-defined satellite peaks in a photoelectron spectrum, will asymmetrically broaden the two main

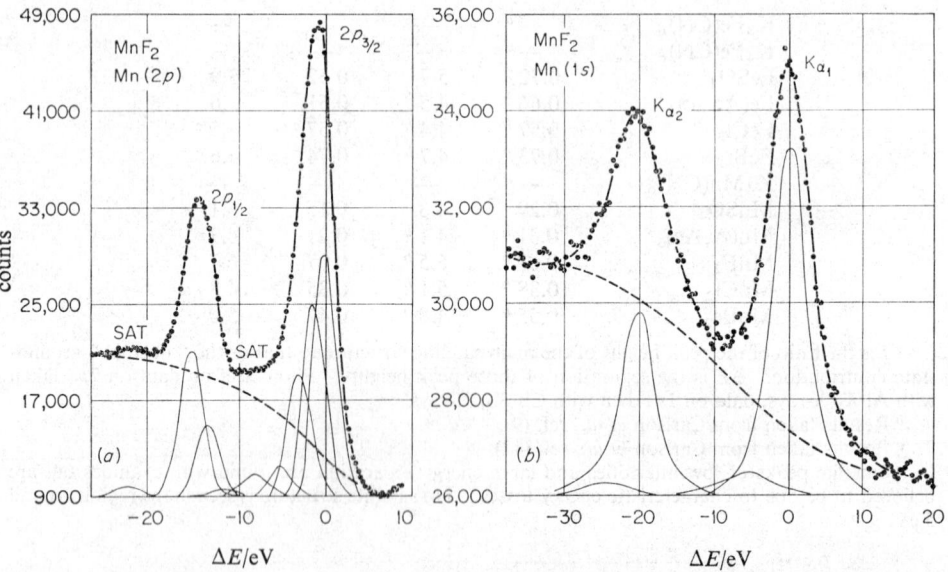

FIG. 3.—Photoelectron spectra of MnF_2 in the $2p$ shell using Al K_α X-ray and in the $1s$ shell using Cu K_α X-rays. The background, which includes consideration of characteristic energy losses, is given by a dotted line. For the $2p$ data the spectrum is deconvoluted into 8 Gaussians using the f.w.h.m. for $F(1s)$ taken in the same spectrum. Data taken from Carlson et al., ref. (9). The deconvolution is qualitatively similar to that given theoretically in fig. 2, with the important exception of the presence of the lowest kinetic energy peak. For the $1s$ data, the spectrum is deconvoluted with two Gaussians with slightly broadened tails corresponding to contributions from both the Cu K_{α_1} and K_{α_2} X-rays. The kinetic energy scale is in eV, arbitrarily set at zero at the highest energy normal peak.

" normal " peaks. The peak associated with the $2p_{\frac{3}{2}}$ subshell is broadened towards lower kinetic energy, while the $2p_{\frac{1}{2}}$ peaks slopes towards higher energy. The net consequence of multiplet splitting also results in the separation of the two main peaks. It has already been suggested [18] that separation and broadening of the $2p$ peaks are caused by multiplet splitting. A small additional satellite peak is also predicted to lie between the two main peaks. These predictions are verified experimentally, particularly by a recent, careful, study of MnF_2 by Kowalczyk et al.[19] (Also see fig. 3). It should, however, be noted that a small satellite peak occurs at 6.5 eV below the $2p_{\frac{1}{2}}$ normal peak that is not predicted by multiplet splitting. I believe this is due to electron shake-up. Satellite structure due to shake-up form the $2p_{\frac{3}{2}}$ subshell is mixed in with structure from multiplet splitting. The results on the K-shell photoionization are consistent with this suggestion, although the intensity of the shake-up structure is too small, combined with the counting rate and resolution available to make the confirmation unambiguous.

Finally, a few words should be written about the possibility of spurious sources of satellite structure. These are (i) the presence of multiple peaks due to more than one chemical form because of impurities or radiation damage, and (ii), the presence of structure due to characteristic energy losses. Contributions to multicomponent structure from different chemical forms can be dismissed on several counts. First, the chemical shifts for the transition metal compounds are generally much smaller than the energy separation evidenced by the main satellite structure. Secondly, if two chemical forms with different core binding energies exist, the intensity and energy separation should be essentially identical for photoionization in each of the different subshells. However, as has been pointed out earlier, the shake-up is often considerably less intense for the $3s$ and $3p$ subshells, and for most of the compounds studied we have been able to verify that this is indeed the situation, although the presence of multiplet splitting in these shells sometimes makes the comparisons difficult. Thirdly, the satellite structure in the $2p$ spectrum is completely reproducible, so if different chemical species are the cause, they must always be in a fixed equilibrium. Also, no changes in the spectra were noted as a function of radiation time.

Characteristic energy losses are always present, and an estimate of their contribution to the background is made in analyzing the data. For non-conducting materials these contributions are more like broad bands than well-defined peaks, with an onset at about 10–20 eV and a long tail towards lower kinetic energy. Such structure will occur with every ejected electron and will be seen in the photoelectron spectra of the ligand atom as well as the metal ion. For cyanide complexes, rather distinct satellite peaks were seen at separations from the normal peak of 12–16 eV. These peaks were, however, also observed arising from the nitrogen and potassium and are believed to be characteristic energy losses rather than electron shake-up contributions.

OTHER METHODS FOR CHARACTERIZING MULTICOMPONENT STRUCTURE, AND THE FUTURE OF K-SHELL PHOTOIONIZATION

In addition to the use of K-shell photoionization for help in characterizing satellite structure, there are other experimental means of distinguishing multiplet splitting from electron shake-up. We have discussed one, namely that in the photoelectron spectrum of the $2p$ shell, electron shake-up will produce similar satellites associated with both the $2p_{\frac{1}{2}}$ and $2p_{\frac{3}{2}}$ peaks, whereas multiplet splitting will generally not. Another method is through the study of K X-rays. The K_α X-rays arise out of transitions from the $2p$ orbitals to the $1s$ shell: $K_{\alpha_1} = 1s - 2p_{\frac{3}{2}}$; $K_{\alpha_2}\ 1s - 2p_{\frac{1}{2}}$. The final states for K fluorescence, with a vacancy in the $2p$ subshell, is the same as that found in photoionization. If multiplet splitting occurs, the same energy spacings for the multiplet lines will be seen as in the $2p$ photoelectron spectrum. (Electron shake-up from the valence shell that might have occurred in creating the initial hole will produce little effect on the K_α X-ray spectrum.[20]) Thus, whereas the study of K shell photoelectron spectra of first-row transition metal compounds will yield results on electron shake-up isolated from multiplet splitting, the study of K_α X-ray spectra will give information on the effects of multiplet splitting without interference from electron shake-up. The two methods are complementary. As examples of its use, Kowalczyk et al.[19] have used Nevedov's X-ray data [21] on MnF_2 to verify their conclusion on the photoelectron spectra, and Asada et al.[22] have used the absence of large satellites in K_α ray data of nickel compounds to demonstrate that the satellite structure seen in the $2p$ photoelectron spectra is due to electron shake-up.

As I have tried to show, the study of K-photoelectron spectra of transition metals has proved to be useful in the interpretation of satellite structure. However, the broadness of the peaks due to the large natural widths prevents the opportunity for studying satellite structure with the detail that one desires. If the X-ray source could be monochromatized, this might be possible. The natural width of the K shell for the first-row transition metals is only about 1.5 eV.[23] A monochromatic source of X-rays will also eliminate the contributions from the K_{α_2} line. Also, a more variable X-ray source is needed. Mn, Fe and Co can be studied with Cu K_α X-rays. Ni may be studied with Cu K_β, Cu with Zn X-rays. Even better would be the use of synchrotron radiation to produce a variable source of X-ray energies with high resolution. A hopeful sign in this direction is the production of an 8 keV X-ray source with resolution of at least 0.2 eV by using synchrotron radiation from the Stanford Storage Ring.[24]

I express my thanks to Dr. J. C. Carver, Dr. G. A. Vernon and Dr. L. J. Saethre for allowing me the use of their data on the 2p photoelectron spectra taken while at Oak Ridge for help in comparing the K-shell spectra. I would also show my appreciation to Dr. W. B. Dress of Oak Ridge National Laboratory for his assistance in processing data taken with the position sensitive detector, and to Dr. S. P. Kowalczyk of the University of California, Berkeley, for making his manuscript available before publication.

[1] A. Rosencwaig, G. K. Wertheim and H. J. Guggenheim, *Phys. Rev. Letters*, 1971, **27**, 479.
[2] D. C. Frost, A. Ishitani and C. A. McDowell, *Mol. Phys.*, 1972, **24**, 861.
[3] L. J. Matienzo, L. I. Yin, S. O. Grim and W. E. Swartz, Jr., *Inorg. Chem.*, 1973, **12**, 2762.
[4] B. Wallbank, C. E. Johnson and I. G. Main, *J. Phys. C, Solid State Phys.*, 1973, **6**, L340, L493.
[5] K. S. Kim, *J. Electron Spectr.*, 1974, **3**, 217.
[6] J. Escard, G. Mavel, J. E. Geurchais and R. Kergoat, *Inorg. Chem.*, 1974, **13**, 695.
[7] D. Briggs and V. A. Gibson, *Chem. Phys. Letters*, 1974, **25**, 493.
[8] D. C. Frost, C. A. McDowell and I. S. Woolsey, *Mol. Phys.*, 1974, **27**, 1473.
[9] T. A. Carlson, J. C. Carver, L. J. Saethre, F. Garcia Santibáñez and G. A. Vernon, *J. Electron Spectr.*, 1974, **5**, 247.
[10] B. Wallbank, I. G. Main and C. E. Johnson, *J. Electron Spectr.*, 1974, **5**, 259.
[11] T. A. Carlson, *Faraday Disc. Chem. Soc.*, 1972, **54**, 292.
[12] T. A. Carlson, J. C. Carver and G. A. Vernon, *J. Chem. Phys.*, 1975, **62**, 932.
[13] B. P. Pullen, T. A. Carlson, W. E. Moddeman, G. K. Schweitzer, W. E. Bull and F. A. Grimm, *J. Chem. Phys.*, 1970, **53**, 768.
[14] C. D. Moak, S. Datz, F. Garcia Santibáñez and T. A. Carlson, *J. Electron Spectr.*, 1975, **6**, 151.
[15] T. A. Carlson, in The *Physics of Electronic and Atomic Collisions*; Invited lecture and Progress Reports, VIII ICPEAC, Beograd, 1973, edited by B. C. Cŏbic and M. V. Kurepa (Institute of Physics, Beograd, Yugoslavia, 1973), p. 205.
[16] D. P. Spears, H. J. Fischbeck and T. A. Carlson, *Phys. Rev. A*, 1974, **9**, 1603.
[17] R. P. Gupta and S. K. Sen, *Phys. Rev. B*, 1974, **10**, 71.
[18] D. C. Frost, C. A. McDowell and I. S. Woolsey, *Chem. Phys. Letters*, 1972, **17**, 320.
[19] S. P. Kowalczyk, L. Ley, F. R. McFeely and D. A. Shirley, *Phys. Rev. B*, 1975, **11**, 1721.
[20] M. O. Krause, in *Proc. Conf. Inner Shell Ionization Phenomena and Future Applications*, ed. R. W. Fink, S. T. Manson, J. M. Palms and P. V. Rao (USAEC, Oak Ridge, Tennessee, 1973) CONF-720404 vol. 3, p. 1586.
[21] V. I Nefedov, *Akad. Nauk. (Phys. Ser.)* 1964, **28**, 724.
[22] S. Asada, C. Satoko and S. Sugano, Tech. Report Inst. for *Solid State Phys.*, A, 671 (1974).
[23] O. Keski-Rahkonen and M. O. Krause, *Atomic Data and Nuclear Data Tables*, 1974, **14**, 139.
[24] I. Lindau, P. Pianetta, S. Doniah and W. E. Spicer, *Nature*, 1974, **250**, 215.

Multiplet Structure in X-ray Photoelectron Spectra of Rare Earth Elements and Their Surface Oxides

By W. C. Lang, B. D. Padalia,† L. M. Watson, D. J. Fabian* and P. R. Norris

Department of Metallurgy, University of Strathclyde,
Colville Building, North Portland Street, Glasgow G1 1XN

Received 13th March, 1975

A systematic study has been made of the valence bands and core-level peaks in X-ray photoelectron spectra of the heavy rare-earth metals and their surface oxides. Multiplet structure observed in the 4d levels is reported here. This structure alters on oxidation, and in addition the 4d peaks broaden and exhibit shifts in energy measured relative to the Fermi energy. For oxidised ytterbium, the appearance of multiplet structure is interpreted as promotion of 4f electrons to bonding orbitals.

INTRODUCTION

Recently, there has been considerable interest in the X-ray photoelectron spectra of the rare earth elements and their compounds.[1-4] These elements are characterised by successive filling of the atomic 4f shell, which in the solid gives rise to spatially localized 4f-electrons near the Fermi level with the well-known consequences for many of the physical properties of these metals.

In general, the X-ray photoelectron spectra are complex, because the final state configuration in the photoemission process contains 4f and 4d holes (configurations of high orbital and spin angular momentum) which couple to give a multiplicity of states. The resulting separation of the terms, and therefore also of photoelectron peaks, is in many cases of 1 eV or less. The use of u.v. rather than X-rays to excite the spectra could be advantageous, on account of the higher resolution attainable, but with u.-v. excitation the intensities are low, so that meaningful measurements are difficult to obtain.

Here we report a systematic investigation of the 4d photoemission for clean and oxidised samples of the heavy lanthanides, using the natural $K\alpha$ lines of Al and Mg for excitation. The results confirm and extend those previously available, and lead to interesting information on the processes of surface oxidation.

EXPERIMENTAL

Photoelectron spectra were recorded using a Vacuum Generators ESCA3 instrument. Clean samples were prepared by repeated evaporation, (pressure of 2×10^{-9} Torr) of spectroscopically pure metals (Johnson and Matthey); details of the method used to ensure uncontaminated surfaces have been given elsewhere.[5] After preparation, samples were transferred without disturbing the vacuum to the analyser chamber, which was held at a base pressure of 6×10^{-10} Torr. At intervals during data collection the degree of surface contamination was monitored by measuring the carbon and oxygen 1s photoemission peaks, and fresh surfaces were evaporated wherever these appeared with observable intensity.

† On leave from Indian Institute of Technology, Bombay.

38   X-RAY PHOTOELECTRON SPECTRA

Controlled oxidation of the metal surface was performed by exposing the evaporated film for measured intervals of time to dry oxygen at pressures $\sim 10^{-9}$-10^{-4} Torr, within the sample preparation chamber. The oxygen, dried in a liquid-nitrogen cold trap, was leaked into the system through a non-bakeable valve.

Facilities are available to excite the samples alternatively with Al $K\alpha$ and Mg $K\alpha$ radiation. Both were employed to examine the possibility of slight changes due to differences of escape depth and photo-ionisation cross-section, and for identification of Auger peaks. Unless otherwise stated, the results presented here are those obtained using the Mg $K\alpha$ line, which was mostly employed for excitation because of its lower natural width. The best resolution obtainable is thus ~ 0.9 eV, which is consistent with the half-widths of the narrowest features observed in the spectra.

All of the results presented are plots of the summed intensity from several successive scans over the same energy interval. This technique enables a fresh film to be re-evaporated more frequently than would otherwise be possible; the data were thus collected in several scans of short duration. Control of the scanning potentials and recording of the data, was effected by running the equipment on line to a PDP8/E computer, which recorded the total electron count at energy intervals of 0.2 eV, which is well within the limits of the instrumental resolution.

RESULTS

Fig. 1-7 show the $4d$ photoelectron spectra of the series of heavy rare earths from terbium to lutetium. In each case the upper curve is the spectrum obtained from the clean metal surface, while the lower is from an oxide layer formed on exposure to oxygen until surface saturation is reached. In the case of Yb a spectrum from a partially oxidised sample is also shown.

The spectra are plotted as total electron intensity against binding energy. The intensity scale is linear in photoelectron current, but the zero point is arbitrarily chosen for convenience; the error bars correspond to a 67 % confidence limit. In

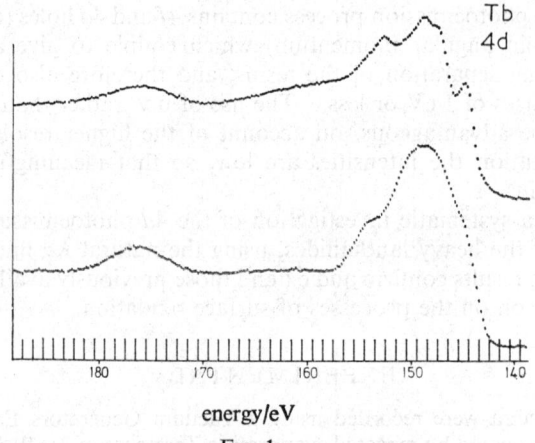

FIG. 1.

FIGS 1-7.—X-ray photoelectron $4d$ core-level multiplets for, respectively: terbium, dysprosium, holmium, erbium, thulium, ytterbium and lutetium, using Mg $K\alpha$ excitation.* Upper spectrum in each case measured for the pure evaporated metal and lower spectrum for the metal film with a saturation layer of oxide formed on exposure to dry oxygen at 10^{-4} Torr pressure. For ytterbium the middle spectrum is that measured with the presence of an intermediate oxide layer (see text). All spectra are the summation of eight successive scans with intermediate re-evaporation of fresh sample where necessary.

* For Tb interference from an Auger peak necessitates the use of Al $K\alpha$ excitation.

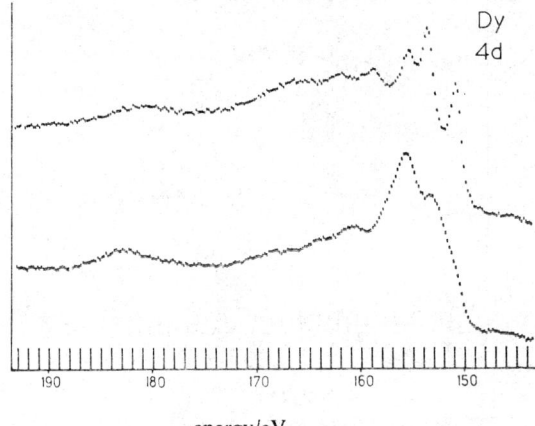

Fig. 2.

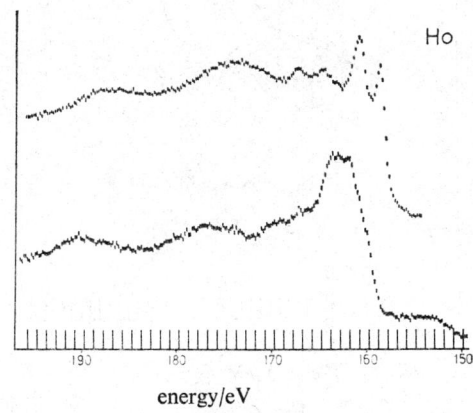

Fig. 3.

Fig. 4.

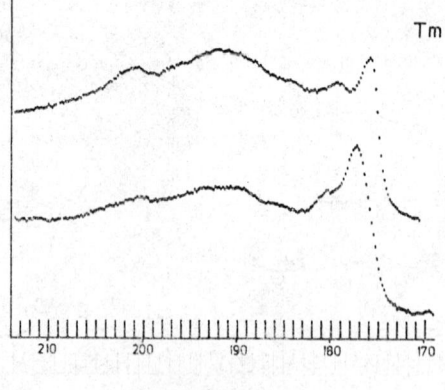

energy/eV

Fig. 5.

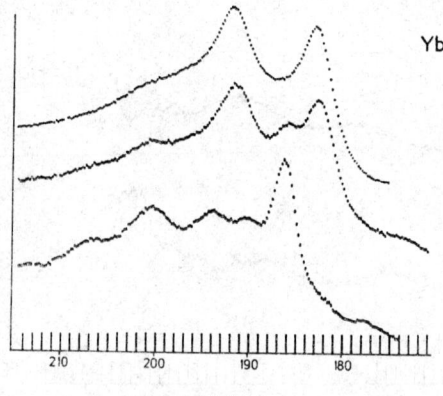

energy/eV

Fig. 6.

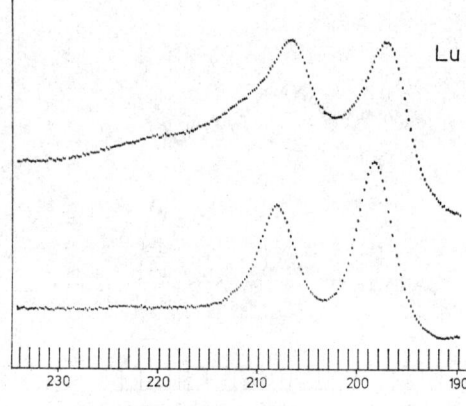

binding energy/eV

Fig. 7.

most cases the position of the Fermi level was determined from actual measurement of the valence band spectra, which were obtained for each of the samples. This procedure could not be followed in all cases (for example, ytterbium and lutetium [6]) because the Fermi edge is masked by the $K\alpha_{3,4}$ satellite of the $4f$ photoemission spectrum; in these cases the binding energy was inferred from the kinetic energy and the previously determined work function of the spectrometer.

In general, the spectra shown in fig. 1-5 all have points of similarity; the $4d$ levels are split into several components, which extend over an energy range of several electron volts. The spectra for clean ytterbium and lutetium are dominated by a double peak, in which the components are separated by 9.0 ± 0.2 and 9.9 ± 0.2 eV respectively, in good agreement with the values of the separation obtained by Kowalczyk et al.[2] (8.9 and 10.0 eV). The oxide spectra consist of broader peaks, none of which occurs at the same energy as any peak in the corresponding spectrum of the clean metal surface. However, for Ho, and Tb a shoulder is evident on the low binding energy side of the $4d$ multiplets, which corresponds to the peak observed in the spectra for the clean samples of these metals. In those cases for which there is no perceptible contribution from the metal substrate, we infer that the oxide layer is, at saturation, of sufficient thickness to prevent the escape of photoelectrons from below the oxide layer.

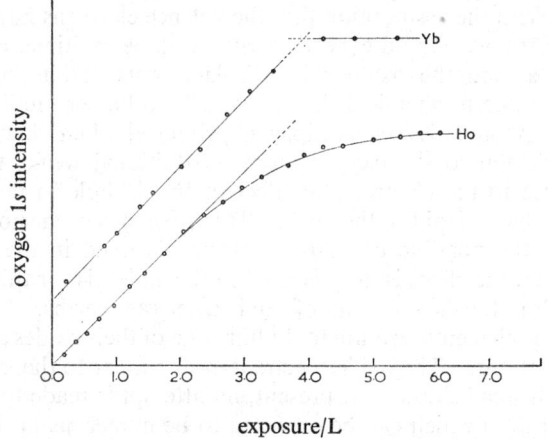

FIG. 8.—Comparison for holmium and ytterbium of the increase of oxygen $1s$ peak intensity with time on exposure in dry oxygen at 10^{-9} Torr pressure. The behaviour illustrated in the case of holmium is typical of all the other heavy lanthanides.

The course of oxidation can be followed more easily in the spectra of ytterbium. In the spectrum from the freshly evaporated surface (upper curve of fig. 6) there is a weak feature at ~ 201 eV and also an indication of further structure between the two components of the doublet. These two features have clearly grown in intensity in the spectrum obtained from a partially oxidised surface (middle curve), and in fact correspond to the two most intense peaks observed for a fully formed oxide film (lower curve). The two weak lines in the spectrum from the pure metal are at the same energies as the strongest lines from the fully oxidised sample, and clearly result from residual oxidation of the clean surface. The growth of the oxygen $1s$ emission during oxidation of the ytterbium sample is indicated in fig. 8. The rate of increase of emission appears to be a constant function of exposure until saturation of the surface is achieved, after which it falls to zero. This behaviour is different from that for the formation of the oxide layer in, for example, holmium (lower curve of fig. 8) for which the rate of growth of the oxygen 1s peak appears to decrease asymptotically to zero with increasing exposure.

Most of the spectra, whether from the clean metal or the oxide, show a broad peak with a maximum about 26 eV higher in binding energy than the most intense features of the 4d complex. This structure is not Auger line, because it is present in spectra obtained using both Mg $K\alpha$ and Al $K\alpha$ excitation. It cannot be interpreted as a plasmon loss satellite of the main 4d peaks, because its energy separation from these does not correspond to the plasmon energy of ytterbium; moreover, no higher order loss peaks have been detected. Interpretation therefore awaits further investigation.

DISCUSSION

The general electronic configuration of the heavy lanthanides in the metallic state is, with the exception of ytterbium, $4f^n 5d^1 6s^2$, and since the 5d and 6s electrons are involved in bonding, the usual ionic configuration is the trivalent form $4f^n$. Thus, the final state produced on emission of a photoelectron from the lower-lying 4d levels will be of the form $4d^9 4f^n$, and the possible final levels will depend on the coupling between the 4d hole and the partially filled 4f shell. Thus complex multiplet splitting is, in general, expected and this is seen in the spectra shown here for Tb, Dy, Ho, Er and Tm. With the assumption that the valence electrons have relatively little effect on the ion cores, we should expect nearly the same multiplet configuration for both the pure metal and the oxide, with displacements in binding energy of the photoelectron peaks due to chemical shifts. A shift to higher binding energy of the components of the photoemission multiplet is produced when charge is transferred from the lanthanide ion to the oxygen atom. Additional weak structure may be expected in response to the electrostatic effect of the 4d hole in the final state and would further shift the centroid of the peak. This effect has been shown by Signorelli and Hayes [4] to be the possible cause of satellites observed in the 3d spectrum of lanthanum; however, the effect is not detected in the 4d levels for this metal and may be considered small in the 4d spectrum of the heavier rare earths.

The spectra of the elements terbium to thulium and of their oxides are in accordance with these interpretations. The oxide spectra appear similar to the metal spectra, but broadened and with small shifts. At present, no attempt is made to characterise the various component lines which can be expected to be numerous in these metals.

The cases of ytterbium and lutetium are rather different. In the metallic state both these metals have a completed 4f shell which thus has a ^1S configuration and cannot contribute to splitting of the 4d photo-ionisation state, which will possess only simple $4d_{\frac{3}{2}}$-$4d_{\frac{5}{2}}$ spin-orbit splitting. This is reflected in the spectrum, which is the simple doublet described previously. However, ytterbium is regarded as trivalent in the oxide so that one of the 4f electrons must be promoted to a bonding orbital, leaving the 4f shell in a ^2F state which can then couple to the ^2D state of the photo-ionisation hole as in the other elements considered. The measured binding energies of the 4d peaks are respectively 186.0, 190.3, 193.7, 200.2 and 206.6±0.2 eV. Early results reported by Hagstrom et al. using Al $K\alpha$ radiation agree reasonably well with these measurements although the resolution and counting statistics of the present work are superior.

Further evidence for anomalous behaviour in ytterbium is given in table 1, which lists the shifts of the 5p, 5s, 4p and 4s core levels of ytterbium and thulium with surface oxidation. The values for thulium are typical of the other heavy lanthanides, while those for ytterbium are significantly greater, which is consistent with the large change in screening parameter accompanying the promotion of a $4f$ electron to a $5d$ orbital. Additionally, the anomalous growth of the oxygen $1s$ emission from the surface of ytterbium suggests that the mechanism of oxygen absorption is radically different from that for the other heavy lanthanides.

TABLE 1.—XPS METAL AND METAL OXIDE CORE LEVELS FOR YTTERBIUM AND THULIUM

	energy level	metal	oxide
ytterbium	$O_3\ 5p_{3/2}$	24.5 ± 0.2	28.0 ± 0.2
	$O_2\ 5p_{1/2}$	30.5 ± 0.2	34.4 ± 0.2
	$O_1\ 5s$	52.7 ± 0.4	56.6 ± 0.4
	$N_3\ 4p_{3/2}$	339.9 ± 0.2	346.5 ± 0.2
	$N_2\ 4p_{1/2}$	388.9 ± 0.4	398.4 ± 0.4
	$N_1\ 4s$	480.9 ± 0.2	488.3 ± 0.4
thulium	$O_3\ 5p_{3/2}$	25.0 ± 0.2	26.9 ± 0.2
	$O_2\ 5p_{1/2}$	31.8 ± 0.2	32.2 ± 0.2
	$O_1\ 5s$	55.7 ± 0.4	54.4 ± 0.2
	$N_3\ 4p_{3/2}$	332.6 ± 0.2	333.5 ± 0.2
	$N_1\ 4s$	470.9 ± 0.4	470.5 ± 0.4

The authors are indebted to the Science Research Council for funds in support of this research, to J. K. Gimzewski and S. Affrossman for their help with the measurements (fig. 8) on rate of oxidation, and to P. Perkins for valuable discussion.

[1] Y. Baer and G. Busch, *J. Electron Spectr.*, 1974, **5**, 611.
[2] S. P. Kowalczyk, N. Edelstein, F. R. McFeely, L. Ley and D. A. Shirley, *Chem. Phys. Letters*, 1974, **29**, 491.
[3] R. L. Cohen, G. K. Wertheim, A. Rosencwaig and H. J. Guggenheim, *Phys. Rev. B*, 1972, **5**, 1037.
[4] A. J. Signorelli and R. G. Hayes, *Phys. Rev. B*, 1973, **8**, 81.
[5] J. C. Fuggle, A. F. Burr, L. M. Watson, D. J. Fabian and W. C. Lang, *J. Phys. F: Metal Phys.*, 1974, **4**, 335.
[6] W. C. Lang, B. D. Padalia, D. J. Fabian and L. M. Watson, *J. Electron Spectr.*, 1974, **5**, 207.
[7] S. B. M. Hagström, P. O. Hedén and H. Löfgren, *Solid State Comm.*, 1970, **8**, 1245.

GENERAL DISCUSSION

Mr. K. Y. Yu (Stanford) *said*: I would like to comment on relaxation or final state effects in the measurement of adsorbate energy levels by UV photoemission ($hv = 21.2$ eV). The adsorption system we have chosen for study is a particularly simple one. It involves physically adsorbing or condensing simple molecules on the cleavage plane of a MoS_2 single crystal, which we found from previous studies to be chemically inert. We have monitored the condensation of seven different gases (H_2O, H_2CO, C_6H_6, C_5H_5N, CH_3OH, C_2H_5OH, NH_3) on MoS_2 ($T = 79$ K).[1] By subtracting the substrate contribution to the spectral emission, it is possible to compare the condensed phase spectrum with the gas phase spectrum (fig. 1). We

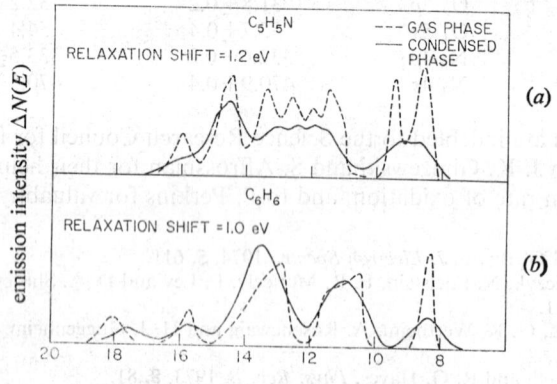

FIG. 1.—Photoemission data at $hv = 21.2$ eV (HeI) comparing the condensed phase spectra of C_6H_6, and C_5H_5N with their respective gas phase spectra. The substrate contribution to the emission (for the condensed phase) has been subtracted, i.e., the condensed phase spectra are in the form of difference curves.

found that there is very good match between the gas phase spectrum and the condensed phase spectrum. The only difference between the two spectra is a uniform reduction in the valence orbitals ionization energies of the condensed molecule (indicated by " relaxation shift " in the figure). This reduction in ionization energy can be understood in terms of a final state screening effect, i.e., the hole produced during photoexcitation polarizes the electron clouds of the surrounding molecules to gain coulombic energy. What is most surprising is that the shift is so orbital independent; the particular orbital can be a π orbital, a σ orbital or a lone pair orbital, and it always suffers the same relaxation shift (within experimental accuracy; 0.3-0.4 eV). Perhaps this is only true for the delocalized valence orbitals, which are the orbitals accessible to HeI radiation. More work at higher photon energy would reveal whether the more localized core levels suffer the same shift as the valence levels.

This work, together with studies on similar adsorption systems, forms the basis for the interpretation of chemisorbed gases spectra. In a chemisorption situation, when one usually observed the relative movement of one group of orbitals relative

[1] K. Y. Yu, J. C. McMenamin and W. E. Spicer, *Surface Sci.*, 1975, **50**, 149.

to the rest (referenced to the gas phase spectra), one can deduce that most of the bonding interaction occurs between that group of orbitals with the substrate.

Dr. I. H. Hillier (*University of Manchester*) said: Kendrick and I have used configuration interaction calculations, similar to those described by Shirley to interpret the satellite spectra of a number of molecules. For N_2, where nine such lines have been observed,[1] the first two peaks correspond to valence-like $\pi \rightarrow \pi^*$ transitions, whilst the remaining peaks correspond to states mainly Rydberg in character, and are predominantly $\sigma \rightarrow \sigma$ in nature.

Dr. C. R. Brundle (*IBM*) (*communicated*): We have also found that the relaxation shifts of condensed molecules are orbital independent, and to answer Yu's question, the same shifts are also found in XPS for the core-levels. These facts and the observed relative movement of IP's in chemisorbed (as opposed to condensed) situations form the basis of much of the analysis in our Paper at this meeting.

Dr. T. A. Carlson (*Oak Ridge*) said: In addition to those examples given in the paper by Kowalczyk et al., two other important cases in which excited states arise in photoionization but are not the result of a simple, single-electron monopole excitation also ought to be mentioned. In one, the exchange or indistinguishability of electrons may alter the roles normally played by the excited and the photo-ejected electron. For example, in the photoionization of He, rather than the normal process of having the photoejected electron go into the continuum with $l = 1$ while the second electron is excited into an s level, it is possible for the photoejected electron to have an angular momentum $l = 0$ while the excited electron is found in a p state. This possibility has been discussed [2] and found to be important, particularly near the ionization threshold. In addition, a satellite peak found in the photoionization of neon [3] that is the result of the final state configuration $1s^1\ 2s^2\ 2p^5\ 3s^1$ can be explained by this mechanism. In the second example, a two electron excitation occurs in the photoionization of the valence shell of argon $3s^2\ 3p^6 \xrightarrow{e^-} [3s^1\ 3p^6] \rightarrow 3s^2\ 3p^4\ 3d^1$. This is found experimentally [4] to be more important than the one electron shake-up process, $3s^2\ 3p^6 \xrightarrow{e^-} 3s^1\ 3p^5\ 4p^1$, because in the former case all three orbitals $3s$, $3p$ and $3d$ have the same principal quantum number, with the advantage of large overlap.

Dr. R. L. Martin (*Berkeley*) said: Carlson pointed out the interesting role which configuration interaction in the initial state plays in determining the intensities of "non-monopole" shake-up satellites (see ref. (4) and (5)). This initial-state "CI" also has important consequences for the intensity of the more conventional satellites. For example, using the final-state wavefunctions discussed in the paper and a well correlated initial-state wavefunction, the intensities for the $(2p \rightarrow 3p)$ lower and $(2p \rightarrow 3p)$ upper peaks are computed to be 2.5% and 2.6% of the main peak, respectively. The increase in intensity relative to a single-determinantal initial-state occurs primarily from the contribution to the overlap from an admixture of the configuration $1s^2\ 2s^2\ 2p^5\ 3p$ into the ground state. Similar increases are observed for the other satellites in the neon spectrum. These examples serve to underline the point that a

[1] U. Gelius, *J. Electron Spectr.*, 1974, **5**, 985.
[2] M. O. Krause and F. Wuilleumier, *J. Phys. B*, 1972, **5**, L143.
[3] U. Gelius, *J. Electron Spectr.*, 1974, **5**, 985.
[4] D. P. Spears, H. J. Fischbeck and T. A. Carlson, *Phys. Rev. A*, 1974, **9**, 1603.

satellite will be observed if there is a component in the ground state which connects it to the final state via the transition operator.

Prof. D. A. Shirley (*Berkeley*) said: Three questions have been raised to which a response may be useful. First, we do not mean to imply from eqn (3.1) and (3.3) that ε_i and $\langle i|V_p^a|i\rangle$ have identical numerical values in free atoms and metals. Thus not all the difference between core-level binding energies in solids can be attributed to extra-atomic relaxation. We believe, however, that most of the difference does in fact arise from this effect. Particularly in the transition metals, where the difference is 10 eV or more, the extra-atomic relaxation effect seems very well established. Unfortunately calculations on metals tend not to be even self-consistent, let alone of Hartree-Fock quality; thus we cannot make accurate estimates of $\varepsilon(\text{atom}) - \varepsilon(\text{metal})$ directly. Indirect estimates based on ε values in diatomic molecules or on cohesive energies suggest that initial-state shifts between atoms and metals are unlikely to exceed ~ 1 eV in most cases.

We raised a small point about the relative magnitude of added relaxation energies between the more- and less-tightly-bound molecular orbitals of adsorbed molecules accompanying adsorption. The benzene spectra of Demuth and Eastman (*Phys. Rev. Letters*, 1974, **32**, 1123) show most clearly the effect of additional relaxation energy in the ~ 20 eV σ orbital peak on chemisorption over that in the other peaks. The same effect is present in other published spectra, although it usually escapes comment. It is of course expected because the inner σ orbital is most core-like.

The subdivision of electron correlation mechanisms into various types is largely a matter of convenience. Bagus has suggested that relaxation accompanying photoemission be divided into a part that can be accommodated within a single determinant (but of relaxed basis functions) and a part that requires additional configurations. This scheme appears to have heuristic value, although it has no observational implications.

Dr. C. R. Brundle (*IBM*) said: Fadley has shown how going to low electron ejection angles increases the surface to bulk sensitivity and has described the additional effects that instrumental parameters and surface roughness can have.

Barrie and I have been examining [1] the effects of changing the ejection angle for the CO/Mo (ribbon) adsorption system in an attempt to establish the magnitude of any chemical shift in the Mo $3d$ level associated with the Mo layer involved in bonding to the adsorbate. Earlier studies [2] at a fixed angle of 45° showed no observable shoulder on the Mo $3d$ line to associate with a chemically shifted surface component. The present studies do indicate, at low angle, a chemically shifted component and possibly a very small shift (*ca.* 0-1 eV) in the peak maximum. Similar results at low angles have been observed in other systems by other authors.[3, 4] A problem arises with the present work however, in that the size of the shifted component at low angle seems incompatible with the non-observance of a shifted component at 45°, if a simple flat surface angle formula is used. Can Fadley think of any physical characteristic of the surface, or instrumental parameter which might explain the apparent exceptional change in surface sensitivity with angle which is observed in this case?

[1] A. Barrie and C. R. Brundle, to be published.
[2] C. R. Brundle and A. F. Carley, *Chem. Phys. Letters*, 1975, **33**, 41.
[3] J. C. Fuggle and D. Menzel, *Chem. Phys. Letters*, 1975, **33**, 46.
[4] A. Barrie and A. Bradshaw, to be published.

Prof. C. S. Fadley (*Hawaii*) said: Brundle has asked if there are any mechanisms which might explain the apparent observation of exceptional surface sensitivity for the CO/Mo-polycrystal system by him and Barrie.

It would help a great deal in answering this question if a more quantitative description of the attempt to fit the data to the flat-surface model were given. How was the surface structure described in terms of the macroscopic model? What value was assumed or derived for the surface layer thickness-to-attenuation length ratio? In any case the starting relationship should be the equivalent of eqn (18) of reference 1 below.

In the absence of further details, it is only possible to state three potentially important effects, all of which lead to *decreased* surface sensitivity at low angle relative to the flat-surface model;
(1) Surface roughness,[1] which causes higher average electron exit angles,
(2) Patching of the surface layer,[1,2] which is predicted to cause lower surface enhancement according to a very simple model,
(3) Electron refraction at the surface,[1] which would be expected to give rise to higher propagation angles inside the surface and thus lower surface sensitivity.

Dr. D. P. Woodruff (*Warwick*) said: Fadley has presented a very clear picture of angular dependence effects in emission from polycrystalline and amorphous surfaces. It is interesting to add, however, that at single crystal surfaces other effects can be extremely important and although these may mask the effects described by Fadley, they may have considerable value in their own right.

One process producing angular dependence of electron emission from atom sites within the crystal surface is the coherent interference of electrons elastically scattered from the surrounding ion cores of the solid. This " diffraction " of the outgoing electrons is also of importance in Auger electron emission so we have performed some calculations to assess the importance of this effect.[3,4] The calculations neglect any angular dependence in the emission prior to diffraction and simply assume an outgoing spherical wave (i.e., initial state effects are neglected). The scattering properties of the surrounding ion cores are properly described using the appropriate phase shifts, but only single scattering is included. The featural similarity of our calculated and experimental Auger emission profiles shows the model is useful but clearly not exact.[3] One particularly interesting aspect of this effect is in the angular dependence of emission from adsorbate species on a surface. In this case the angular dependence is very sensitive to the adsorbate site location on the surface. The figure shows some calculations of intensity versus polar emission angle in the ⟨100⟩ azimuth for 145 eV electrons emitted from an adsorbate atom located in three different sites on a Ni(100) surface. ((*a*), (*b*) and (*c*) correspond to 1-fold, 4-fold and 4-fold adsorption sites at 2.2 Å, 1.7 Å and 1.3 Å respectively above the top Ni surface layer) and shows this sensitivity to site location.[4] While this calculation is intended to relate to a strong Auger electron emission from sulphur it is equally relevant to XPS from any adsorbate where the kinetic energy of the photoelectron is 145 eV outside the crystal. Although in most XPS experiments kinetic energies are higher than this, the effect should be preserved. However, for adsorbate emission, calculations indicate that at much

---

[1] C. S. Fadley, R. J. Baird, W. Siekhaus, T. Novakov and S. Å. L. Bergström, *J. Electron Spectr.*, 1974, **4**, 93.
[2] J. Brunner and H. Zogg, *J. Electron Spectr.*, 1974, **5**, 811.
[3] L. McDonnell, D. P. Woodruff and B. W. Holland, *Surface Sci.*, 1975, **51**, 249.
[4] D. P. Woodruff, Proceedings of *Interdisciplinary Surface Science*, Warwick, England, March 1975; to be published in *Surface Sci.*

higher energies ($\gtrsim 500$ eV) temperature effects will largely remove the observed angular dependence at room temperature. The same arguments do not apply to emission from substrate species where the angular dependence contains much structure from forward scattering with small associated Debye-Waller factors. In this case marked angular dependence is to be expected at room temperature throughout the

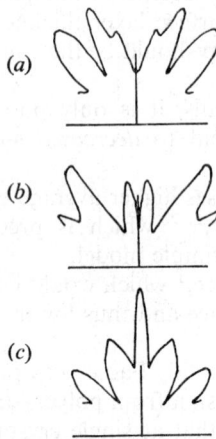

usual XPS kinetic energy range; indeed such effects have been observed, by Fadley [1] amongst others. In the case of substrate species emission it is not clear that this will give useful structural information, although it will certainly interfere with depth analysis of the kind described by Fadley, and will also affect the quantification of XPS as a surface composition analysis technique where small collection angle analysers are used at fixed orientation to the crystal surface.

Prof. C. S. Fadley (*Hawaii*) said: Woodruff has suggested that, in XPS studies on single crystals, diffraction or channelling effects may significantly alter the overall form of photoelectron angular distributions such that qualitative or quantitative depth profiling of atomic species may have to be done within a different theoretical framework.

A definite answer to this comment is not possible at the present time, and other experimental work on very well-defined single crystal surfaces is needed. However, our earlier work on gold single crystals with disordered carbon-containing overlayers,[2] as well as more recent gold studies with thinner carbon overlayers ($C1s/Au4f$ as low as 0.04),[3] appear to indicate that, for the high-energy photoelectrons in XPS, single-crystal effects cause only relatively small-amplitude deviations of intensity (up to $\sim 30\%$) away from angular distribution curves whose shapes are at least qualitatively predicted by the polycrystalline models discussed in this paper and elsewhere.[2,4] The presence of any kind of amorphous surface layer (for example, oxide on metal) I would also expect to generate diffuse scattering that should tend to make observed angular distributions follow more closely a polycrystalline model.

Thus, based on the presently available data, we have tentatively suggested [2] that

[1] C. S. Fadley and S. Å. L. Bergström, *Phys. Letters*, 1971, **35A**, 375.
[2] C. S. Fadley, R. J. Baird, W. Siekhaus, T. Novakov and S. Å. L. Bergström, *J. Electron Spectr.*, 1974, **4**, 93, and earlier references therein.
[3] R. J. Baird, C. S. Fadley and L. F. Wagner, to be published.
[4] W. A. Fraser, J. V. Florio, W. N. Delgass and W. D. Robertson, *Surface Sci.*, 1973, **36**, 661.

it may be possible to average over single-crystal structure and obtain angular distributions which can be analyzed in terms of simpler macroscopic models to yield information such as depth profiles.

Dr. R. W. Joyner (*University of Bradford*) said: Can Fadley comment on what he would expect to be the effect of progressively increasing surface roughness on the absolute intensities of counts observed in XPS? This point is of interest since it is observed experimentally in both XPS and Auger electron spectroscopy that very rough samples show unexpectedly low total count rates.

Prof. C. S. Fadley (*Hawaii*) said: The effects of roughness on absolute intensities in XPS have been considered previously,[1] and they can be divided into two categories:

(*a*) *Shading of portions of the surface from X-ray incidence.* This will occur only for roughness contours with average dimensions greater than or approximately equal to the X-ray attenuation length, and will act to reduce total intensities *provided* that the areas shaded are also visible to the electron analyzer from the point of view of electron escape. Ebel, Ebel and Hillbrand[2] have also performed an experimental study which appears to indicate some effects due to X-ray shading.

(*b*) *Shading of portions of the surface for electron escape into the analyzer.* If the effects of X-ray shading are negligible (as, for example, for a surface with roughness contours small with respect to the X-ray attenuation length), then it is a straightforward matter to show that roughness will have *no* effect on the total intensities observed from a homogeneous specimen with an atomically clean surface.[1] A note of caution in interpreting this result is worthwhile, however, because, as judged relative to a flat surface, surface roughness will generally tend to decrease the average value of the true electron exit angles for high θ values ($\theta \simeq 90°$), and conversely will act to increase the average electron exit angle for low θ values ($\theta \approx 0°$). Thus, the degree of surface enhancement possible with a rough surface may be greater or lesser than that for an equivalent flat surface, as is shown in the grating data presented in this paper and elsewhere.[3] Therefore, although total intensities may be equal between clean rough- and flat-surfaces, the surface-atom contributions to these intensities may be quite different in the two cases. Extending this argument to specimens consisting of homogeneous substrates with covering overlayers means that surface-atom- or substrate-atom-intensities may be higher or lower for certain orientations of a rough surface as compared to equivalent orientations or a flat surface. A more quantitative discussion of such effects for model periodic surfaces is given in ref. (1).

Although a general formalism for dealing quantitatively with the simultaneous effects of X-ray- and electron-shading has been presented previously,[1] no numerical results have been obtained to date.

Prof. C. S. Fadley (*Hawaii*) said: Landman has asked if there might not be differential cross section variations with electron emission angle that would tend to make less useful the simple definition of the instrument response function $R(\theta)$ suggested in this paper.

For polycrystalline specimens and an experimental geometry with the angle between X-ray incidence and electron exit fixed (as we have assumed) such effects

---

[1] R. J. Baird, C. S. Fadley, W. Siekhaus, T. Novakov and S. Å. L. Bergström, *J. Electron Spectr.*, 1974, **4**, 93.
[2] H. Ebel, M. F. Ebel and E. Hillbrand, *J. Electron Spectr.*, 1973, **2**, 277.
[3] R. J. Baird, C. S. Fadley, S. Kawamoto and M. Mehta, *Chem. Phys. Letters*, 1975, **34**, 49.

should be negligible. The only possible problem in this case would be for an electron spectrometer with extended solid angle, as has been mentioned in the discussion following eqn (3).

For single-crystal specimens, diffraction or channelling effects may cause $d\sigma_k/d\Omega$ for either core or valence levels to exhibit a great deal of fine structure that will in general rotate with the crystal, even if the X-ray/electron angle is held fixed.[1-4] In this case, peak intensities would be proportional to a more complex quantity

$$R(\theta)_k = \sin\theta \iint_A I(\theta, x, y)[d\sigma_k/d\Omega]\Omega(\theta, x, y)\,dx\,dy,$$

a response function characteristic of a given level k, and lacking the simplicity of dependence only on instrumental parameters. Only if the single-crystal fine structure in $R(\theta)_k$ were averaged over in some way would it be possible to carry out an analysis in terms of the simpler $R(\theta)$ defined in eqn (1) of this paper. The validity of such averaging has been discussed in my reply to Woodruff's comment, and so far it seems to be at least qualitatively correct.

Dr. I. H. Hillier (*University of Manchester*) said: Darko, Kendrick and I have performed configuration interaction calculations to interpret the Mn 1s satellite observed by Carlson in MnF_2. Such a calculation on the $^7A_{1g}$ Mn 1s hole state of the cluster MnF_6^{4-} yields a weak satellite, 10 eV from the main ionization peak, which corresponds to $F(t_{2g}, e_g) \to Mn(t_{2g}, e_g)$ excitations.

[1] K. Siegbahn, U. Gelius, H. Siegbahn and E. Olsen, *Phys. Letters*, 1970, **32A**, 221.
[2] C. S. Fadley and S. Å. L. Bergström, *Phys. Letters*, 1971, **35A**, 375.
[3] C. S. Fadley and S. Å. L. Bergström, in *Electron Spectroscopy*, ed. D. A. Shirley, (North Holland, Amsterdam, 1972) p. 233.
[4] R. J. Baird, C. S. Fadley and L. F. Wagner, to be published.

XPS and UPS Studies of the Adsorption of Small Molecules on Polycrystalline Ni Films

By C. R. Brundle* and A. F. Carley

School of Chemistry, University of Bradford

Received 14th April, 1975

The interactions of O_2, H_2O, CO, CO_2, NO, N_2O, NO_2, H_2S and SO_2 with evaporated nickel films have been studied over temperatures of 77 to 400 K and adsorbate pressures of 10^{-8} Torr to several Torr. The reactions range from reversible adsorption at 77 K only (CO_2) to complex molecular and dissociative adsorptions leading to surface phases of a few monolayers thickness (O_2, NO, NO_2). With the aid of gas-phase data and the spectra of some of the molecules adsorbed on gold at 77 K it is possible to separate non-dissociative and dissociative adsorptions. Proposals concerning the products of dissociation can be made, though for NO, NO_2 and SO_2 the reactions are very complex.

In a previous study [1] on the polycrystalline Ni/O_2 system three distinct adsorption species were identified from the XPS $O(1s)$ spectrum, their relative intensities depending on oxygen exposure and substrate temperature. The spectra were interpreted in terms of the formation of oxygen atoms in an oxide-like electronic environment (the lowest B.E. peak at 529.5 eV) with the features at higher B.E. (531.4, 533.2 eV) representing chemisorbed oxygen (and possibly defect "Ni_2O_3") bonded in an electronically different fashion, probably involving high d-electron contributions from the Ni. The present work surveys the adsorption of a series of oxygen-containing small molecules, plus H_2S and N_2, using both XPS core-level and HeI and HeII valence level spectra. The Ni/O_2 results will not be discussed again here. As was found previously for the Mo/O_2, CO, H_2O, CO_2 systems [2] it is useful to have available XPS and UPS data for the adsorbate molecules in the gas phase, and in their weakest surface interaction situations (physical adsorption), to provide reference spectra for the stronger interactions in terms of chemical shifts, relative peak intensities and valence level "orbital" assignment. Therefore many of the above molecules were studied, condensed onto an Au substrate at 77 K, and interaction with Ni at 77 K was always followed.

It should be stressed that this large collection of data is intended as a survey only since each individual system deserves a study in greater detail using single crystal surfaces and LEED characterization.

EXPERIMENTAL

All spectra were recorded on the Vacuum Generators prototype ESCA III spectrometer.[9] Base pressures varied from 5×10^{-10} to 2×10^{-9} Torr with the X-ray source (Al $K\alpha$ at 1486.6 eV) in operation at 10 kV, 50 mA. Nickel films were evaoprated from 0.5 mm wire (Johnson-Matthey, 99–999%) in the preparation chamber. Adsorbates were admitted to the sample in the analysis position from a separate UHV line using appropriate cold traps. Pressures were measured (nude ion-gauge) at the inlet position and previous calibration [4] indicates that the true pressure near the sample is a factor of about 5 lower. Purity was

* present address: I.B.M. Research Laboratory, San José, California 95193, U.S.A.

checked with a Vacuum Generators Q7 quadrupole. All core-level B.E. are referenced to Au at 83.7 eV,[3] and all HeI and HeII spectra to E_F. Normal procedure was to cool a clean sample (Ni($2p$) : O($1s$) >130 : 1, Ni($2p$) : C($1s$) >260 : 1 by peak height. Most initial films had less than half this amount of contamination) to 77 K by circulation of liquid N_2 through the probe, let in the adsorbate at pressures between 10^{-8} and 10^{-5} Torr either to saturation, or several monolayers coverage when the pressure exceeded the saturation vapour pressure, while taking XPS and UPS spectra sequentially without moving the sample. The chamber was then evacuated, the sample warmed to 300 K, and the spectra re-taken. More adsorbate was then admitted at 300 K, sometimes to pressures of many Torr. In many cases the system was also studied with the entire reactions taking place at 300 K, sometimes followed by cooling to 77 K for subsequent further adsorption, or heating to 400 K in the presence of the gas-phase adsorbate. Unfortunately, though the initial films were clean, a small amount of CO was always adsorbed from the ambient during the cooling period. Except

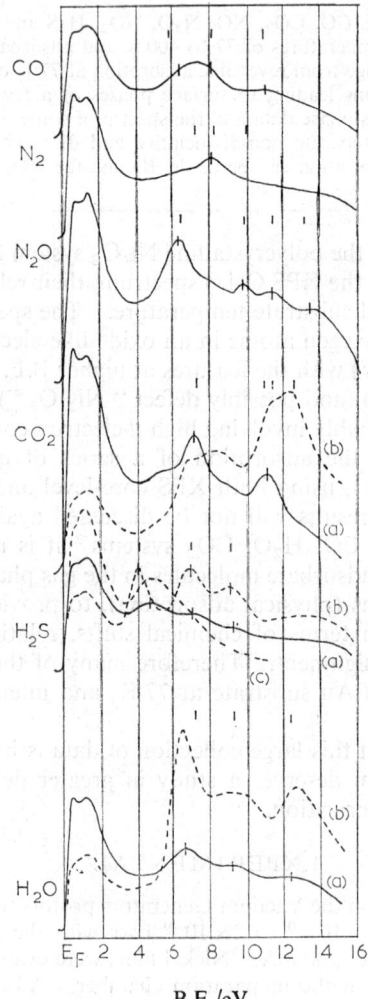

FIG. 1.—HeI spectra of CO, N_2, N_2O, CO_2 and H_2O adsorbed on Ni at 77 K. CO_2, trace (*a*) below s.v.p.; trace (*b*) multilayer condensation. H_2S, trace (*a*) initial adsorption; trace (*b*) multilayer adsorption; trace (*c*) (*b*) warmed to 300 K. H_2O, trace (*a*) initial adsorption; trace (*b*) multilayer adsorption.

for the case of N_2O where only one run has been made on a badly contaminated film (ca. 25% coverage of CO present), the CO coverage at the commencement of adsorption was always less than 10%.

RESULTS

Fig. 1-4 show the HeI (21·2 eV) and HeII (40.8 eV) spectra for all the systems concerned. Representative spectra of the original clean Ni surfaces are given in fig. 3(a) and 4(a). Page and Williams [5] have shown that such spectra are consistent with a weighted sum of (100), (110), and (111) crystal faces. The broad weak hump

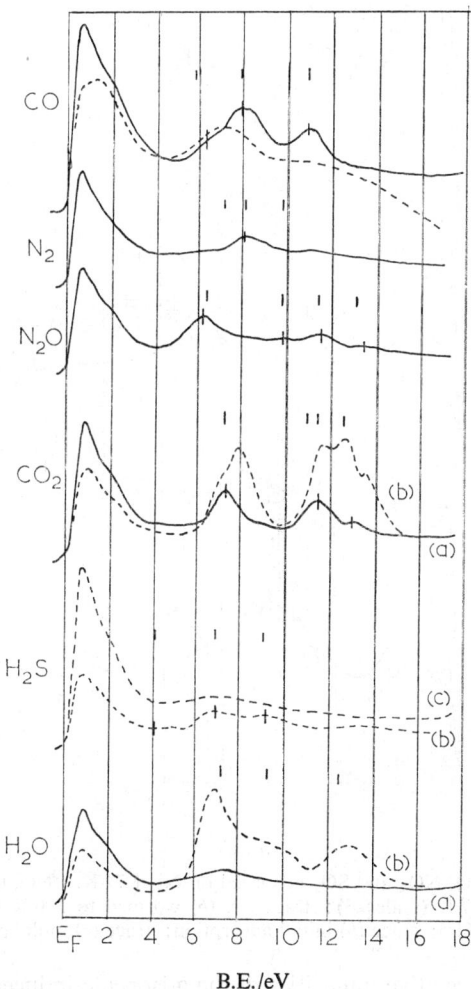

FIG. 2.—HeII spectra of CO, N_2, N_2O, CO_2 and H_2O adsorbed on Ni at 77 K. CO, dashed curve shows HeI spectrum for comparison. CO_2, H_2O, H_2S, labelled traces as for fig. 1.

centred at about 6.5 eV in the HeI and HeII spectra in considered to be a genuine part of the Ni spectrum. Its possible origin has been discussed before and will not concern us here. The presence of carbon or oxygen contamination on the evaporated film increases the spectral intensity in the 4–7 eV region, as well as increasing the intensity of the Ni *d*-band feature at 1.2 eV relative to that at 0.4 eV in the HeI spectrum. These changes are at least as sensitive a monitor of the cleanliness of the initial film as the XPS core-levels. The relative intensities of the spectra shown in the figures have no significance, depending on the conditions of operation of the

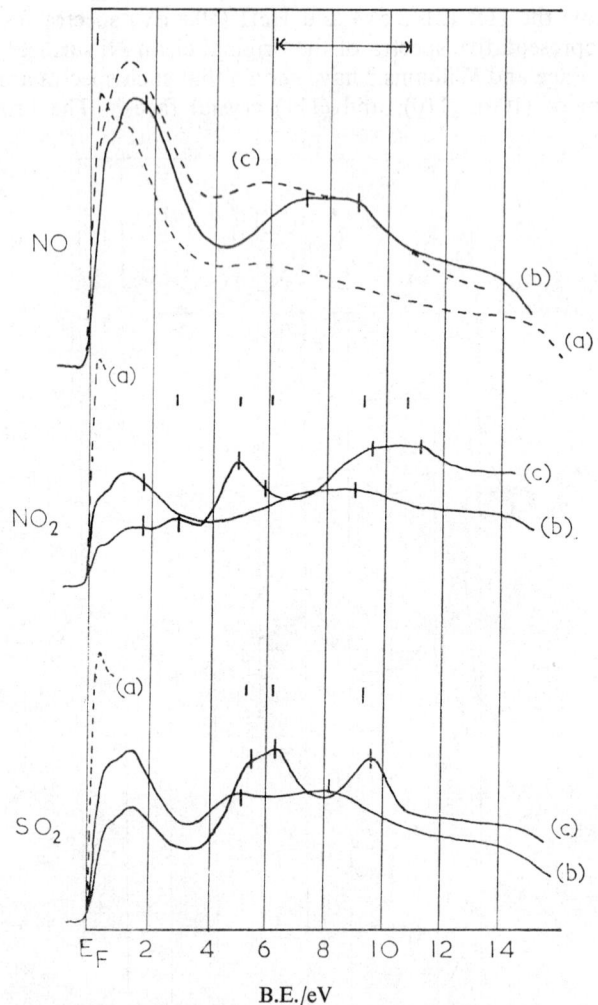

Fig. 3.—HeI spectra of NO, NO_2 and SO_2 adsorbed on Ni at 77 K. NO, trace (*a*), clean Ni film; trace (*b*), adsorption at 77 K (scale × 3); trace (*c*), (*b*) wormed to 300 K (scale × 3). NO_2, SO_2 trace (*a*) clean Ni film; trace (*b*) initial adsorption; trace (*c*) multilayer adsorption.

discharge lamp. The relative intensities of the adsorbate induced features to the Ni *d*-band features do have significance, reflecting both the extent of interaction (as does also the attenuation of the 0.4 eV Ni *d*-band feature, examples of which are shown in fig. 3 and 4 for NO, NO_2 and SO_2) and the relative ionization cross-sections of the levels involved. Likewise changes in the relative intensities of the features in a given spectrum on going from HeI to HeII reflect differences in escape depths and changes in ionization cross-section (see, e.g., the H_2S spectra, fig. 1 and 2).

For CO, N_2 and NO the spectra shown represent maximum coverage at about

5×10^{-6} Torr pressure (below the saturation vapour pressure at 77 K for these molecules). All the other molecules concerned may be condensed as multilayers at 77 K, and spectra are shown both in the early stages of adsorption and in the multilayer regime (except for N_2O where data was only taken at one coverage).

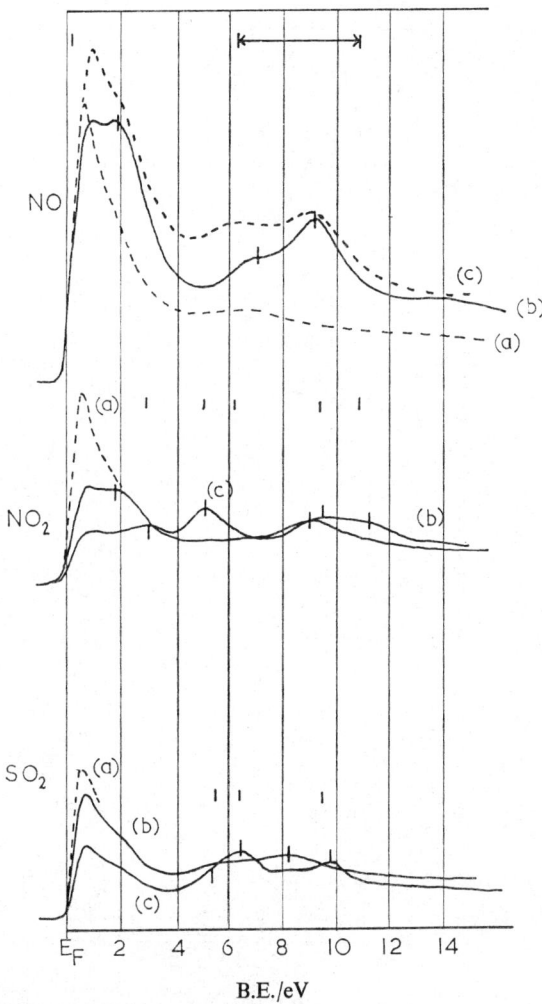

FIG. 4.—HeII spectra of NO, NO_2 and SO_2 adsorbed on Ni at 77 K; data as for fig. 3.

The centres of the adsorbate-induced features are marked in the spectra. The relative positions of the orbitals of the gas-phase adsorbate molecules are also marked. The gaseous values are lined up with the adsorbed state spectra simply by sliding the scales until a best fit is obtained with all or some of the orbitals (see discussion). UPS binding energies for the adsorbed and gas phase situations are listed in table 1.

Fig. 5–9 show the XPS O(1s), C(1s), N(1s) and S(2p) regions for the adsorbate molecules for many of the same reaction situations as recorded in fig. 1–4. The data has been collected over a period of about 18 months and though signal strengths are much more reproducible than for UPS, there are fluctuations. The relative intensities of the core-levels from system to system may therefore be only approximately judged from the spectra shown (some of the spectra are taken at an analyzing energy of 50 eV and some at 100 eV, depending on signal strength. The Au $4f_{1\frac{4}{2}}$ peak has a half-width of 1.6 eV at 50 eV, 2.0 eV at 100 eV and the intensity increases by a factor of 2.7 by height and 3.7 by area on changing from 50 eV to 100 eV). A representative clean Ni $2p_{\frac{3}{2}}$ spectrum is shown in fig. 10.

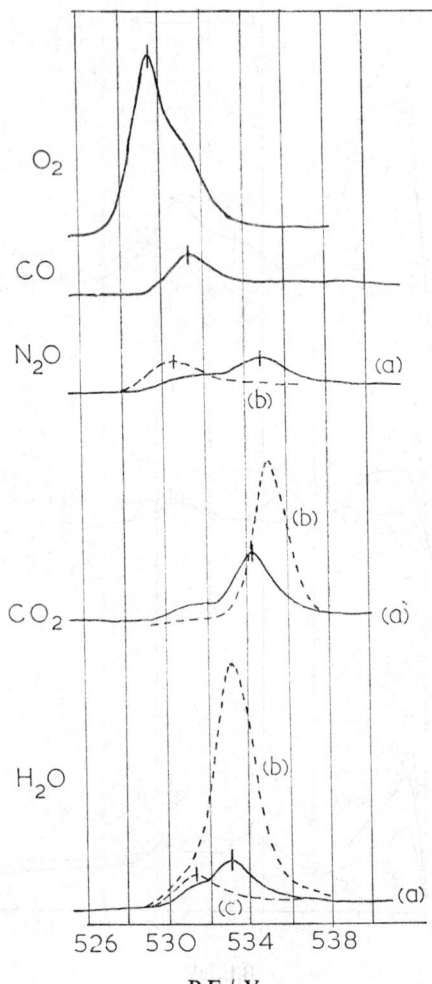

Fig. 5.—XPS O(1s) spectra for the adsorption of O_2, CO, N_2O, CO_2 and H_2O on Ni (All data at 50 eV analyzing energy).[2] O_2, adsorption at 300 K; CO, adsorption at 77 K; N_2O, trace (a) adsorption at 77 K; trace (b) warmed to 300 K; CO_2, trace (a) initial adsorption at 77 K; (b) multilayer adsorption at 77 K (scale÷3); H_2O trace (a) initial adsorption at 77 K; (b) multilayer adsorption at 77 K; (c) warmed to 300 K, plus further H_2O at 300 K.

The position of the Ni $2p_{\frac{3}{2}}$ peak does not change for most of the adsorption situations discussed here, but merely suffers a decrease in intensity and a slight broadening (fig. 10). Only in the later stages of interaction with NO and NO_2 (well beyond the monolayer stage) does one get clearly observed chemical shifts of the substrate peak (fig. 10). This lack of effect on the substrate levels at low coverage is well-documented for a number of systems,[7,8] though if one goes to grazing electron

TABLE 1.—UPS VALENCE LEVEL ENERGIES (eV) FOR FREE MOLECULE [a] AND ADSORBED STATES (77 K)

CO	assignment		5σ	1π	4σ	
	free mol.		14.0	16.8	19.7	
	Ni$_{ads}$.	HeI	6.3	8.1	—	
		HeII	6.3	8.1	11.1	
N$_2$	assignment		5σ	1π	4σ	
	free mol.		15.6	16.7	18.8	
	Ni$_{ads}$.	HeI		8.0	11.8?	
		HeII		8.0	11.1?	
N$_2$O	assignment		2π	4σ	1π	3σ
	free mol.		12.9	16.4	18.2	20.1
	Ni$_{cond}$.	HeI	6.3	9.7	11.3	13.5
		HeII	6.3	9.8	11.5	13.3
	Au$_{cond}$.	HeI	—	9.5	11.3	13.2
		HeII	6.1	9.5	11.3	13.3
CO$_2$	assignment		$1\pi_g$	$1\pi_u$	$2\sigma_u$	$2\sigma_g$
	free mol.		13.7	17.6	18.0	19.4
	Ni$_{ads}$.	HeI	7.2		11.2	?
		HeII	7.4		11.4	12.9
	Ni$_{cond}$.	HeI	all ca. 1–1.5 eV higher than for Ni$_{ads}$			
		HeII				
	Au$_{ads}$.[c]	HeI	6.9	9.6	11.0	13.1
		HeII	6.8	9.3	11.0	12.5
	Au$_{cond}$.[c]	HeI	8.3	11.6	12.3	13.6
		HeII	8.2	11.4	12.4	13.7
H$_2$O	assignment		$1b_1$	$3a_1$	$1b_2$	
	free mol.		12.6	14.7	18.3	
	Ni$_{cond}$.	HeI	6.5	9.2	13.0	
		HeII	6.5	9.2	13.0	
	Au$_{cond}$.[c]	HeII, HeI	6.3	10.2	12.6	
H$_2$S	assignment		$1b_1$	$3a_1$	$1b_2$	
	free mol.		10.5	13.3	15.4	
	Ni$_{ads}$.	HeI	4.7	7.0	9.2	
		HeII	—	—	—	
	Ni$_{cond}$.	HeI	4.0	7.0	9.0	
		HeII	—	6.9	9.1	
	Au$_{cond}$.	HeI	4.2	7.4	9.4	
		HeII	—	7.2	9.3	
NO	assignment		$\pi_g 2p$	$\pi_u 2p, \sigma_g 2p$		
	free mol.		9.5	15.5–20		
	Ni$_{ads}$.	HeI	1.9	7.3, 9.1		
		HeII	1.9	7.1, 9.2		
NO$_2$	assignment		$4a_1$	$1a_2\ 3b_2$	$1b_1\ 3a_1$	$2b_2$
	free mol.[b]		11.2	13.0, 13.6, (t); 14.1, 14.5(s)	17.6	19.0
	Ni$_{cond}$.	HeI	3.1	←5.2→	←6.3→ 9.7	11.3
		HeII	3.1	←5.3→	? 9.5	11.3
	Au$_{cond}$.	HeI, HeII	very similar to Ni$_{cond}$.			
SO$_2$	assignment		$4a_1$	$1a_2\ 3b_2$	$1b_1\ 3a_1$	
	free mol.		12.5	13.4	16.5	
	Ni$_{cond}$.	HeI	5.4	6.4	9.8	
		HeII	5.4	6.5	9.9	
	Au$_{cond}$.	HeI, HeII	5.6	6.7	10.1	

a ref. (6), *b* ref. (26), *c* Au$_{ads}$ data from ref. (2), *t* triplet states, *s* singlet states.

ejection angles it has now become clear that very small shifts (generally ≤ 0.3 eV) can be observed.[9, 10] The measured binding energies of the adsorbate core-levels, calibrated against Ni $2p_{\frac{3}{2}}$ at 852.5 eV,[8] are collected in table 2. The separation between well-resolved features in any one spectrum is considered accurate to ± 0.1 eV. The absolute positions are likely to be less accurate giving a realistic error limit for one adsorption system relative to another of ± 0.3 eV, though the accuracy for intense sharp peaks is better. Also included in table 2, are the gas phase values for the adsorbates,[11] and those found for condensation on Au at 77 K.

TABLE 2.—XPS BINDING ENERGIES (eV) [a] OF ADSORBATES [b]

molecule	condition		C(1s)	O(1s)	N(1s)	S(2p)
CO	gaseous		296.2	542.3		
	Ni$_{ads.}$	77 K	285.6	531.4		
		300 K	285.6	531.4		
N$_2$	gaseous				409.9	
	Ni$_{ads.}$	77 K			400.6, 405.7	
		300 K			none	
N$_2$O	gaseous			541.2	408.5, 412.5	
	Ni$_{cond.}$	77 K		534.8	402.1, 406.1	
		300 K		530.5	none	
	Au$_{cond.}$	77 K		534.5	401.6, 405.6	
CO$_2$	gaseous		297.7	541.3		
	Ni$_{ads.}$	77 K	291.1	534.4		
	Ni$_{cond.}$	77 K	ca. 292	535.5		
		300 K	none	none		
	Au$_{ads.}$	77 K	290.9	534.3		
	Au$_{cond.}$	300 K	292.3	535.6		
H$_2$O	gaseous			539.7		
	Ni$_{cond.}$	77 K		533.1		
		300 K		531.4		
	Au$_{cond.}$	77 K		532.7		
H$_2$S	gaseous					170.1
	Ni$_{ads.}$	77 K				162.4
	Ni$_{cond.}$	77 K				163.9
		300 K				162.4
	Au$_{cond.}$	77 K				163.8
NO	gaseous			543.3	410.3	
	Ni$_{ads.}$	77 K		530.9	399.9	
		300 K		530.2	397.8, 399.9	
	+further NO, 300 K			529.5, 531.7	397.8, 403.0	
NO$_2$	gaseous			541.3	412.4	
	Ni$_{ads.}$	77 K		531.9	401.1	
	Ni$_{cond.}$	77 K		532.9	400.3, 404.2	
		300 K		529.6, 531.0	397.7, 400.3	
	+further NO$_2$ 300 K			529.5, 531.9	402.9	
	Au$_{ads.}$	77 K		531.4	402.4	
	Au$_{cond.}$	77 K		532.5	404.7	
SO$_2$	gaseous			539.6		174.8
	Ni$_{ads.}$	77 K		531.2		165.8
	Ni$_{cond.}$	77 K		532.5		168.0
		300 K		531.2		162.9, 166.9
	+further SO$_2$, 300 K			531.2		162.9, 166.9
	Au$_{cond.}$	77 K		532.5		168.4

[a] referenced to Ni $2p_{\frac{3}{2}}$ at 825.5 eV; [b] gaseous values taken from ref. (11).

DISCUSSION

PHILOSOPHY AND PHENOMENOLOGY OF THE INTERPRETATIONS

Certain patterns have emerged from these and other metal surfaces studies which allow a framework for interpretation to be developed. This involves *assumptions* which are seen to be reasonable in situations where the interpretation is not in doubt, and are then carried over into unknown situations.

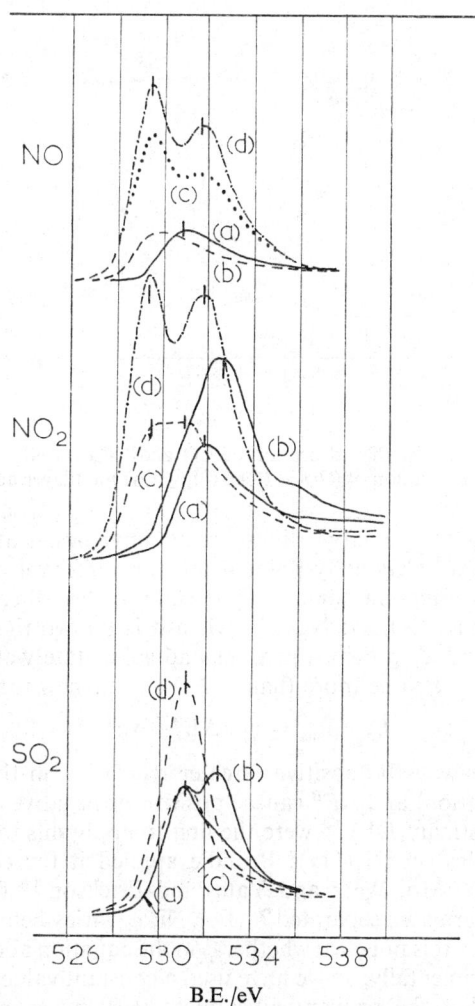

FIG. 6.—XPS O(1s) spectra for the adsorption of NO, NO_2 and SO_2 on Ni (all data at 50 eV analyzing energy). NO, trace (*a*) initial adsorption at 77 K; trace (*b*), warmed to 300 K; trace (*c*) interaction at 10^{-2} Torr at 300 K; trace (*d*) further interaction at 10^{-2} Torr. NO_2, SO_2, trace (*a*) initial adsorption at 77 K; trace (*b*), multilayer condensation at 77 K; trace (*c*), warmed to 300 K; trace (*d*), further interaction at 300 K.

Throughout the discussion emphasis is placed upon comparison to gas-phase B.E. values for the adsorbate molecules. Such comparisons were originally made for UPS data by simply adding a work-function correction ϕ to the adsorbed state adsorbate orbital energies E_{ads} to reference the values to the same level (vacuum level) as used for the free molecule values, E_q. $E_g - E_{ads}$ is then made up of two terms, $\phi + \Delta E^B$, where ΔE^B represents the initial state chemical shift dictated by bonding differences. In addition to the difficulty of knowing the appropriate ϕ value to add for a molecule sitting in the surface dipole, it is now realized that this procedure is

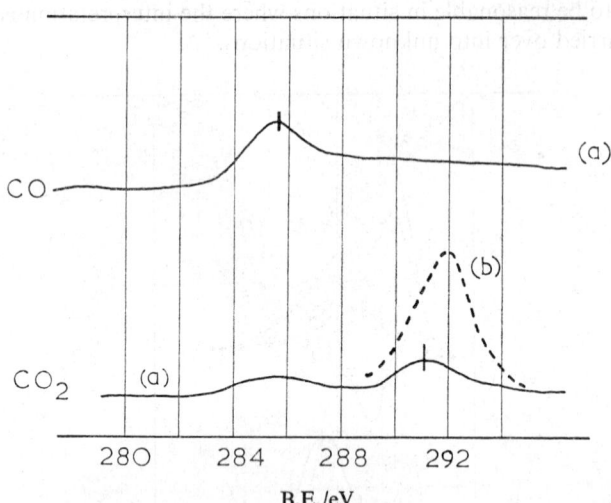

FIG. 7.—XPS C(1s) spectra for the adsorption of CO and CO_2 on Ni (analyzing energy 100 eV). Trace (a) initial adsorption at 77 K; trace (b), (CO_2) multilayer adsorption at 77 K.

inappropriate owing to the large difference in final hole-state relaxation energies ΔE^R in the two cases. Originally pointed out for core-level measurements of bulk materials,[12, 13] it is apparent for adsorbed state values where $E_g - E_{ads}$ can be as large as 11–12 eV (e.g., CO/Ni; NO/Ni) when it is known that ϕ can only account for 5–6 eV of this and all previous gaseous and solid state work would suggest that the ΔE^B term ought not to be more than 2–3 eV. The new model is therefore

$$E_g - E_{ads} = \phi + \Delta E^R + \Delta E^B, \quad (1)$$

in which ΔE^R will always be positive (greater relaxation in the adsorbed situation, e.g., solid-state situation) and ΔE^B can be positive or negative.

Demuth and Eastman (DE)[14] were the first to apply this to valence level orbitals for organic molecules on Ni (111); Brundle applied it for core and valence level measurements of CO/Mo, W[15]; and Yates and Erickson[16] for Xe/W. How well can the individual terms be separated? For Ni, ϕ varies between about 5–6 eV for the systems studied. It is not clear whether ϕ in the equation above should be identical to ϕ measured experimentally, so we have used a constant value of 5.5 eV throughout, and acknowledge that the proper value might vary by as much as ± 1 eV. DE showed[14] that when organic molecules were *condensed* at 77 K on a Ni (111) surface the $E_g - E_{ads.}$ values for all the valence levels were the same for a given molecule but larger than ϕ by 1 or 2 eV. They made the reasonable assumption the ΔE^B should be near zero for condensed organic molecules (i.e., no strong molecule-molecule forces) so

$$E_g - E_{ads.} \simeq \phi + \Delta E^R.$$

ΔE^R is therefore constant for the valence levels and has a value of 1 or 2 eV. We shall see here that in all the cases where we are certain to have condensed inorganic species at 77 K (H_2O, H_2S, CO_2, SO_2) we find the same argument to hold, not only for the valence levels but also for the core-levels. Near constancy of $E_g - E_{ads.}$ for all orbitals with a value only 1 or 2 eV greater than ϕ is therefore a diagnostic test of

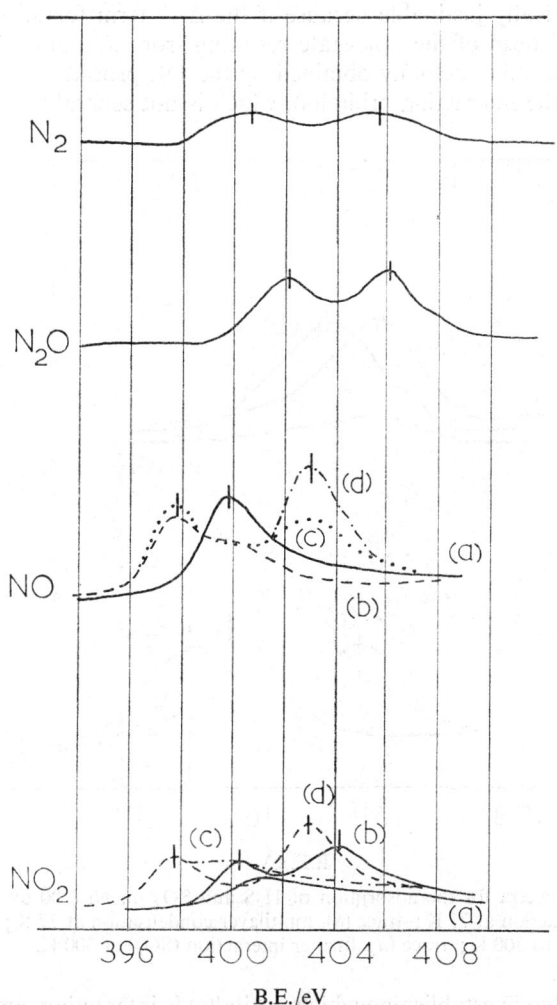

FIG. 8.—XPS N(1s) spectra for the adsorption of N_2, N_2O, NO and NO_2 on Ni. N_2, N_2O 100 eV analyzing energy, adsorption at 77 K. NO 50 eV analyzing energy scale × 3. Trace (a), adsorption at 77 K; trace (b), warmed to 300 K and pumped on for several hours; trace (c), interaction at 10^{-2} Torr; trace (d), further interaction at 10^{-2} Torr. NO_2, 50 eV analyzing energy. Trace (a), initial adsorption at 77 K; trace (b), multilayer condensation at 77 K; trace (c), warmed to 300 K; trace (d), further interaction at 300 K.

a non-interacting surface species (condensed; weak physical adsorption). Implicit in this also is the ready detection of all the free molecule orbital levels in the adsorbed situation for a condensed molecule.

For more strongly interacting molecular cases it can be assumed that ΔE^B is still near zero for those valence orbitals which are not involved in the surface bonding (such as the σ non-bonding levels in benzene). With this assumption it was demonstrated [14] that ΔE^R increased for these orbitals on going from condensed to chemisorbed situations (because $E_g - E_{ads.}$ increased). DE then estimated ΔE^B for those orbitals which were involved in the surface bonding by making the additional assumption that this same new ΔE^R applied to them also and substituting in eqn (1). This is not theoretically justifiable because if the ΔE^B term for an orbital becomes significant the ΔE^R term of the hole-state resulting from ionisation is also likely to be affected. The number actually obtained by the DE procedure is a composite of $\Delta E^B + \Delta(\Delta E^R)$ for the interacting orbital(s), which is not separable. Nevertheless, it

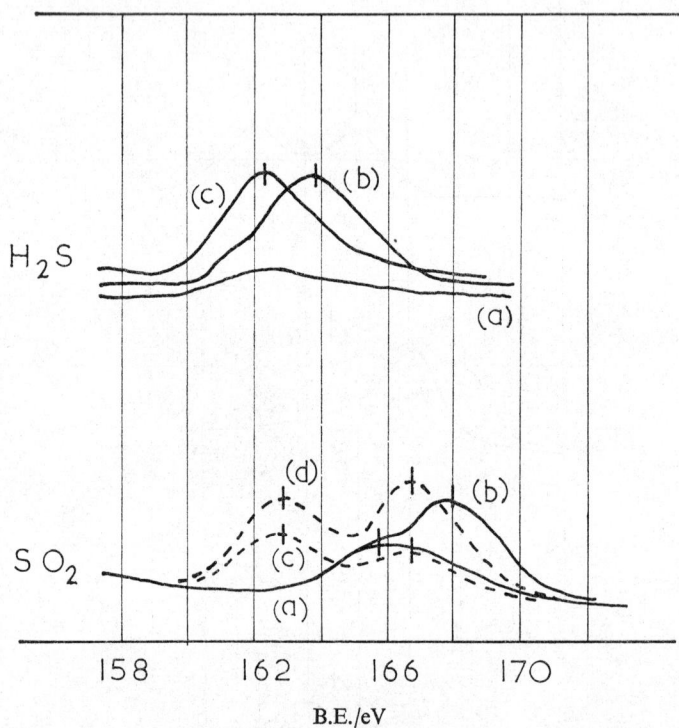

FIG. 9.—XPS S(2p) spectra for the adsorption of H_2S and SO_2 on Ni (100 eV analyzing energy). Trace (a), initial interaction at 77 K; trace (b), multilayer condensation at 77 K; trace (c), warmed to 300 K; trace (d), further interaction (SO_2) at 300 K.

is of diagnostic value in establishing which orbital(s) is interacting, and is used during the present discussions, including sometimes the non-justifiable assumption that $\Delta(\Delta E^R) \simeq 0$. For core-levels it is expected that substantial values of ΔE^B somewhere amongst the valence level orbitals would produce a measurable ΔE^B_{core}, but again since such an effect would also be expected to alter ΔE^R_{core} it is not possible to separate the terms. Neither is it to be generally expected that $\Delta E^B_{core} + \Delta E^R_{core}$ will approximate any of the $\Delta E^B_{valence} + \Delta E^R_{valence}$ values.

Dissociation reactions reactions occur for several of the systems studied here. The characteristics of a dissociation are complete loss of the correct valence level m.o. structure and shifts in the core-levels to values known (or assumed) to correspond to known dissociation products. A study of a series of small molecules containing the same elements is an obvious advantage here.

(a) CO

Several UPS studies of CO+Ni have been made at 300 K,[5, 17, 18] and there is one previous XPS/UPS study.[19] Adsorption at 77 K (this work) produced nearly identical UPS spectra to 300 K. Two adsorbate-induced features appear in the HeI spectrum (fig. 6) one at 7.2 eV and a weak one at 11.1 eV (more clearly seen in a

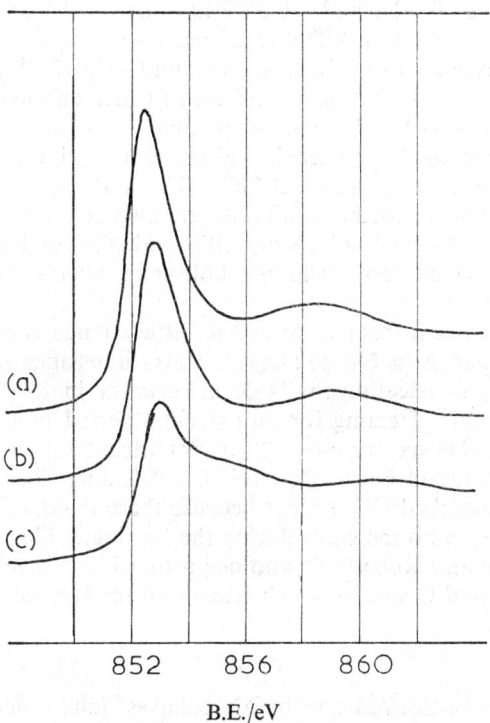

FIG. 10.—XPS Ni($2p_{\frac{3}{2}}$) spectra (20 eV analyzing energy): (a) clean Ni; (b) adsorption of ≤ monolayer of any of the adsorbates; (c) after interaction at 10^{-2} Torr with NO_2.

difference spectrum). The original assignments [17] were to the 5σ and 1π orbitals of CO respectively, indicating molecular adsorption. It has since been suggested,[20, 21] and is now generally recognised that the 7.2 eV feature includes both 5σ and 1π levels, and that the 11.1 eV feature represents the missing 4σ level. A careful comparison of the HeI and HeII spectra (fig. 2a) reveals clearly that the 7.2 eV band consists of two features at 6.3 and 8.1 eV which have equal intensities in the HeI spectrum, but unequal (cross-section variation) in the HeII. The adsorbed state core-level values, (table 2) give $E_g - E_{ads.}$ values of 10.9 and 10.6 eV for O(1s) and C(1s) respectively. As stated in the previous section, ϕ can only account for about 5.5 eV of this, and it is inconceivable that ΔE^B could be as large as +5.5 eV (calculations on Ni(CO)$_4$ and CO indicate *negative* ΔE^B values of no more than *ca.* 2 eV between the C(1s) and O(1s) values in these two environments).[22] On the arguments developed in the previous section, therefore, this state of CO is strongly interacting with the surface and it is not to be expected that core $E_g - E_{ads.}$ will equal the valence level values. With the new orbital assignment of the latter one obtains (table 1) $E_g - E_{ads.}$ values of 7.7,

8.7, and 8.6 eV for the 5σ, 1π, and 4σ levels respectively. If one assumes that the bonding concerns only the 5σ orbital (C " lone pair "), and therefore $\Delta E^B_{1\pi,\,4\sigma} \simeq 0$, one obtains $\Delta E^R_{1\pi,\,4\sigma} = 8.6 - \phi \simeq 3.7$ eV. Making the DE assumption that $\Delta E^R_{5\sigma}$ also equals 3.7 eV, one obtains $\Delta E^B_{5\sigma} = -1.0$ eV, i.e., the bonding between free CO and Ni metal has caused a shift of 5σ of CO to 1 eV higher BE, as schematically indicated on fig. 1 and 2. The justification for making the DE assumption is that the result is that reasonably expected from an end-on interaction of CO through the 5σ orbital (ΔE^B values of -1.1 and -0.3 eV have been calculated for the 5σ and 1σ levels of CO).[23] The values of ΔE^B and E^R are also of similar magnitude to those found by the same treatment for C_2H_2, C_2H_4, and C_6H_6 on Ni (111).[14]

Attempting to constrain the unknowns even further and assuming ΔE^R_{core} to be $+3.7$ eV also, produces ΔE^B C(1s) and O(1s) values of $+1.4$ and $+1.7$ eV, implying a net charge transfer to the CO molecule and therefore an important contribution to the bonding via back-bonding into CO $2\pi^*$. The authors do not believe that the discussion can be advanced further until data are available for (a) CO condensed as multilayers; (b) gas phase and solid state UPS and XPS of Ni(CO)$_4$; (c) accurate calculations on Ni/CO clusters, including hole-state calculations to estimate ΔE^R values.

On warming the system from 77 to 300 K, little change is observed in the UPS spectrum, but the change in Ni(2p) : O(1s) : C(1s) intensities in the XPS indicate about 10% desorption. Heating to 400 K *in vacuo* or in 10^{-4} Torr CO produced only further desorption. Heating for an extended period of time produced a low BE shoulder in the C(1s) spectrum (*ca.* 283.4 eV) attributable to diffusion of carbon to the surface of the nickel film. That this C(1s) feature does not arise from any dissociation of the adsorbed CO is clear because there is no shift in the O(1s) level and the C:O intensity ratio increases during the heating. This is in contrast to the conclusion of Joyner and Roberts[19] who claim that CO dissociation does occur at 400 K to produce C and O species which remain on the surface.

(b) N_2

Nitrogen, though isoelectronic with CO, behaves quite differently. Adsorption occurs at 77 K, but complete desorption is obtained on warming to 300 K. Two broad peaks are observed in the N(1s) spectrum (fig. 8), of approximately equal intensity. $E_g - E_{ads.}$ for the highest B.E. feature is only 4.2 eV (table 2), the lowest observed for any state of any molecule here and accountable for entirely by ϕ, suggesting a very weakly adsorbed state with little electronic interaction with the surface. The second peak has $E_g - E_{ads.} = 9.3$ eV. It is unlikely to represent a state where the molecule is dissociated because the results for NO and NO_2 suggest that the N(1s) for atomic nitrogen (at least as nitride) comes 2.6 eV lower at 397.8 eV. We attempted to make a nitride surface by reaction with N_2 at 1 atmosphere but none was observed. This reaction does occur with Fe, however,[24] and produces an N(1s) value 2–3 eV lower than the reversible adsorption value, confirming this as the region for nitride. The binding energy of bulk VN is also quoted as 397.2 eV.[25] The possible assignments of the two peaks are therefore: (a) two different adsorbed states, or (b) since the signal intensities are equal, two distinguishable atoms of a single state. The separation between the peaks (5.1 eV) is surprisingly large for either assignment, but for (b) the size of ΔE^R may well be dominated simply by the distance of the adsorbed atom from the surface (a final-state image-charge potential effect) so that an end on N_2 molecule weakly bonded with negligible ΔE^B could have different experimental B.E. for the two nitrogens. (The fact that C(1s) – O(1s) for the Ni/CO system—where it is thought that an end-bonding mechanism pertains—is the same as for free CO provides a counter argument however).

Either one, or two, clear m.o. patterns in the UPS would resolve the argument, but no clear-cut orbital pattern is observed. A clear feature at 8.0 eV is observed (fig. 1 and 2) with the possibility of a very weak feature 2.2 eV higher (which could also be accounted for by co-adsorption of CO—clearly observable from the XPS $O(1s)$ and $C(1s)$ regions). The logical (but tentative) assignment is to correlate the 8.0 eV feature with $5\sigma+1\pi$, as for CO, with 4σ 2.2 eV higher. $E_g - E_{ads. (1\pi, 4\sigma)}$ would then be 8.7 eV, also identical to that for CO. A shift of 5σ relative to 1π of 1 eV would bring them into exact coincidence since 5σ and 1π are only 1 eV apart in free N_2. This is consistent with the narrowness of the feature correlated with $5\sigma+1\pi$ for N_2, with no evidence of a shoulder in the HeII spectrum (see fig. 2). Having suggested these tentative assignments we must admit that the UPS data are the least satisfactory of all that are presented here, since they represent only one experiment and no spectra were taken after N_2 desorption at 300 K to ascertain what features were due to co-adsorbed CO, which would remain on the surface at that temperature.

(c) N_2O, CO_2

The adsorption of these isoelectronic molecules of Ni at 77 K reveals the clearest examples of molecular orbitals which are directly correlatable with the gas phase (fig. 1 and 2). Desorption occurs on warming to 300 K, but with N_2O a small amount of oxygen remains at the surface. CO_2 adsorption above saturation vapour pressure gave multilayer condensation with a second set of m.o. appearing at ca. 1–1.5 eV higher B.E. in the UPS spectra and with $O(1s)$ values also about 1 eV higher (cf. the same phenomenon on Au).[2] N_2O was not studied in the multilayer regime. The difference in ΔE^B between physically adsorbed and condensed CO_2 should be very small (as should ΔE^B itself), and therefore the experimentally observed shift is probably due to a difference in ΔE^R. $E_g - E_{ads.}$ values of 6.7 and 6.4 eV are obtained for the valence and core levels of N_2O; 6.3, 6.4, and 6.9 eV for the valence, $C(1s)$ and $O(1s)$ levels of physically adsorbed CO_2; and ca. 5.1 and 5.4 eV for the valence and core-levels of condensed CO_2. The condensed values can be accounted for by ϕ alone (cf. N_2) and if ΔE^B is near zero as expected, ΔE^R increases from near zero to about 1–1.5 eV on going from the condensed to the physical state.

The small $O(1s)$ feature remaining in warming N_2O/Ni to 300 K falls at about 1 eV higher B.E. than the normal " oxide, O^{2-} " position [1] and presumably therefore represents atoms which remain isolated at the surface. Further exposure to N_2O or CO_2 at 300 K produces no further reaction. It is worth noting that the feature identified at 534.4 eV in the previous Ni/O_2 work [1] at 77 K is shown by the present work to represent CO_2 adsorption and this is confirmed by the presence in that work of a small $C(1s)$ feature at 291.1 eV.

(d) H_2O and H_2S

H_2O and H_2S have been studied in the initial and multilayer stages at 77 K (since pressures are always above saturation vapour pressure the two cases are not distinguishable in the same way as for CO_2). On warming to 300 K desorption occurs leaving only chemisorbed species on the surface. The three m.o. of water can be easily seen in the HeI spectrum (fig. 1) in the early adsorption stages, but are too weak to be distinguishable in the HeII (fig. 2). $E_g - E_{ads.}$ for the $3a_1$ and $1b_2$ orbitals is 5.5 eV, and that for the $1b_1$ O lone pair orbital 6.1 eV, figures which can largely be accounted for by the ϕ term. The relative shift of $1b_1$ to the other orbitals of

0.6 eV is in the opposite direction to that proposed for the lowest B.E. orbital for CO and N_2. For multilayer coverage the orbitals are clear in both the HeI and HeII spectra (fig. 1 and 2), and remain at the same positions. H_2O is the only molecule studied here where a significant relative shift of one of the orbitals, compared to the free molecule, is clearly observed in the multilayer condensation regime. This is probably more connected with the strong affinity of H_2O for H_2O (H-bonding) than with any interaction with the surface, as is the lack of any distinction between initial and multilayer spectra. $E_g - E_{ads.}$ for $O(1s)$ is 6.6 eV at 77 K, which, being significantly different from the valence levels, also suggests a significant H-bonding effect on $O(1s)$. On warming to 300 K, a small $O(1s)$ signal remains (fig. 5) at 531.4 eV. It is too large to represent only co-adsorbed CO (certainly present to some extent) as judged by the intensity of $C(1s)$ and therefore represents dissociated H_2O as either OH or O atoms at the surface (cf. the 531.4 eV $O(1s)$ feature in the Ni_4O_2 system).[2] The HeI spectrum at 300 K is consistent with this interpretation of dissociation, having lost all correlation with an H_2O m.o. pattern and showing a small broad feature centred at 6.0 eV, characteristic of oxygen atoms (fig. 1). The knowledge that this feature represents dissociated H_2O raises the possibility that in the early stages of H_2O adsorption at 77 K partial dissociation is also occurring and the 6.5 eV peak assigned to $1b_1$ represents a mixture of the 6 eV dissociated feature plus $1b_1$. If so, a 531.4 eV $O(1s)$ peak from dissociated H_2O should be present at that time, but it is not possible to confirm this from the present results because of the interfering $O(1s)$ peak from co-adsorbed CO.

For H_2S, initial adsorption reveals the expected three orbital structure in HeI (fig. 1) but at first sight it appears that there is a relative shift of the $1b_1$ orbital (S lone pair) to high B.E. ($E_g - E_{ads.}$ for $1b_2$ = 5.8 eV; for $3a_1$ and $1b_2$ = 6.4 eV), i.e., in the opposite directions to that for the $1b_1$ orbital (O lone pair) of H_2O. At high coverage the $1b_1$ orbital moves to its expected position. Examination of the $S(2p)$ region during this sequence, and both the $S(2p)$ region and HeI on warming to 300 K shows that the apparent displacement of $1b_1$ at low coverage at 77 K is due to partial dissociation. The initial adsorption $S(2p)$ shows a broad peak centred at ca. 162.5 eV with a long high B.E. tail (fig. 9). As coverage increases a distinct peak at 163.9 eV emerges and grows, representing molecular condensed H_2S ($E_g - E_{ads.}$ = 6.3 eV, identical to that of the valence levels). On warming to 300 K, $S(2p)$ moves to 162.4 eV. Reaction entirely at 300 K produces only the 162.4 eV feature. In the HeI spectrum the H_2S orbitals disappear at 300 K, leaving a feature at 4.9 eV. (fig. 1). These observations are consistent with a dissociation reaction taking place at 300 K, and in the initial stages at 77 K. We believe the species remaining to be S rather than SH because the 162.4 eV $S(2p)$ feature is very close to that observed for SO_2 interaction at 300 K (fig. 8) a condition under which SH cannot be formed. Thus the "$1b_1$" feature at 4.7 eV in the HeI spectrum at 77 K represents a combination of dissociated H_2S (4.9 eV) and molecular $1b_1$ (4.0 eV), which is not shifted from its expected position.

The HeII spectrum of H_2S (fig. 2) demonstrates the lack of photoionization cross-section of the second-row element at 40.8 eV photon energy. For high-coverage adsorption the $3a_1$ and $1b_2$ orbitals are revealed in the HeII spectrum, but the pure $S(2p)$ character of $1b_1$ leads to a complete absence of this orbital. For dissociation at 300 K the HeII spectrum is indistinguishable from that of clean Ni, except for an attenuation of the Ni d-bands. This is an additional reason for believing the dissociation product to be S rather than SH, since any $H(1s)$ character in the valence region should produce intensity in the HeII spectrum.

(e) NO, NO$_2$

The 77 K NO UPS spectra of fig. 3 and 4 represent saturation coverage below the saturation vapour pressure, whereas for NO$_2$ the multilayer condensation region can also be studied. Features centred at 1.8, 7.2 and 9.2 eV are identified as adsorbate orbitals in both HeI and HeII spectra. When the system is warmed to 300 K some subtle changes are observed. There is a relative loss of intensity in the 1.8 and 9.2 eV bonds, a new broad band centred about 6 eV appears, and there is possibly also some intensity increase around 3 eV. It is not possible to say whether the 7.2 eV feature has decreased because of the new overlapping 6 eV band. This behaviour is consistent with partial dissociation occurring at 300 K, the 6 eV band representing atomic oxygen and nitrogen. Two alternative interpretations of the 77 K spectra are possible.

(i) Two states of NO co-exist, the major one, a physically adsorbed state (A), represented by the 1.8 and 9.2 eV bands, the other, (B), represented by the 7.2 eV band, being a more strongly chemisorbed state. On warming to 300 K, (A) partly desorbs and (B) dissociates. For this assignment the 1.8 eV band corresponds to the gaseous $(\pi g 2p)^1$ electron ($E_g - E_{ads.} = 7.7$ eV) and the broad 9.2 eV feature to the complex mixture of states (singlet and triplet) resulting from ionization of the $(\pi_u 2p)^4$ and $(\sigma_g 2p)^2$ electrons ($E_g - E_{ads.} \simeq 8.1$ eV). Constraining ΔE^B to zero or near-zero for this state (cf. a condensed molecule) implies a ΔE^R of around 2.5 eV for the valence levels. The 7.2 eV band of (B) would have to be assigned to the $\pi_u 2p$, $\sigma_g 2p$ combination ($E_g - E_{ads.} \simeq 10.1$ eV), with the $(\pi_g 2p)^1$ electron lost in the Ni d-band region (possibly also at 1.8 eV). For a state other than one approximating a non-interacting condensed one, there is no justification in the case of NO for constraining ΔE^B to zero for either the $\pi_g 2p$ or $\sigma_g 2p$ levels since both could be involved in an end-on bond so we can only state that $E_g - E_{ads.} \simeq 10.1$ eV for the 7.3 eV experimental level.

(ii) The second interpretation is that at 77 K only one state exists with the 1.8 eV feature representing $(\pi_g 2p)^1$ ($E_g - E_{ads.} = 7.7$ eV) and both the 7.3 and 9.2 eV features representing the $\pi_u 2p$, $\sigma_g 2p$ combination. This is sensible since in the free molecule the I.P. from these orbitals spread over 5.5 eV and exhibit more than one maximum.[6] $E_g - E_{ads.}$ would be approximately 9 eV for these orbitals. The difference in $E_g - E_{ads.}$ values (7.7 and 9 eV) suggests a state more strongly adsorbed than a condensed or physically adsorbed molecule with the $(\pi 2_g p)^1$ electron perhaps more involved in the bonding than the other orbitals. It would, in fact, be remarkably similar to CO or N$_2$ if ΔE^B for $\pi_u 2p$, $\sigma_g 2p$ is contrained to zero.

In the XPS spectra only single O(1s) and N(1s) features are observed (fig. 6 and 8) though each has a high B.E. tail which could correspond to a smaller concentration of an additional state. After warming to 300 K and pumping for several hours, the XPS spectra confirm the dissociation reaction suggested by the UPS results. The O(1s) peak broadens and its centre moves to 530.2 eV (considered to be made up of unresolved components from 529.5 and 530.7 eV). The 529.5 eV region is characteristic of O^{2-}.[1] The N(1s) behaviour is similar but the shifts are larger. The 399.9 eV peak decreases, the intensity being transferred into a peak at 397.8 eV, the nitride position (see § (b)). The XPS results are not compatible with interpretation (1) for the UPS because the N(1s) and O(1s) parameters are not compatible with a predominate presence of a physically adsorbed state at 77 K. $E_g - E_{ads.}$ N(1s) = 10.4 eV and $E_g - E_{ads.}$ O(1s) = 12.4 eV. These are high values, representing a $\Delta E^R + \Delta E^B$ term of some 5 eV for N(1s) and 7 eV for O(1s), much higher values than expected for a physically adsorbed or condensed molecule and different (and higher) than those proposed for the valence levels of the physical state of assignment (i). In addition, the O(1s)−N(1s) separation value of 131.0 eV is 2 eV smaller than the gas-phase value suggesting a strong chemical effect on the adsorbed molecule. Assignment (ii) therefore seems to be correct, though the presence of high B.E. tails in the XPS spectra both at 77 and 300 K suggest a low concentration second state which is more weakly bonded, and is not observed in the UPS. Demuth and Eastman [23] have briefly reported the HeI spectrum of NO/Ni (111) at 300 K and though they do not consider it, partial dissociation may also be occurring there.

Further reaction at high pressure (10^{-2} Torr) produces a large increase in the oxide O(1s) peak, a peak at 531.7 eV, a small increase in the nitride N(1s) and a large increase at 402.9 eV. The new peaks are similar to those observed for extensive reaction with NO_2, and this, together with the fact that the 402.9 eV feature is at so much higher B.E. than the other NO states, but is close to that observed for K/NO_2 (403.2 eV)[11] suggests an NO_2^- assignment. The large increase in the oxide O(1s) could result from NO reactions producing O^{2-}, but may also come from oxygen present in the NO at these high pressures.

In the multilayer condensation regime for NO_2 at 77 K, all the valence orbitals can be identified as present in the UPS spectra, though there are so many IP for free NO_2 [26] that some unresolved broad bands are formed. $E_g - E_{ads}$ values vary between 7.8 and 8.2 eV for the assigned orbitals (table 1), and for the core levels are 8.4 and 8.2 eV respectively for O(1s) and N(1s). The core-levels are also close to those observed for NO_2 condensed on Au (table 2) and the valence level patterns are the same. For initial adsorption at 77 K the UPS spectra are quite different and are somewhat similar to those for NO at 77 K, plus possibly a little molecular NO_2 and some " O ". It appears then from the UPS that NO_2 may dissociate and then multilayers of NO_2 are formed on the covered surface. The XPS data, however, are not entirely compatible with this suggestion. The O(1s) and N(1s) peaks (fig. 6 and 8) are both centred at about 1 eV higher B.E. than for initial adsorption of NO at 77 K. The O: N intensity ratio is consistent with NO_2 remaining present, being about twice that for NO, so no oxygen is lost from the surface. Either a strongly perturbed NO_2 adsorbed state (for which $E_g - E_{ads}$ would be 9.4 and 11.5 eV for O(1s) and N(1s)), or dissociation to NO+O with NO in a different adsorbed state from that formed by initial NO adsorption (possibly owing to the influence of the remaining O), are the probable interpretations. More careful investigation is required before further speculation. The condensed values of N(1s) 404.1 eV and O(1s) 532.6 eV confirm the suggestion made for NO adsorption that the major species at 77 K is a rather strongly chemisorbed state, since by comparison with the gas-phase values a physical or condensed state of NO O(1s) should be 2 eV higher than for NO_2 condensed at ca. 534.6 eV, and N(1s) should be 2 eV lower at ca. 402 eV.

On warming to 300 K the condensed NO_2 desorbs and also reacts further with the surface. The N(1s) region shows a nitride peak (397.9 eV), an NO region (ca. 400 eV) and a long high B.E. tail. The O(1s) region shows O^{2-} (529.6 eV), NO (531.0 eV), and a short B.E. tail. The reaction may be represented as:

$$NO_2 \text{ (77 K)} \xrightarrow{\text{warm}} NO+O \xrightarrow{\text{partial}} N+O \text{ (300 K)}.$$

Further exposure to NO_2 at 10^{-3} Torr increases the O^{2-} peak and replaces the 531 eV peak by a larger one at 531.9 eV. The nitride peak remains unchanged, but the 400 eV peak is replaced by a large 403.1 eV peak. This is compatible with the production of NO_2^- species, as in the later stages of NO treatment. Any remaining NO_{ads} is removed in the process.

(f) SO_2

The HeI and HeII spectra of condensed multilayers of SO_2 reveal all the free molecule orbital features (fig. 3 and 4, table 1) $E_g - E_{ads}$ is the same for all the orbitals within experimental error (*ca.* 6.8 eV), and the core-level values are also the same, as expected. The initial interaction at 77 K is quite different, however, producing broad features between 5 and 9 eV in the UPS, on $O(1s)$ value of 531.2 eV, and an $S(2p)$ of *ca.* 165.8 eV. This may represent either a dissociation reaction to $SO+O$ (but not S^{2-}, since there is nothing in the 162.4 eV region) or a strongly chemisorbed SO_2 molecular species ($E_g - E_{ads}$. $S(2p) = 9.0$ eV; $O(1s) = 8.4$ eV). Without having

TABLE 3.—B.E. (eV) OF SOME BULK Cu,[1] AND Ni COMPOUNDS

CuO		Cu_2S		CuSO		$\Delta(S(2p))$	$O(1s)-S(2p)$ for $CuSO_4$
$O(1s)$		$S(2p)$		$O(1s)$	$S(2p)$		
529.7		161.3		530.8	168.0	6.7	362.8
NiO [b]		Ni sulphide [c]		Ni sulphate [c]		$\Delta(S(2p))$	$O(1s)-S(2p)$ for Ni sulphate
$O(1s)$		$S(2p)$		$O(1s)$	$S(2p)$		
529.4		162.4		531.2	166.9	4.5	364.2

a values taken from ref. (11) have all been reduced by 3.2 eV so that $O(1s)$ CuO coincides with several other determinations of CuO, including one by the present authors, *b* ref. (1), *c* present work-suggested surface species.

the spectra of a known SO species it is not possible to speculate further. On warming to 300 K the condensed SO_2 desorbs and also interacts further with the surface leaving an $O(1s)$ peak at 531.2 eV (broader than the original), a sulphide $S(2p)$ peak at 162.9 eV, and another $S(2p)$ peak at 166.8 eV. Further interaction at 10^{-3} Torr increases all three features. Since the interaction is producing one $O(1s)$ feature but two $S(2p)$ features, a disproportionation reaction is likely:

$$2SO_2 + 4e \rightarrow S^{2-} + SO_4^{2-}.$$

There are no data on bulk $NiSO_4$ in the literature. There are, however, figures for $CuSO_4$, CuO and Cu_2S which may be used for comparison.[11] These are given in table 3, together with the present surface data for Ni. Considering that the calibration of the Cu data is suspect, the agreement is close enough to support the suggestion of SO_4^{2-} species. Measurements on the bulk nickel sulphides and sulphates are required to confirm the analysis.

CONCLUSIONS

The simple scheme of comparison with gas phase UPS and XPS data works well for the identification of very weakly interacting molecular species. It provides information on the orbitals involved in the stronger adsorption process, on whether dissociation occurs, and what the products are. Solid-state bulk-compound data is also useful, particularly for comparison to the later adsorption stages. The presence of O^{2-}, S^{2-}, NO_2^- and SO_4^{2-} species are suggested by such comparisons. The small size of the experimental B.E. shifts involved and the problem of co-adsorption of CO prohibit definitive identification in some cases. Higher resolution (monochromator, deconvolution) might help, but the excessive width of some of the XPS peaks probably represents a number of slightly different adsorption states. Whether this is a property of the polycrystalline nature of the surface remains to be seen. Reliable model calculations are badly needed to help interpretation of binding energies, particularly for estimates of relaxation energies involved.

Note added in proof: Subsequent analysis [27] of the relative intensities of the N(1s) and O(1s) core-levels during the adsorption and desorption sequences for the NO/Ni and NO_2/Ni interactions have lead us to the conclusion that the assignment of the N(1s) peak at 402.9 eV as an NO_2^- state is incorrect. The intensity analysis indicates that a nitrosyl assignment (NO) is appropriate. The N(1s) B.E. range for bulk nitrosyls stretches from 399.5 eV to 403.5 eV, overlapping the NO_2^- bulk position and leading to the ambiguous assignment based on B.E.'s only. Independent work by Ertl et al.,[28] confirms, by flash desorption, that an NO state is formed after extensive interaction of NO with Ni(100).

[1] C. R. Brundle and A. F. Carley, *Chem. Phys. Letters*, 1975, **31**, 423.
[2] S. J. Atkinson, C. R. Brundle and M. W. Roberts, *Faraday Disc.*, 1974, **58**.
[3] C. R. Brundle, M. W. Roberts, D. Latham and K. Yates, *J. Electron. Spectr.*, 1974, **3**, 241.
[4] S. J. Atkinson, *Ph.D. Thesis*, (Bradford University, 1974).
[5] P. J. Page and P. M. Williams, *Faraday Disc.*, 1974, **58**.
[6] D. W. Turner, A. D. Baker, C. Baker, and C. R. Brundle, *Molecular Photoelectron Spectroscopy* (Wiley and Sons, London 1970).
[7] C. R. Brundle, *J. Vac. Sci. Tech.*, 1974, **11**, 212.
[8] C. R. Brundle and A. F. Carley, *Chem. Phys. Letters*, 1975, **33**, 41.
[9] J. C. Fuggle and D. Menzel, *Chem. Phys. Letters*, 1975, **33**, 46; *Surface Sci.*, to be published.
[10] A. Barrie and C. R. Brundle, to be published.
[11] *Handbook of Spectroscopy*, vol. 1, ed. J. W. Robinson, (Chemical Rubber Company, 1974).
[12] P. H. Citrin and J. D. Thomas, *J. Chem. Phys.* 1972, **57**, 4466.
[13] D. A. Shirley, *Chem. Phys. Letters*, 1972, **16**, 220.
[14] J. E. Demuth and D. E. Eastman, *Phys. Rev. Letters*, 1974, **32**, 1123.
[15] C. R. Brundle, *Surface Sci.*, 1975, **48**, 99.
[16] J. T. Yates and N. E. Erickson, *Surface Sci.*, 1974, **44**, 489.
[17] D. E. Eastman, and J. K. Cashion, *Phys. Rev. Letters*, 1971, **27**, 1520.
[18] D. E. Eastman in Electron Spectroscopy, ed. D. A. Shirley, (North-Holland, Amsterdam), 1972).
[19] R. W. Joyner and M. W. Roberts, *J.C.S. Faraday Trans. I*, 1974, **70**, 1974.
[20] D. R. Lloyd, *Faraday Disc.*, 1974, **58**.
[21] G. Blyholder, *J. Vac. Sci. Tech.*, 1974, **11**, 865; I. P. Batra and P. S. Bagus, *Solid State Comm.* 1975, **16**, 1097.
[22] M. Barber, J. A. Connor, M. F. Guest, M. B. Hall, I. H. Hillier and W. N. E. Meredith, *Faraday Disc.*, 1972, **54**, 219.
[23] J. E. Demuth and D. E. Eastham, *Jap. J. Appl. Phys. (Suppl. 2)*, 1974, **2**, 827.
[24] K. Kishi, personal communication.
[25] D. N. Hendrickson, J. M. Hollander and W. L. Jolly, *Inorg. Chem.*, 1969, **8**, 2642.
[26] C. R. Brundle, D. Neumann, W. C. Price, D. Evans, A. W. Potts and D. G. Streets, *J. Chem. Phys.*, 53, **705**, 1970.
[27] C. R. Brundle, *J. Vac. Sci. Tech.*, to be published.
[28] G. Ertl, private communication.

Photoelectron Spectroscopic Study of the Interaction of Nickel and Oxygen

By P. R. Norton* and R. L. Tapping

Physical Chemistry Branch, Chalk River Nuclear Laboratories,
Atomic Energy of Canada Limited, Chalk River, Ontario,
Canada, KOJ 1JO

Received 18th February, 1975

The adsorption of oxygen on nickel foils is characterized by two kinetic regions. In the first (coverage, $\lesssim 0.4$ monolayers) it is established that chemisorption occurs with little consequent change in the core and valence spectra which indicates that the surface metal atoms retain their metallic character. In the range $0.5 \lesssim \theta \lesssim 2.0$ the core and valence level spectra reflect the occurrence of oxide nucleation and finally the formation of a passive film of NiO.

Holloway and Hudson [1a, b] have shown that the initial stages of oxidation (coverage, $\theta, \lesssim 2$ monolayers of oxygen) of nickel (100) and (111) surfaces are characterized by two distinct kinetic regions. In the region below 0.5 monolayers, oxygen forms ordered chemisorbed arrays while above ~ 0.5 monolayers the LEED data [1a, b] indicate that nickel oxide (NiO) nucleates, finally producing a passive film two layers thick. Similar behaviour has been observed on polycrystalline nickel films.[2]

The Ni-O_2 system is thus ideal for investigating the effects of chemisorption and oxidation on the photoelectron spectra of core and valence levels of both adsorbent and adsorbate. An improved understanding of both the information provided by ultra-violet and X-ray photoelectron spectroscopy (UPS and XPS respectively) and the oxidation process itself should come from such a study.

Of particular interest is the behaviour of the core levels of the adsorbent in the chemisorption region in view of the observed lack of effect on the Pt$N7$, $N6$ levels in the Pt-O_2 and Pt-CO systems.[3, 4] In the present study we clearly delineate, for the first time, the coverage at which the core and valence level shifts are observed in the Ni-O_2 system.

EXPERIMENTAL

Spectrometer calibration and temperature control have been discussed previously.[3] The binding energies (B.E., ± 0.1 eV) reported are all referred to the Fermi level of nickel. The polycrystalline nickel foils were mounted in the usual manner [3] and were cleaned by outgassing (1300 K, several hours), argon ion sputtering at high temperature (1300 K, 5 keV Ar$^+$, 30 μA cm^{-2}, 3.6 ks), and annealing (1300 K, 1–4 ks) at pressures $< 6 \times 10^{-8}$ Pa. This process effectively removed sulphur, the main impurity to diffuse from the bulk. The surfaces then contained no more than 1–2 % of a monolayer of residual impurities (chiefly carbon) as determined by XPS. After oxidation, a clean surface was regenerated by reduction in H$_2$, 4–8 $\times 10^{-4}$ Pa, 1300 K, 600 s) followed by flashing to 1100 K to remove adsorbed hydrogen.

All XPS spectra reported were obtained using Al Kα X-rays and all spectra of the valence band were obtained using HeII radiation (40.8 eV). The spectrometer resolution was set to ~ 0.4 eV in most of the experiments, yielding an overall resolution in the XPS mode of

~1.2 eV. All oxygen pressures were measured by reference to a calibrated ion gauge [5] and the oxygen purity was monitored by a VG Q7B quadrupole mass spectrometer.

RESULTS

T = 295 K

Fig. 1 shows the variation of the oxygen 1s (O 1s) signal at 529.9 eV binding energy with oxygen exposure in Langmuirs (1 L = 10^{-6} Torr s ≡ 1.33×10^{-4} Pa s). The insert shows the low-coverage region at higher sensitivity.

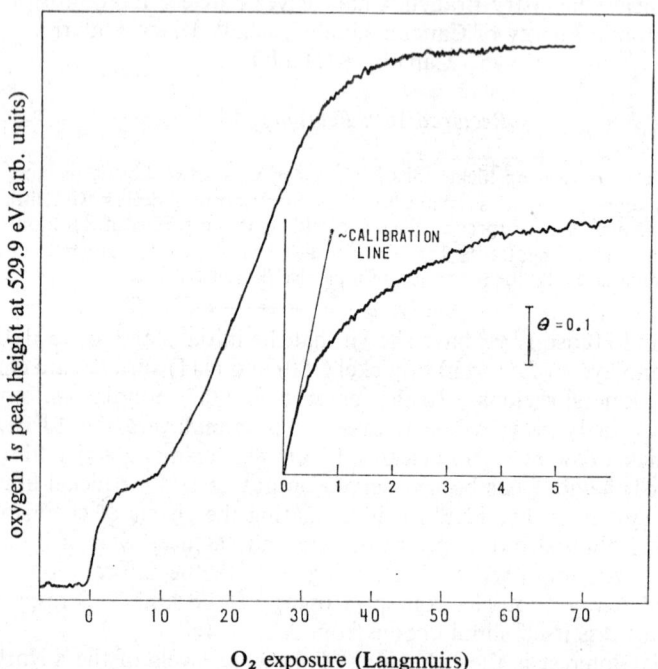

FIG. 1.—Exposure dependence of the oxygen 1s signal at 529.9 eV of oxygen adsorbed on nickel foil at 295 K. *Insert*: Initial region at higher sensitivity showing calibration line for sticking coefficient, $S = 1$.

The exposure calculation was based upon the reasonable assumption of unit sticking coefficient S at zero coverage.[2] The ion gauge indication also gave a value of $S \sim 1$. The O 1s line width and position both change with coverage over the range studied, but these effects are not sufficiently large to invalidate the qualitative characteristics of the kinetic plot of fig. 1. There are clearly two kinetic regions. In the range of exposure 0 to ~ 5 L, S decreases with increasing coverage. Above ~ 5 L, S increases again to a broad maximum value at ~ 20 L decreasing to near zero at > 70 L exposure. The results are similar to those of Holloway and Hudson [1a, b] and Horgan and King.[2] As discussed below and in ref. (1(a), (b)), a reasonable interpretation of the results is that oxygen chemisorption takes place below ~ 5 L, and above 10 L nickel oxide (NiO) nucleation occurs resulting finally in a passive film of two NiO layers.

The exposures shown in the following figures correlate with those in fig. 1. The coverage calibrations shown are based upon the areas of the O 1s peaks (taking attenuation into account) and the assumption that the saturation coverage of oxygen at 295 K corresponds to 2 layers of NiO.[1a, b] This will be more fully discussed below.

Fig. 2 shows the spectra of the valence band region of nickel (2a); the Ni $2p_{\frac{3}{2}}$ level (2b) and the oxygen 1s (O 1s) region (2c), in the exposure range 0-5 L. The O 1s peaks are slightly asymmetric, the maximum shifts from 530.1 eV (0.6 L and 1.4 L) to 529.9 eV (5.1 L) and the FWHM decreases from 2.5 to 2.2 eV under our resolution conditions.

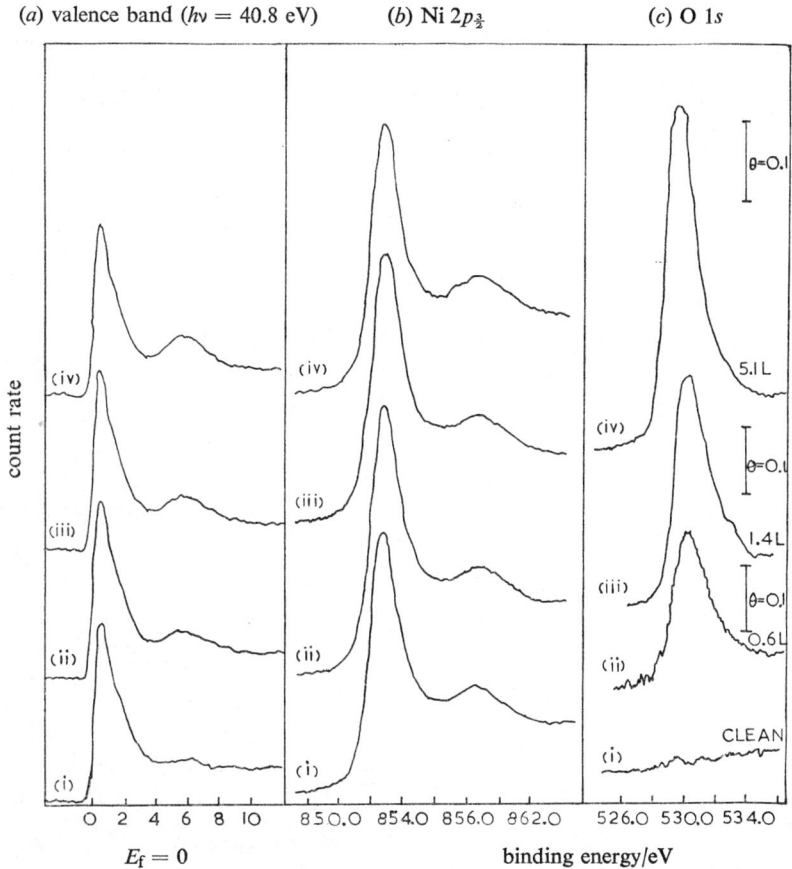

FIG. 2.—Variation of spectra of valence band (2a), nickel $2p_{\frac{3}{2}}$ (2b) and oxygen 1s (2c) regions with increasing oxygen exposure in range 0-5 L (shown at right). T = 295 K. Numbers (i-iv) apply to spectra obtained at corresponding oxygen exposure.

The spectra of the valence band (fig. 2a) show little change except an increase in the intensity in the 5-6 eV region which has been observed before.[6, 7] The integrated area in the 0-10 eV region is thus increasing. The clean surface spectrum is characterized by an intense peak at ~ 0.5 eV, a shoulder at 1.8 eV and a broad maximum centred at 5.6 eV, as reported previously.[6] Of particular importance is the present observation that there is little attenuation (~ 5 %) even after a 5 L exposure (θ ~ 0.4). This was confirmed by monitoring the peak at 0.5 eV during oxygen exposure. No change in intensity was observed until $\theta \gtrsim 0.4$. No changes were observed in the Ni $2p_{\frac{3}{2}}$ region in this coverage range.

Fig. 3 shows results obtained in the second kinetic region at exposures of 12, 18, 29 and > 100 L. The O 1s peak has shifted to 529.7 eV and the FWHM decreases further from 2.1 eV at 12 L ($\theta = 0.65$) to ~ 1.9 eV at 18 L ($\theta = 1.15$), then remaining constant up to $\theta = 2.0$. Under our highest resolution the FWHM of the O 1s line at $\theta = 2.0$ was ~ 1.6 eV.

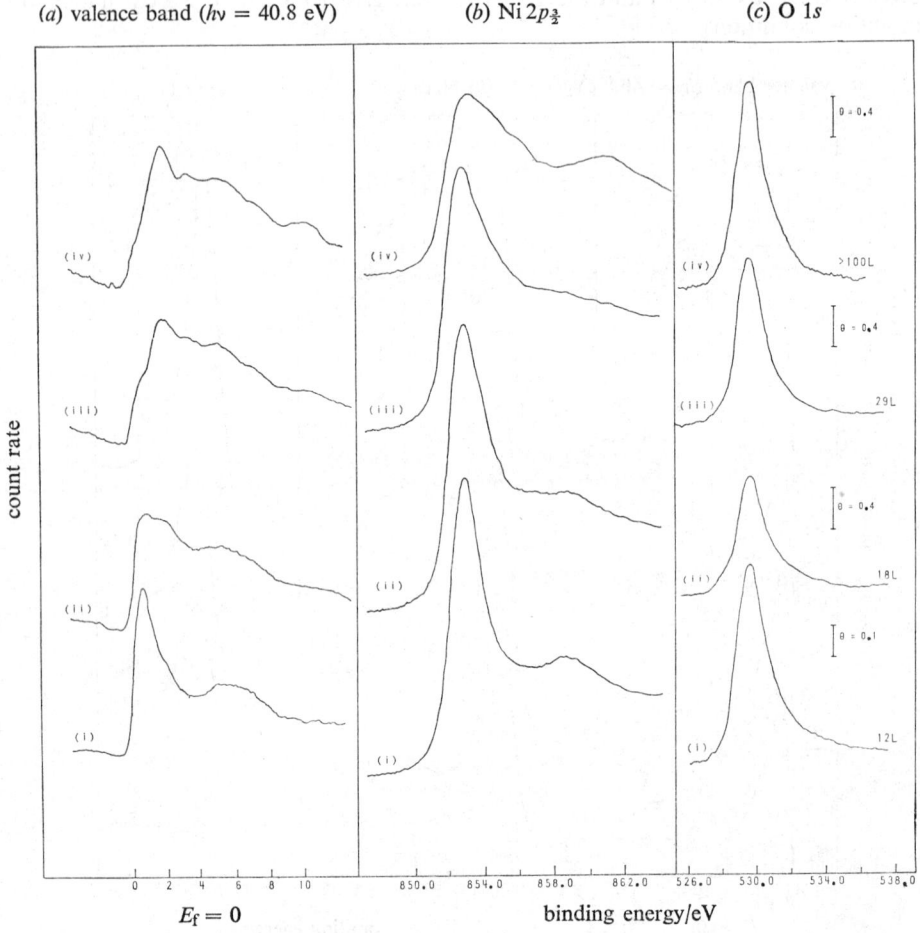

FIG. 3.—Variation of spectra of valence band (2a), nickel $2p_{\frac{3}{2}}$ (2b) and oxygen 1s regions (2c) with increasing oxygen exposure in the range 10–100 L (shown at right); $T = 295$ K. Numbers (i-iv) apply to spectra obtained at corresponding oxygen exposure.

Fig. 3b(i-iv) show the Ni $2p_{\frac{3}{2}}$ region. Almost no change has occurred in either the main or the shake-up peaks up to 12 L but by 18 L ($\theta \sim 1.15$) the Ni $2p_{\frac{3}{2}}$ peak (852.8 eV) is broadened and attenuated ~ 5 % compared to clean nickel and the shake-up peak (858.6 eV) is markedly attenuated. At $\theta \sim 1.60$ (29 L) the intensity at 852.8 eV is attenuated 17 % and the satellite has almost disappeared. At saturation (2.0 monolayers; exposure > 100 L) the signal at 852.8 eV is attenuated 43 % and the shake-up line has shifted to 861.4 eV. Just visible are two additional shoulders at ~ 854.4 and 856.1 eV.

The spectra of the valence band (fig. 3a) change rapidly in this coverage region. At $\theta \sim 0.64$ (12 L) the intensity at 0.5 eV is down 15 % while that at ~ 5.6 eV is further increased. As the exposure is increased through 18 L, 29 L and 100 L the peak near the Fermi edge attenuates further and a new peak grows in at 1.8 eV and structure continuously develops at 3.3, 5.5, 7 and 10 eV. The *decrease* in intensity of the 0.5 eV peak with exposure in the range 10–100 L exactly parallels the observed *increase* in the oxygen signal in this region. At $\theta \sim 2.0$ the peak at 0.5 eV is attenuated approximately 80 %. Presumably this residual signal originates in the underlying metal.

Fig. 4(i) shows the clean nickel $2p_{\frac{3}{2},\frac{1}{2}}$ lines. Fig. 4(ii) shows the effect on the $2p_{\frac{3}{2},\frac{1}{2}}$ lines of heating the sample to 473 K in oxygen until the coverage θ was ~ 2.2 (based upon peak height). The O 1s line was similar in shape to that observed in experiments at 450 K discussed below. In fig. 4(ii) the shoulder at 854.4 eV in fig. 3b(iv) has increased in intensity and the new shoulder at 852.8 eV must originate in the underlying nickel. This indicates that the nickel metal signal is attenuated approximately 60 % at $\theta \sim 2.2$.

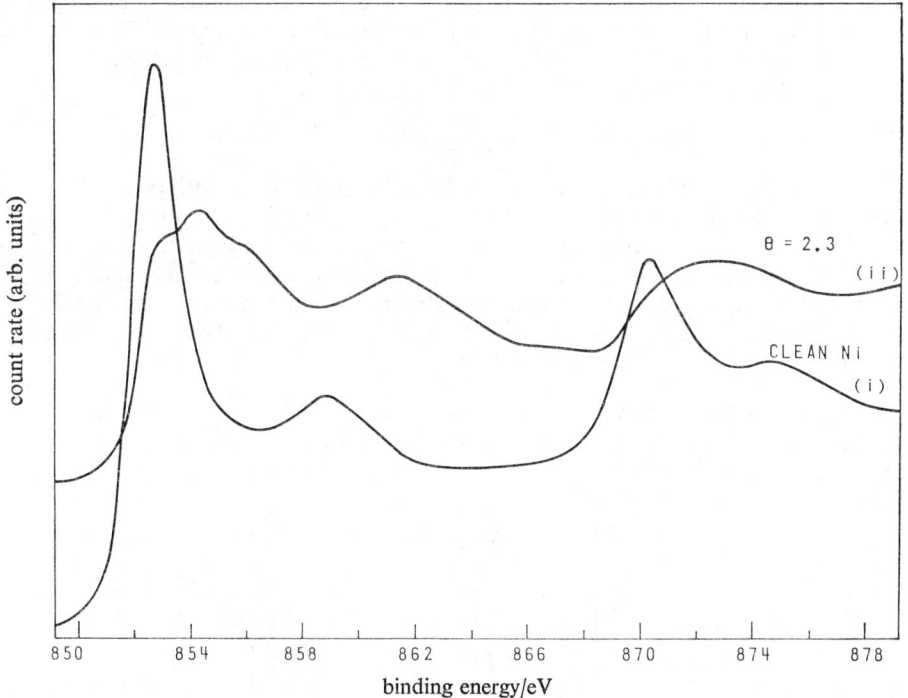

FIG. 4.—Spectra of Ni $2p_{\frac{3}{2},\frac{1}{2}}$ region before and after exposure to oxygen: (i) clean nickel foil, (ii) same foil after > 200 L exposure at ~473 K; $\theta = 2.3$.

COVERAGE CALIBRATION

The data in fig. 3 and 4 may be used to construct an approximate coverage calibration to calculate the attenuation of the 957 eV (K.E.) O 1s electrons in the region in which the oxide nuclei form and grow.

The method is as follows. We assume that the area of the O 1s peak at saturation (295 K) corresponds to 2.0 monolayers [1, 8] and therefore the coverage of oxygen that produced the 60 % attenuation of the Ni $2p_{\frac{3}{2}}$ metal line in fig. 4(ii) was ~ 2.2. Using an electron take-off angle of 45° and averaging over the interplanar spacing of the (100), (110) and (111) planes of NiO we calculate a mean free path λ for the 634 eV electrons of ~ 11 Å in NiO. A similar calculation from the UPS data in fig. 3a(iv), where the shoulder near the Fermi edge is assumed to originate from the underlying nickel metal, yields a value of λ of ~ 6 Å for the 40 eV electrons. These compare well with literature values.[9] If we assume the energy dependence of λ shown in ref. (9), we estimate a mean free path for the 957 eV electrons of ~ 14 Å. By a process of iteration this may be used to correct the original estimate of the oxygen coverage of 2.2 layers.

The peak areas in fig. 3 for $\theta > 0.5$ may then be approximately corrected for attenuation by using this value of λ and an average escape depth of half the final ($\theta = 2.0$) film thickness for the second layer. On this basis the coverage at 5 L exposure is calculated to be 0.43 monolayers. This may be compared to the oxygen coverages at the onset of oxide nucleation on Ni (100) and (111) surface (0.4 and 0.34 monolayers respectively [1a, b]). The external consistency of this calibration may be checked by using the measured oxygen O 1s peak areas at $\theta = 0.43$ to compute the saturation coverage of CO at 295 K (from the experimentally determined carbon monoxide O 1s peak areas). On this basis the saturation coverage of CO was 0.48, close to the value of ~ 0.5 deduced by Brennan for polycrystalline nickel.[10]

Some preliminary results at 90 and 450 K are now reported.

$$T = 90 \text{ K}$$

The adsorption of O_2 to $\theta \sim 1.4$ resulted in a broad O 1s line with peak maximum at 530.0 eV and a shoulder at ~ 531.4 eV (fig. 5c). The Ni $2p_{\frac{3}{2}}$ region at $\theta = 0$ and 1.4 is shown in fig. 5b, indicating that the attenuation and broadening are much less than at an equivalent coverage produced by adsorption at 295 K. The spectrum of the valence band (fig. 5a) also shows less structure than the corresponding 295 K exposure spectrum (cf. fig. 3a (ii, iii)). The peak at 1.8 eV is not evident and there is a featureless maximum in the range 4–8 eV.

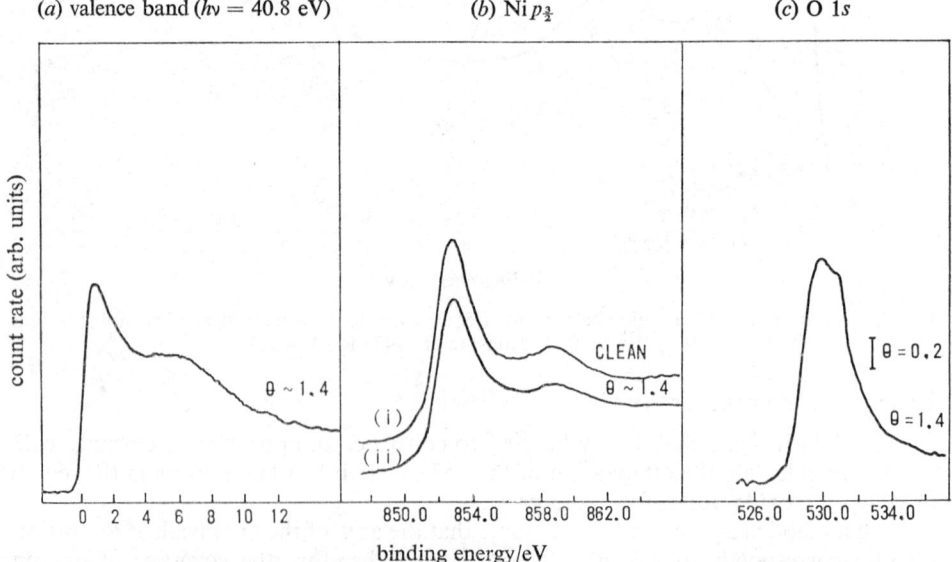

FIG. 5.—Adsorption of oxygen at 90 K. (a) Valence band region $\theta = 1.4$. (b) (i) clean nickel foil; Ni $2p_{\frac{3}{2}}$ region, (ii) same foil, $\theta = 1.4$; Ni $2p_{\frac{3}{2}}$ region. (c) Oxygen 1s region, $\theta = 1.4$.

The spectra were unchanged by warming the sample to 295 K in vacuum, but exposure to > 100 L O_2 at 295 K produced valence and core level spectra that were very similar to those shown at saturation in fig. 3, except that the O 1s line is broader, has a maximum at 529.7 eV and a shoulder at 531.5 eV. This transformation was observed even when $\theta \sim 1.8$ at 90 K before exposure to O_2 at 295 K.

$$T = 450 \text{ K}$$

The kinetics of oxide nucleation at 450 K are slower than at 295 K (in agreement with Holloway and Hudson [1a, b]) and the O 1s line is broader with a FWHM of 2.4 eV at $\theta \sim 1.8$ (fig. 6b). The peak maximum is at 529.7 eV but a shoulder has developed at 531.5 eV. The nickel $2p_{\frac{3}{2}}$ region shown in fig. 6a, is similar to fig. 3b(iv). The valence band spectrum was similar to fig. 3a(iv).

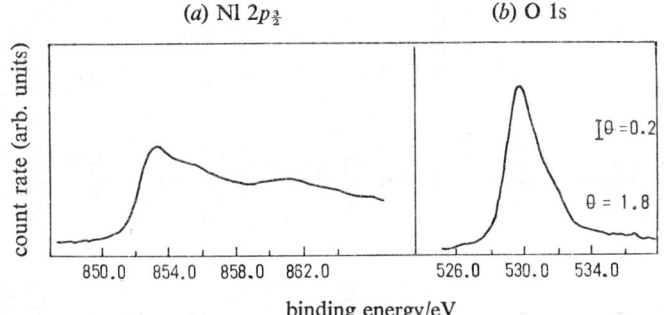

FIG. 6.—Adsorption of oxygen at 450 K. (a) Ni $2p_{\frac{3}{2}}$ region, $\theta = 1.8$. (b) Oxygen 1s region, $\theta = 1.8$.

DISCUSSION

The data are summarized in table 1.

A. NICKEL CORE AND VALENCE LEVELS

(i) $T = 295$ K, $\theta \gtrsim 0.5$—Oxide Region

This coverage region is discussed first as the changes in the spectra are understandable in terms of the formation of nickel oxide (NiO).

At zero coverage and at saturation, the XPS line positions agree well with the data of Kim and Winograd [11] for clean nickel and bulk NiO respectively. In fig. 4(ii) the shoulder at 852.8 originates from the Ni $2p_{\frac{3}{2}}$ line of the underlying nickel, while the line at 854.4 eV comes from Ni^{2+}. The satellites at 856.1 and 861.4 eV have been discussed previously.[11-13]

The UPS data confirm that nickel oxide is formed in that the levels at 1.8, 5.4 and 10 eV have been attributed to NiO.[12] We also observe the levels at 3.3 and 7 eV reported by Hagstrum and Becker.[14] Kim[12] attributed these to Ni_2O_3. He also assigns an O 1s line position of 531.7 eV to Ni_2O_3. This is close to the line position observed during low- or high-temperature oxygen adsorption. We have also observed an oxygen line at ~ 531.5 eV on evaporated nickel films, with no increased formation of the 3.3 and 7 eV levels. The peak at 1.8 eV presumably originates from the Ni^{2+} d electrons, shifted 1.8 eV from Ni^0. This may be compared to the observed shift of the $Ni^0 (2p_{\frac{3}{2}}) \to Ni^{2+} (2p_{\frac{3}{2}})$ of 1.6 eV.

TABLE 1.—XPS AND UPS DATA ON Ni+O$_2$ SYSTEM
Zero Energy = Fermi Level of Ni

adsorption temperature/ K	coverage θ	O 1s B.E./ eV	O 1st FWHM/ eV	Ni 2p$\frac{3}{2}$ Region B.E./eV	valence Region B.E./eV	comments
295	0–0.3	530.1	2.5	852.8 858.6	0.5 1.8 5.6	no attenuation of core or valence levels growth of level at 5.6 eV
295	0.4$_3$	529.9	2.2	852.8 858.6	0.5 1.8 5.6	slight attenuation of 0.5 eV peak growth of 5.6 eV level no detectable change in Ni 2p region
295	0.6$_8$	529.7	2.1	852.8 858.6	0.5 1.8 5.6	15 % attenuation of 0.5 eV peak growth of 5.6 eV level little change in Ni 2p region
295	1.2–2.0	529.7	1.9	not clearly defined in this region; see $\theta = 2.3$	~0.5 1.8 3.3 5.4 10	levels given for $\theta = 2.0$ Ni 2p levels attenuate and broaden 858.6 level decreases and shifts to higher B.E. O 1s FWHM constant
473	2.3	529.7 ~531.5	~2.3	852.8 854.4 856.1 861.4	as $\theta = 2.0$	O 1s level broadened after oxidation at 473 K; probably a doublet
90	1.4	~530.0 ~531.4	~3.1	852.8 858.6	~1.0 ~5.6	O 1s probably a doublet little change in Ni 2p region
450	1.8	529.7 ~531.5	~2.4	not clearly developed	as $\theta = 2.0$	O 1s probably a doublet

† experimental FWHM, not unfolded from spectrometer resolution and X-ray line width.

(ii) $T = 295$ K, $\theta \lesssim 0.4$—Chemisorption Region

The absence of attenuation in either the core or valence levels in this coverage region is puzzling and is an effect that has been tentatively identified before.[3, 4, 15] The calculated escape depths suggest that if a surface layer of nickel oxide were formed at $\theta \sim 0.4$ the Ni $2p_{\frac{3}{2}}$ line and 0.5 eV peak in the Ni d band should be attenuated 12 and 25 % respectively. The absence of attenuation indicates that no nickel oxide (Ni^{2+}) is present in the chemisorbed layer and that the attenuation of the photoelectrons by oxygen atoms is very small. It therefore seems that in the outer layer the nickel atoms retain their metallic nature below $\theta \sim 0.4$. This may be understood if the Ni-O bond is formed from the itinerant electrons; i.e., it is delocalized. The existence of an Ni^{2+} ion at the surface of the metal (a conductor) would imply a difference in electrostatic potential between two points in the conductor. This is clearly impossible unless the state so formed is not in electrical equilibrium with the rest of the solid—i.e., it is a surface state. When the oxide nucleates in the surface region, changing it from a conductor to a semi-conductor or insulator, then a difference in potential can exist between one point and another, as charge can no longer flow to neutralize the charge difference. Similar ideas to these have been advanced previously.[3] We should therefore perhaps not expect to see a chemical shift in the chemisorption region unless a localized surface state is formed.

The absence of attenuation of the peak near the Fermi edge is in distinct contrast to the behaviour of the Pt-O_2 system[3] where 0.4 monolayers of oxygen attenuated a peak at the Fermi edge by ~ 20 %. Certainly, if electron transfer from levels near the Fermi edge is important in bonding oxygen on nickel one should see some effect in this region, bearing in mind the extremely short escape depth (~ 6 Å). The occurrence or absence of attenuation of the peak near the Fermi edge depends on more subtle considerations than electron transfer since the adsorption of CO on our samples attenuated the 0.5 eV peak in a manner similar to CO on Pt.[16]

B. OXYGEN $1s$ LINES $T = 295$ K

Fig. 2 and 3 show that the oxygen line narrows and shifts to lower binding energy as θ increases above ~ 0.6. The greater width below $\theta \sim 0.4$ may be related to site heterogeneity which is removed in the nickel oxide. The decrease in binding energy is small and remains unexplained. At high coverage the line position agrees well with the data of Kim for NiO.[11]

C. ADSORPTION AT 90 AND 450 K

Holloway and Hudson[1a, b] found that oxygen adsorption at 147 K on (111) and (100) nickel produced surfaces on which the NiO LEED patterns were barely detectable even though the total coverage was ~ 2 monolayers. This agrees with the lack of effect on the Ni $2p_{\frac{3}{2}}$ lines after adsorption at 90 K, i.e., very little NiO is present. That this layer is stable in vacuum but not in oxygen indicates that the energy released upon adsorption may be sufficient to activate the reordering of the surface to NiO.

The presence of an O $1s$ peak at ~ 531.5 eV ($T = 90$; 450 K) for oxygen on evaporated films $T = 295$ K) may reflect disorder in the surface layer; the lack of chemical shift at 90 K suggests that the layer is still a conductor. This shoulder almost coincides with an O $1s$ line in Ni_2O_3 which Kim[11] regards as a gross defect structure of NiO.

CONCLUSIONS

Adsorption of oxygen on nickel to coverages less than that required to nucleate NiO produces little change in the valence spectra and no detectable change in the core level spectra. This is interpreted as indicating that the bonding is delocalized and that the surface metal atoms do not become positively charged, because of requirements of electrical equilibrium in a conductor. This may be a result of general importance in XPS studies of adsorption. The observations at higher coverage are consistent with the growth of a 2-layer-thick nickel oxide film. Chemical shifts occur in this coverage region because charge differences can exist between the oxide and metal. More detailed studies on single crystal surfaces are in progress and confirm many of the above ideas.

The authors gratefully acknowledge helpful discussions with Dr. D. Mitchell and Dr. P. Sewell.

[1] (a) P. H. Holloway and J. B. Hudson, *Surface Sci.*, 1974, **43**, 123. (b) P. H. Holloway and J. B. Hudson, *Surface Sci.*, 1974, **43**, 141.
[2] A. M. Horgan and D. A. King, *Surface Sci.*, 1970, **23**, 259.
[3] P. R. Norton, *Proc. 3rd Int. Symp. Surface Physics* (Utrecht, The Netherlands, June 1974); *Surface Sci.*, 1975, **47**, 98.
[4] P. R. Norton and R. L. Tapping, to be published.
[5] P. R. Norton and P. J. Richards, *Surface Sci.*, 1974, **41**, 293.
[6] D. E. Eastman and J. K. Cashion, *Phys. Rev. Letters*, 1971, **27**, 1520.
[7] P. J. Page, D. L. Trimm and P. M. Williams, *J.C.S. Faraday I*, 1974, **9**, 1769.
[8] D. R. Mitchell and P. B. Sewell, private communication.
[9] I. Lindau and W. E. Spicer, *J. Electron Spectr.*, 1974, **3**, 409.
[10] D. Brennan and F. H. Hayes, *Phil. Trans. Roy. Soc. A*, 1965, **258**, 347.
[11] K. S. Kim and N. Winograd, *Surface Sci.*, 1974, **43**, 625.
[12] K. S. Kim, *Chem. Phys. Letters*, 1974, **26**, 234.
[13] S. Hüfner and G. K. Wertheim, *Phys. Rev. B*, 1973, **8**, 4857.
[14] H. D. Hagstrum and G. E. Becker, *Proc. Roy. Soc. A*, 1972, **331**, 394.
[15] C. R. Brundle, *Proc. Int. Conf. Electron Spectr. J. Electron Spectr.*, 1974, **5**, 291.
[16] P. R. Norton and P. J. Richards, *Surface Sci.*, 1975, **49**, 567.

Note added in proof:

The points raised by Somorjai and Landman on pp. 146-7 are well taken. Our interest in the nickel/oxygen system was however dependent upon the fact that the chemisorption and oxide regimes are clearly distinguishable on the basis of the kinetics. It therefore seemed an ideal system in which to investigate qualitatively the two types of interaction by photoelectron spectroscopy. There also is the considerable interest in the corrosion behaviour of polycrystalline nickel which prompted this investigation. We presented (111) Ni data at this meeting and have since also completed a study of the (100) Ni/O_2 system with the result that the qualitative behaviour on polycrystalline nickel (e.g., absence of effect on the nickel d band in the chemisorption region) is exactly duplicated on (111) and (100) nickel. Differences of the order of a factor of 2 to 3 exist in the nucleation kinetics and the coverage at which nucleation starts differs on the two surfaces. There are also slight differences in the $O1s$ behaviour on the two surfaces that appear to be correlated with the structure of the final oxide, but again our qualitative observations stand.

Electron Spectroscopic Studies of the Initial Oxidation of Zinc

By D. Briggs

Imperial Chemical Industries Limited, Corporate Laboratory, The Heath, Runcorn, Cheshire

Received 24th March, 1975

The initial stages of oxidation of evaporated zinc films at room temperature have been studied by means of photoelectron spectroscopy (HeI, HeII and MgK_α) and Auger electron spectroscopy (MgK_α excited). Two oxygen species (BE's 530 and 532 eV) are found to coexist even at the lowest coverages whilst the formation of zinc oxide has been detected at exposures of 100 L oxygen. The results of experiments in which the angle of electron emission from the sample surface was varied provide extra information about the nature of the two oxygen species and reveal a measureable chemical shift between the $Zn2p$ levels in the metal and its oxide.

The combination of X-ray and u.-v. photoelectron spectroscopy (XPS and UPS) is now probably accepted to be the most powerful tool for the investigation of chemisorption phenomena to emerge from the newly established group of surface-sensitive techniques. However, recent studies of the interaction of small molecules with metal surfaces at the " monolayer " level, especially oxidation studies, are disappointing in that substrate core-level binding energy shifts are not generally observed. Furthermore, multiple adsorbate peaks are frequently observed in XPS and, relaxation effects apart, it is difficult to explore the possibility that these may represent identical species at different sites in or on the surface, besides representing different species.

As has been previously pointed out,[2] X-ray photoelectron spectra also contain high resolution Auger electron peaks which tend to be neglected even though containing potentially useful information about substrate atom environments. Also, variation of the angle of electron emission enables some depth resolution to be made within the electron escape depth.[3, 4] Reports of the use of this technique are as yet very limited, possibly because not all commercial instruments have the required sample geometry.

The object of the present work was to apply all these techniques to a metal + oxygen system which could serve as a prototype for later catalytic oxidation studies. Zinc was chosen because it avoids the problem of variable oxidation state, which can only complicate spectrum interpretation, and because it has not previously been studied under controlled reaction conditions by these techniques.

EXPERIMENTAL

Measurements were made on an AEI ES 200B electron spectrometer at room temperature, with an analyser transmission energy of 65 eV. High purity zinc was evaporated from a tungsten filament onto the stainless steel backing of a translatable probe tip inside a sample preparation chamber (base pressures *ca.* 5×10^{-10} Torr). The source chamber (base pressure *ca.* 4×10^{-11} Torr) housed the X-ray (MgK_α) and u.-v. (HeI and II, NeI and II) sources mounted at 135° to each other and at 90° to the analyser entrance slit. The sample geometry allowed sequential excitation by either source to produce an emergent photo-electron beam at any desired angle to the sample surface ($\theta = 0$–90°). Oxygen (BOC Grade X) was admitted directly from a cylinder via a bakeable UHV manifold.

During u.-v. measurements a small negative bias was applied to the sample to increase sensitivity and to alow the observation of a sharp electron emission threshold for the determination of work functions.

The instrument was calibrated as follows. The difference between the KE of the Au$4f_{\frac{7}{2}}$ peak when excited by Al$K_{\alpha 1,2}$ and Mg$K_{\alpha 1,2}$ was set to 233 eV (there is sufficient Al impurity in the Mg target used to give rise to detectable AlK_α excited spectrum). Comparison of the Au conduction band excited by MgK_α and HeI/II then allowed the BE of the Au$4f_{\frac{7}{2}}$ peak relative to the Fermi level to be measured (Mg$K_{\alpha 3,4}$ satellites obscure the direct observation of the Fermilevel in the Mg$K_{\alpha 1,2}$ spectrum). This value was 84.0 eV. Binding energies in this work were referred directly to the Fermi level (E_F). This was located in the UP spectrum to be 9.8 eV below the core-like Zn$3d_{\frac{5}{2}}$ peak. This peak can also be observed in the XP spectrum and hence serves as intermediate reference point. Deconvolution of overlapping peaks to give relative areas was performed on a Du Pont 310 curve resolver.

By rotation of the sample probe, the angle of electron emission to the surface (θ) could be varied between 0 and 90°.

The position $\theta = 0°/90°$ had to be estimated visually—the accuracy here can only be to within 5°. Changes in angle could be measured directly from a graduated scale on the rotation mechanism to $\pm 1°$.

RESULTS

1. UPS AND WORK FUNCTION MEASUREMENTS

Oxidation was found to proceed at a convenient rate at ca. 10^{-7} Torr; exposure for some minutes at pressures $\leqslant 10^{-8}$ Torr had no obvious effect on the spectrum of the clean surface. Fig. 1 shows the reaction sequence using HeI (21.2 eV) radiation.

The linear growth of an O$2p$-like peak some 5 eV below E_F is observed up to ca. 200 L exposure, with concomitant decrease in the Zn$3d$ intensity. Thereafter, the

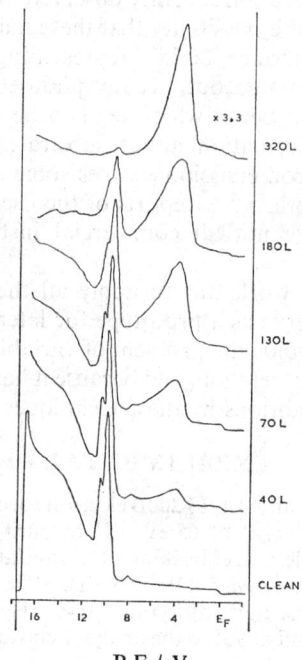

Fig. 1—HeI (21.22 eV) spectra at increasing exposures of Zn to O$_2$ at ca. 10^{-7} Torr. The feature at 8 eV is the HeIβ satellite of the Zn$3d$ peak. Count rate = 10^5 c s^{-1} f.s.d.

overall intensity begins to decrease as the " O2p " peak continues to increase in size relative to the Zn3d. The Zn3d$_{\frac{3}{2},\frac{5}{2}}$ components are unresolvable by 250 L exposure and the peak has almost disappeared by 320 L exposure. Weak structure is observed between E_F and the onset of the " O2p " peak after 250 L and this increases in relative intensity on further exposure.

Pumping on a film exposed to 320 L O$_2$ for 16 h at 10^{-10} Torr caused the UPS spectrum to revert to a form similar to that produced by *ca.* 200 L exposure.

Although the HeII spectra were not continuously recorded there is some evidence that these contain more structure in the " O2p " region. A more thorough study is required before these spectra can be discussed.

Simultaneous measurement of the threshold energy for electron emission (E_T) and the Fermi energy (E_F) in a HeI spectrum allows the sample work function ($\bar{\phi}$) to be computed from:

$$\bar{\phi} = h\nu - (E_T - E_F) \text{ where } h\nu = 21.2 \text{ eV}$$

or, more simply, if E_T is expressed as a BE relative to $E_F = 0$

$$\bar{\phi} = 21.2 - E_T \text{ (eV)}$$

$\bar{\phi}$ for a freshly evaporated Zn film was found to be 4.2 ± 0.05 eV in excellent agreement with the two reliable values reported previously.[5] During oxidation $\bar{\phi}$ decreased monotonically, reaching a value of 3.9 eV after *ca.* 40 L exposure and finally 3.45 eV after 320 L.

2. XPS AND X-AES MEASUREMENTS

After evaporation of a new film, a small O1s contamination peak was detected in the XPS spectrum. Since great difficulty had been experienced in thoroughly

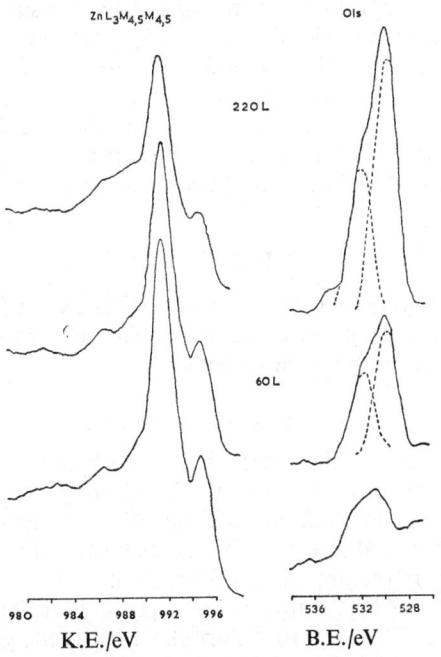

FIG. 2.—Selected XPS (MgK$_\alpha$) spectra at increasing exposures of Zn to O$_2$ at *ca.* 10^{-7} Torr. Note the original oxygen contamination of the " clean " film (est. 20 % of a monolayer). Count rates = 10^4 c s^{-1} f.s.d. (Zn *LMM*) and 10^3 c s^{-1} f.s.d. (O 1s).

de-gassing Zn and then evaporating in a controlled way (because of its low m.p.) it was decided to continue an oxidation run. During exposure the oxygen pressure was 10^{-7} Torr and the regions of the spectrum scanned covered the O1s and Zn $L_3M_{4,5}M_{4,5}$ peaks. The Zn2p region was not continuously recorded, since preliminary experiments confirmed the general finding that there is a negligible shift between peaks for the clean and oxidised surface.

From the lowest exposures two oxygen peaks were discernible at 529.8 and 532.0 (± 0.3) eV. During oxidation both peaks increased in intensity—the 532 eV peak at a steady linear rate, initially only one third of the growth rate of the 530 peak. By ca. 150 L exposure, however, the 530 peak growth started to tail off rapidly so that total growth of the O1s peak almost "plateaued" at 220 L exposure. Fig. 2 shows three of the ten spectra recorded.

The most prominent Auger peak, Zn $L_3M_{4,5}M_{4,5}$ at 992.3 eV decreased steadily in intensity during the run. Resolution of the low KE features was noticeably degraded by 100 L exposure, while at saturation a significant increase in intensity in this region was apparent (see fig. 2). Relative intensities of the main peak to the shoulder at 4 eV higher BE for the "clean" surface and after 100 L and 220 L exposure are approximately 1:0.22, 0.27 and 0.30 respectively. Evacuation at $< 10^{-10}$ Torr for 16 h did not change the spectra recorded after 220 L exposure.

3. ANGULAR VARIATION EXPERIMENTS (Zn + 220 L O_2)

Fig. 3 shows the rather dramatic effects of varying θ between 16 and 75° while fig. 4(a) quantifies the data. It seems probable that the overall drop in intensity between $\theta = 60$ and 75° is due to rotation slightly off-axis leading to a drop in incident X-ray flux. This does not affect the curves in fig. 4(b) which show the angular variation of peak ratios. The "ZnO" data were derived by measuring the intensity of the shoulder 4 eV to higher BE of the Zn $L_3M_{4,5}M_{4,5}$ peak and subtracting an amount calculated from the intensity ratio of this shoulder to that of the main peak measured from the spectrum of the clean surface. This assignment is discussed below. Measurements of the Zn$2p_{\frac{3}{2}}$ peak at $\theta = 15$ and 75° revealed a 75 % increase in intensity, a slight increase in FWHM and a shift in peak maximum of 0.3 eV to higher BE at $\theta = 15°$ relative to $\theta = 75°$. This effect was reproduced in a second oxidation run.

DISCUSSION

The data will be discussed in terms of two questions: what do they reveal about the mechanism of the initial stages of oxidation of Zn and how do they relate to other studies of metal + oxygen adsorption systems?

1. ZINC OXIDATION

It is worth commenting at the outset that the only other report of the controlled oxidation of a clean Zn surface is an A.C.S. Meeting Abstract [6] which discusses LEED/Auger studies of the oxidation of Zn(0001). At temperatures up to 425 K log (Okll intensity) increased linearly with exposure to saturation at 10^3 Torr s and LEED showed no superstructure pattern formation.

We believe that a number of features in our spectra indicate the formation of oxide during exposures up to 200 L (2×10^{-4} Torr s). First is the growth of a broad band in the region 0–5 eV to high BE of the Zn $L_3M_{4,5}M_{4,5}$ peak. Well oxidised Zn surfaces show [7,8] two sets of Auger peaks separated by 4 eV, those to higher BE are nearly twice as broad and can be correlated directly with those from bulk ZnO.[7]

We have independently reproduced these results in our system and conclude that this broad band represents the emergence of a chemically shifted " ZnO " peak superimposed on the existing structure to high BE of the Zn $L_3M_{4,5}M_{4,5}$ peak. Second is the appearance of the UP spectra. HeI and NeI spectra of a cleaved ZnO crystal

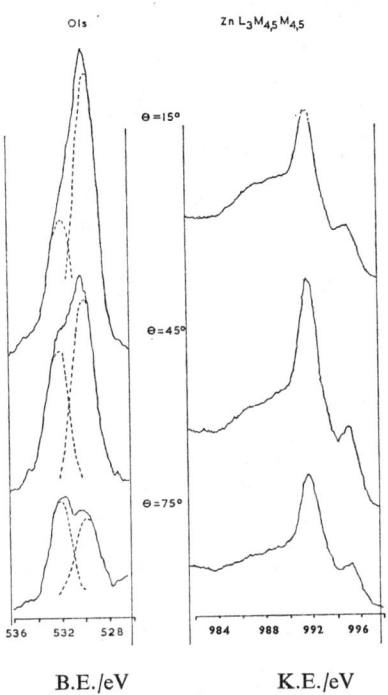

FIG. 3.—Variation of angle of electron emission (Θ) for Zn+220 L O_2. Count rates as in fig. 2.

have been reported [9] and these are virtually identical to those shown in fig. 5. In our experiments the relative intensity of the low KE-scattered electron background changed appreciably during oxidation and became markedly dependent on the applied bias voltage. It is therefore fortuitous that the overall EDC's are so close to those in ref. (9); but this does not affect the relative intensities of the principal features above the background. Third is the evidence from angular variation that the $Zn2p_{\frac{3}{2}}$ peak consists of two components after oxidation, the new component having a BE at least 0.3 eV higher than the metal component consistent with cation formation.

We now consider the identification of the two oxygen species indicated by the O1s peaks at 530 and 532 eV respectively. The relative magnitudes at saturation and the adsorption kinetics suggest that the peak at 530 eV is due to oxide, and this is consistent with the BE of the O1s peak in bulk ZnO.[7] Thus the higher BE peak would be consistent with some form of " chemisorbed " oxygen with a lower formal negative charge than O^{2-}.

The angular variation data allows this picture to be viewed in more detail. Immediately, it is clear from the (I_θ, θ) curves (fig. 4(a)) that the O (530 eV) and the ZnO peaks are linked and so are the O (532 eV) and Zn peaks. This confirms the assignment of the two oxygen species. Using the model of Fraser et al.[3] for a layer of ZnO (thickness δ) on Zn, the variation in the ratio (R_θ) of the ZnO/Zn peak heights (fig. 4(b)) gives a value of $\delta/\lambda \simeq 0.25$ where λ is the electron escape depth. For the Zn

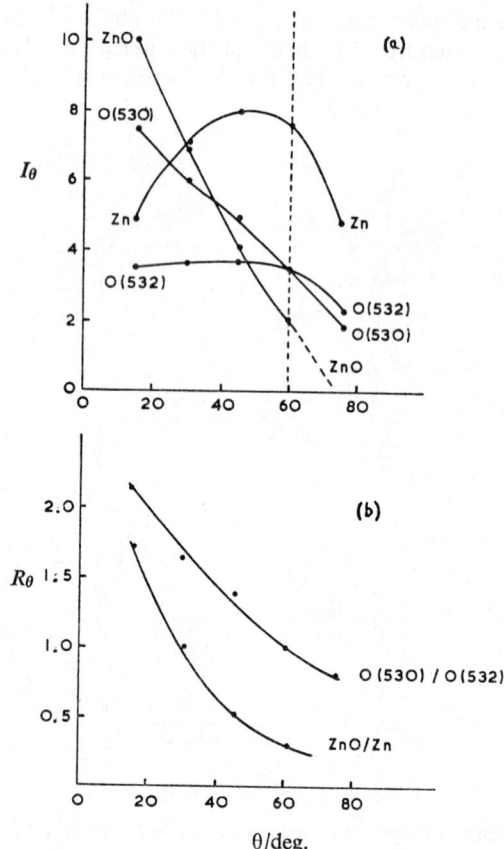

FIG 4.—(a) Variation of peak intensities (I_θ) with angle of electron emission ($\theta = 15°–75°$, see text) (b) Variation of peak intensity ratios (R_θ) with $\theta = 15°–75°$.

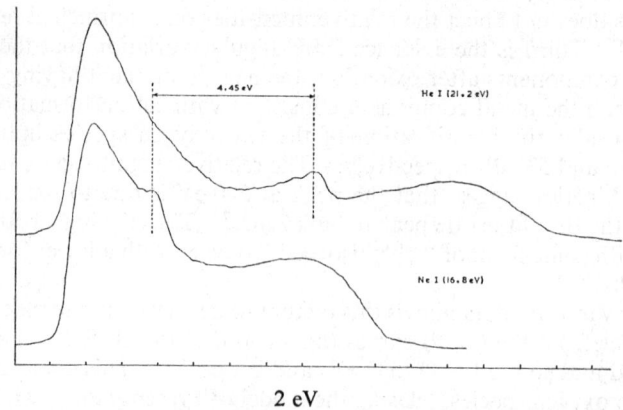

FIG. 5.—HeI (21.22 eV) and NeI (16.67, 16.85 eV) spectra of $Zn+220$ L O_2. Count rate $= 10^4$ c s^{-1} f.s.d.

LMM peak we used $\lambda \simeq 14$ Å, from which $\delta \simeq 3.5$ Å. Using the Zn$2p_{\frac{3}{2}}$ and O$1s$ peaks the calculated [9] Zn/O atomic ratio is 2.5 ($\theta = 75°$) and 1.6 ($\theta = 15°$) where $\lambda \simeq 7$–10 Å and 2–3 Å respectively for these lower KE electrons. Hence, within the value of δ the Zn/O ratio is > 1 which suggests that the ZnO layer is in the form of islands leaving Zn exposed. Moreover $\delta = 3.5$ Å corresponds only to a monolayer, which does not constitute a ZnO "lattice".

The argument may now be taken a step further. The angular variation of the Zn LMM peak is consistent with the model of the Zn substrate being covered by islands of ZnO. However, the O (532 eV) peak shows an invariance to change in Θ similar to that expected from a clean surface. This result would be expected from a situation in which the O atoms were embedded *in* the exposed Zn patches, rather than adsorbed *on* them, in other words the "chemisorbed" oxygen leads to a reconstructed Zn surface. A more detailed treatment of this surface using a patched surface model [4] will be given elsewhere [11] in an attempt to rationalise the relative intensities of the two O$1s$ signals. This model would also explain why the "ZnO" LMM peak is a broad hump rather than a distinct peak (as observed in well oxidised Zn surfaces); the reconstructed Zn layer contains Zn atoms perturbed by oxygen "atoms" to a lesser extent than in ZnO (Zn/O ratio > 1) giving rise to a range of smaller shifts between the Zn and ZnO peaks.

It would appear from this data, therefore, that upon adsorption on Zn, oxygen is dissociated into atoms which reconstruct the surface and which may or may not be disposed towards nucleation of ZnO islands (at least double layers). Both processes seem to occur from the start of chemisorption at 300 K. This mechanism is not inconsistent with the observed decrease in average work function ($\bar{\phi}$).[12]

COMPARISONS WITH OTHER SYSTEMS

We restrict ourselves here to comparisons with XPS work on initial oxidation of metals since the UPS features observed here are very broad and do not allow meaningful comparison with those systems in which narrow resonances occur (e.g., ref. (13)).

Unfortunately, published studies of oxygen chemisorption on Ag, Au and (especially relevant) Cu [14] were carried out in relatively poor vacuum conditions (10^{-8} Torr base pressure) with etched metal films and rather high exposures (10^{-4} Torr), with few details of relative peak intensities. Comparable studies of the Ni+O$_2$ system have been carried out by Brundle [15] (which also cover the low temperature (77 K) adsorption) and (in less detail) by Page et al.[16] These are especially interesting since the Ni+O$_2$ system has been much studied by a variety of techniques, whereas the Zn/O$_2$ system has not.

The results from both these studies (constrained by the apparatus to $\theta = 45°$) at ca. 300 K show similar results to the present work in that two O$1s$ peaks are seen at saturation with relative intensities of ca. 2 : 1 (more of the lower BE peak). Page et al. do not quote BE's, those of Brundle, when corrected to Au$4f_{\frac{7}{2}} = 84.0$ eV, agree closely with the data for Zn+O$_2$. However, Page et al. note that relative intensities of the O$1s$ doublet do not change between 6 L and 56 L exposure while Brundle's reaction sequence of saturation at 77 K, warming to 300 and resaturation at 10^{-4} Torr suggests a growth sequence for these two peaks similar to that observed here. This casts doubt on the comparability of the oxygen exposures quoted. In this work, two pressure gauges in pumping throats close to, and removed from, the sample area and both some distance from the diffuse point of gas entry to the chamber gave the same reading, giving rise to same confidence in the quoted exposures. Page et al. conclude either that (a) the two oxygen species correspond to oxide (lower BE peak) and oxygen adsorbed on that oxide or (b) the O1s doublet is due to multiplet splitting. Brundle suggests the 530 eV peak to be due to O^{2-} incorporated into the surface and the 532 eV peak to represent more covalently adsorbed oxygen on the surface, and that the data support an island growth model for NiO.[17]

Assuming some correspondence between the two systems the present data for O_2/Zn would not support either of the suggestions of Page et al.[16] On the other hand, and here the angular variation data is crucial, they largely agree with Brundle's suggestion [15] except that all chemisorbed oxygen seems to be incorporated, there being a BE difference between this and nucleated oxide. The observation [15] of an even higher BE ($\sim$ 533 eV) precursor state, stable only at low temperature, adds weight to this interpretation; this species must then be truly chemisorbed oxygen *on* the Ni surface.

In conclusion, we have demonstrated how X-AES data can provide useful additional information in XPS on chemisorption and that the angular variation experiment literally adds an extra dimension to the picture obtained. In this dissociative system UPS data makes little contribution to understanding the mechanism, but should come into its own when the work is extended to low temperatures. There seem to be strong similarities between the initial oxidation stages of Zn and Ni at $\sim$ 300 K.

The author acknowledges the experimental contributions of V. A. Gibson and Miss S. J. Cross and wishes to thank Dr. D. T. Clark for provision of curve resolution facilities and Dr. C. R. Brundle for access to data prior to publication.

[1] C. R. Brundle, *J. Electr. Spectr.*, 1974, **5**, 291.
[2] D. Briggs, *Faraday Disc. Chem. Soc.*, 1974, **58**.
[3] W. A. Fraser, J. V. Florio, W. N. Delgass and W. D. Robertson, *Surface Sci.*, 1973, **36**, 661.
[4] C. S. Fadley, *J. Electr. Spectr.*, 1974, **5**, 725.
[5] J. C. Riviere in *Solid State Surface Science*, ed. M. Green (Dekker, New York, 1969).
[6] W. N. Unertl and J. M. Blakely, *Abs. Pap. ACS.*, 1974, 113.
[7] E. Schön, *J. Electr. Spectr.*, 1973, **2**, 75.
[8] J. E. Castle and D. Epler, *Proc. Roy. Soc. A*, 1974, **339**, 49.
[9] R. A. Powell, W. E. Spicer and J. C. McMenain, *Phys. Rev. B*, 1972, **6**, 3056.
[10] W. J. Carter, G. K. Schweitzer and T. A. Carlson *J. Electr. Spectr.*, 1974, **5**, 827.
[11] D. Briggs, to be published.
[12] M. W. Roberts and B. R. Wells, *Trans. Faraday Soc.*, 1966, **62**, 1608.
[13] A. M. Bradshaw, D. Menzel and M. Steinkilberg, *Faraday Disc. Chem. Soc.*, 1974, **58**.
[14] S. Evans, E. L. Evans, D. E. Parry, M. J. Tricker, M. J. Walters and J. M. Thomas, *Faraday Disc. Chem. Soc.*, 1974, **58**.
[15] C. R. Brundle, *Chem. Phys. Letters*, 1975, **31**, 423.
[16] P. J. Page, D. L. Trimm and P. M. Williams, *J.C.S. Faraday I*, 1974, **69**, 1769.
[17] R. H. Holloway and J. B. Hudson, *Surface Sci.*, 1974, **43**, 123, 141

Interaction of Oxygen with Cu (100) Studied by Low Energy Electron Diffraction (LEED) and X-Ray Photoelectron Spectroscopy (XPS)

By Malcolm J. Braithwaite, Richard W. Joyner and M. Wyn Roberts

School of Chemistry, University of Bradford, Bradford BD7 1DP

Received 12th April, 1975

At 80 K, exposure of the Cu(100) surface to oxygen results in a disordered (2×2) structure, followed by a ($\sqrt{2}\times\sqrt{2}$) R45° mesh at higher coverage. The FWHM value of the O(1s) peak at 80 K is large (~4.5 eV) indicating a range of metal-oxygen bond strengths. On warming to 290 K about half of the oxygen adsorbed at 80 K is desorbed below 150 K and there is a narrowing of the O(1s) peak to 2.3 eV. The ($\sqrt{2}\times\sqrt{2}$) R45° mesh is, however, retained but with apparently improved ordering of the surface structure. A model for the adlayer is suggested.

At 290 K, with increasing exposure to oxygen at low pressure (10^{-6} Torr), three diffraction patterns are observed; an oblique "4 spot" mesh, followed by a ($\sqrt{2}\times\sqrt{2}$) R45° and finally a ($\sqrt{2}\times2\sqrt{2}$) R45°. We suggest that the ($\sqrt{2}\times\sqrt{2}$) R45° structure, the stable structure at 80 K, involves only chemisorbed oxygen while the ($\sqrt{2}\times2\sqrt{2}$) R45° reflects oxygen incorporation.

At 290 K, exposure of the Cu(100) surface to 5 Torr of oxygen leads to substantial oxygen uptake, a disordered diffraction pattern, the copper 2p peaks showing for the first time the satellites known to be characteristic of copper (II). Heating to 520 K *in vacuo* causes the reappearance of the ($\sqrt{2}\times2\sqrt{2}$) R45° mesh and the disappearance of the satellites. Heating to 570 K in 0.5 Torr of oxygen and subsequent cooling in oxygen results in the appearance of diffraction rings which are compatible with the formation of CuO, which XPS shows to be present.

Interpretation is facilitated by recourse to thermodynamic data and to results already published for polycrystalline copper surfaces.

There have been a number of previous structural studies [1-9] of the interaction of oxygen with Cu(100) surfaces; these are summarised in table 1. There is some disagreement both on the sequence of diffraction patterns observed and on the patterns themselves, e.g., in the most recent study by McDonnell and Woodruff [8] a (2×2) structure at high coverage is reported for the first time.

Earlier studies on polycrystalline copper surfaces [10] by work function and adsorption techniques have clearly indicated that lattice penetration occurs at 290 K and that this is facilitated at higher oxygen pressures. In this respect copper [11] is more similar to nickel than to molybdenum, but very different from iron where even at 80 K there is convincing evidence for lattice penetration.[10] With molybdenum, surface coverage and work function data argue against lattice penetration even at 290 K.

The question of surface re-arrangement has also been considered in relation to recent X-ray photoelectron spectroscopic (XPS) data, particular attention being given to the absolute O(1s) binding energy value and the presence or absence of satellite peaks. The 530.3 eV value frequently observed for chemisorbed oxygen on metals has been discussed [12] and, although direct experimental evidence is slight, in one case where oxygen is known in the chemisorbed stage to be located below the surface [13] the O(1s) value is appreciably different (528.7 eV).

It thus seemed appropriate to examine the interaction of oxygen with a number of single crystal planes of copper using a combined *in situ* Low Energy Electron Diffraction (LEED)-XPS System over the temperature range 80 to 500 K. We have started with the Cu(100) surface and will have very much in mind (*a*) the variation in surface structure as revealed by the LEED pattern; (*b*) whether there is evidence for disorder superimposed on the ordered structure; (*c*) the O(1*s*) binding energy, peak profile and peak energy; (*d*) the onset of surface reconstruction and (*e*) the influence of substrate temperature, oxygen exposure and oxygen pressure on (*a*), (*b*), (*c*) and (*d*). We have discussed elsewhere the relationship between defects in the adlayer and imperfect LEED patterns [14] and the dependence of O(1*s*) binding energy on the strength and nature of the metal-adsorbate bond [15, 16] (e.g., H_2O, CO, CO_2).

TABLE 1.—SUMMARY OF PREVIOUS STUDIES OF OXYGEN ADSORPTION ON Cu(100): 290 K

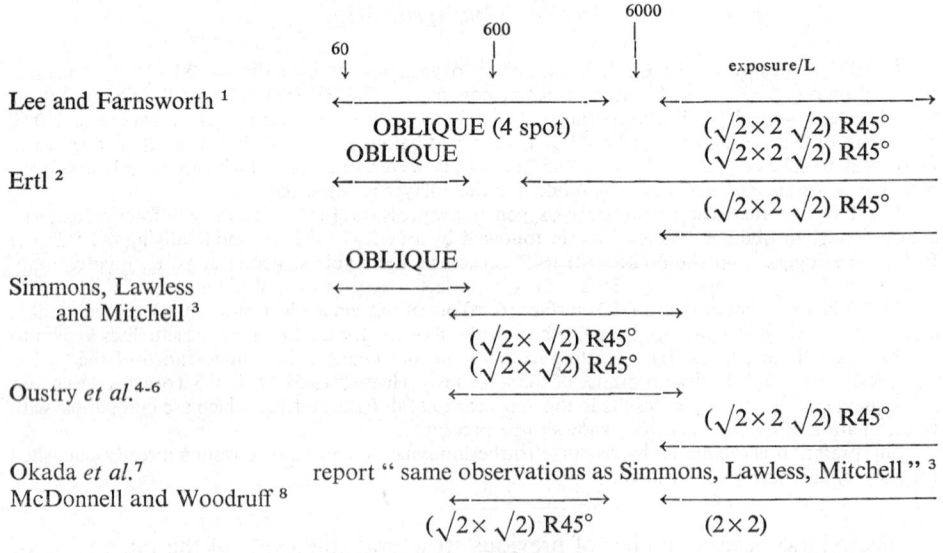

EXPERIMENTAL

The experiments were preformed in a stainless steel U.H.V. system built to our specification by Vacuum Generators Ltd., (East Grinstead, Sussex). A schematic diagram of the apparatus is shown in fig. 1; it contained 4-grid LEED optics and a hemispherical electron analyzer as well as a twin $MgK\alpha/AlK\alpha$ X-ray anode capable of up to 500 W X-ray power. After 12 h bakeout at 450 K the ambient pressure was 4×10^{-11} Torr and the " working pressure " with the X-ray source on etc. was 8×10^{-11} Torr. The titanium sublimation pump could be isolated from the main experimental chamber during exposure of the substrate to active gases such as oxygen.

The copper single crystal was cut by spark erosion from a larger crystal obtained from Metals Research Ltd., and was oriented parallel to the (100) plane within 1°. Before insertion into the vacuum chamber it was polished to a mirror finish using the procedure of Ahern, Monaghan and Mitchell.[17] In the chamber an atomically clean surface was generated by cycles of argon ion bombardment and annealing or argon bombardment at 600 K; spectroscopically pure gases were obtained from B.O.C. Ltd., manipulated in a subsidiary U.H.V. systems and introduced into the main chamber through a metal leak valve.

The electron analyzer was that developed [18] for the ESCA3 spectrometer, and generated 5×10^4 c s^{-1} at 500 W with the Al anode and a resolution for the gold $4f_{\frac{7}{2}}$ peak of 1.2 eV. Digital data were acquired using a Digico Micro 16V mini-computer, sampling the spectrometer output every 500 μs, and storing the averaged value every 0.5 s.

Some Auger data obtained in a separate retarding field LEED/Auger System are also reported.

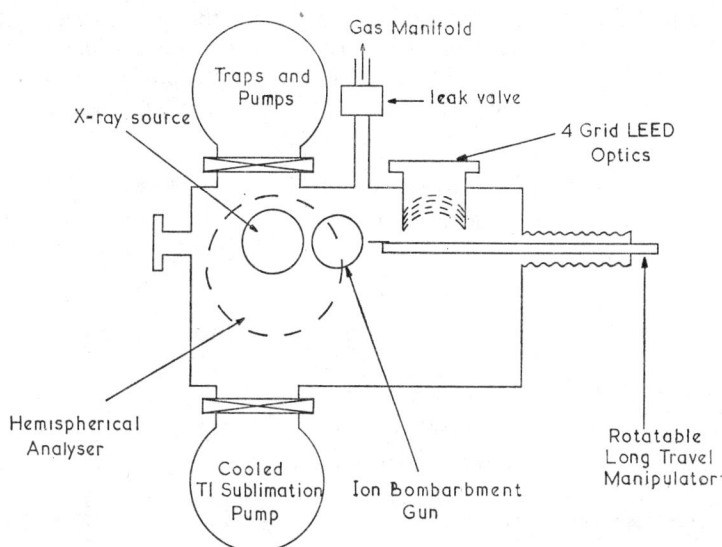

FIG. 1.—The combination LEED/XPS ultra-high vacuum system.

RESULTS

ADSORPTION AT 290 K

Cu(100) surfaces showing well ordered diffraction patterns were generated by ion bombardment, (3 μA, at 1000 eV) while the crystal was held at 600 K. A typical LEED Pattern is shown in fig. 2a. Since the X-ray photoelectron spectrum, (XPS) showed no peaks other than those due to copper, the surface is considered to be atomically clean and well ordered. This surface was exposed to oxygen, flowing at 5×10^{-8} Torr (1 Torr = 133.3 Pa); the titanium sublimation pump was closed off from the experimental chamber during exposure to oxygen in order to reduce regurgitation of carbon monoxide and water. After an exposure of only 20 L (1 L = 10^{-6} Torr s) weak diffraction features in the vicinity of the ($\frac{1}{2}$, $\frac{1}{2}$) positions were observed at about 27 eV electron energy. As the exposure was increased to $\sim$ 300 L. this pattern increased in intensity and became observable at higher electron energies, (fig. 2b and 2c). Between 500 and 800 L exposure this unusual " 4 spot " pattern gave way to a ($\sqrt{2} \times \sqrt{2}$) R45° structure, (this structure is often but inappropriately described as the $c(2 \times 2)$ mesh) shown in fig. 2d and 2e. The oxygen 1s photoelectron peaks associated with the LEED patterns are shown in fig. 3a, and the plots of oxygen 1s peak height and peak area versus exposure are shown in fig. 3b. Also given are the oxygen KL_2L_2 Auger derivative peak heights observed in the separate LEED/ Auger system. Above 2000 L exposure, at an oxygen pressure of 1×10^{-5} Torr, a new mesh indexed as Cu(100)—($\sqrt{2} \times 2 \sqrt{2}$) R45°—O was formed, (fig. 2f) which did not change even after an oxygen exposure of 5×10^6 L at 10^{-4} Torr. During these high exposures the intensity of the copper photoelectron peaks (Cu $2p_{\frac{3}{2}}$) were attenuated, although there was no obvious change in peak shape. Computer analysis of the time-averaged digital data, suggests that the Cu $2p_{\frac{3}{2}}$ peak may broaden slightly on oxygen adsorption, to a value of 1.35 eV, which compares with 1.30 eV for the clean surface. A more detailed examination of this is in progress.

During the development of the chemisorbed layer the oxygen (1s) binding energy changes by 0.6 eV (table 2). Our earlier suggestion [12a] that O(1s) binding energies may be invariant during the growth of such a layer is thus not correct in this case. A recent study by Fuggle and Menzel [12b] also shows a variation (0.3 eV) in oxygen binding energy with coverage; what is significant, however, is that the variation, when it does occur, is relatively small and this must be of considerable theoretical interest.

TABLE 2.—SUMMARY OF LEED AND XPS DATA

T/K	exposure Langmuir*	maximum pressure/Torr†	diffraction pattern	B.E. (F)/eV	oxygen (1s) F.W.H.M. /eV	O (1s)/Cu ($2p_{\frac{3}{2}}$) area ratio
80	20	5×10^{-8}	Cu(100)–(2×2)–O (weakly ordered)	530.5	4.7	0.011
80	300	1×10^{-6}	–($\sqrt{2}\times\sqrt{2}$) R45°	531.7	4.3	0.023
80	30 000	5×10^{-5}	–($\sqrt{2}\times\sqrt{2}$) R45°	532.0→533.6	4.4	0.033
	as above, warmed to 290 K		($\sqrt{2}\times\sqrt{2}$)R45° (increased ordering)	530.8	2.3	0.014
290	20–600	1×10^{-6}	"4 spot" (4 domains)	530.2→530.7	2.05→2.2	0.017 at 300 L
290	800–2000	5×10^{-6}	($\sqrt{2}\times\sqrt{2}$) R45°	530.8	2.4	0.020 at 1000 L
290	2000–5×10^6	1×10^{-4}	($\sqrt{2}\times 2\sqrt{2}$) R45° (2 domains)	531.1	2.4	0.023 at 2000 L
290	16 h at 5 Torr		disordered	531.0	2.8	0.099
	as above, heated to 520 K		($\sqrt{2}\times 2\sqrt{2}$) R45°	531.0	2.2	0.059
600	30 mn in 0.5 Torr followed by cooling in oxygen		diffraction rings (see text)	531.0	2.4	0.144

* 1 L = 10^{-6} Torr s.
† 1 Torr = 133.3 Pa.

The ($\sqrt{2}\times 2\sqrt{2}$) R45° surface was exposed to 5 Torr of oxygen in a closed system for 16 h. The resulting XPS spectrum showed no features other than those due to copper and oxygen and the diffraction pattern, although having a high overall intensity, showed no observable diffraction spots. The oxygen 1s peak, which was symmetrical is shown in fig. 3a. The resulting Cu $2p_{\frac{3}{2}}$ peak and X-ray induced $L_{23}M_{45}M_{45}$ Auger peak, which are significantly different to those of the clean metal, are shown in fig. 4a and 4b, together with the spectra for Cu(100) and copper (II) oxide powder.

Heating this heavily "oxidised" Cu(100) surface to 520 K in high vacuum, (5×10^{-9} Torr) caused the reappearance of the ($\sqrt{2}\times 2\sqrt{2}$) R45° mesh which was now very highly ordered, (fig. 2g). This was accompanied by desorption of oxygen, reflected by the pressure rising to 2×10^{-8} Torr, and the oxygen/copper peak area ratio decreasing. Also, as a result of the heating the Cu $2p_{\frac{3}{2}}$ peak shape reverted to that for clean copper.

The surface was further exposed to 0.5 Torr of oxygen during heating to 600 K for 20 min and then cooled in oxygen. After evacuation the diffraction rings shown in fig. 2h were observed; the profiles of the spectra were similar to those shown in fig. 4.

Table 2 summarises the experimental data: substrate temperature, oxygen exposure LEED pattern and O(1s) binding energy and FWHM.

ADSORPTION OF OXYGEN BETWEEN 80 K AND 290 K

Exposure of the clean surface after cooling to ~80 K to 20 L of oxygen at a pressure of 5×10^{-8} Torr generated the rather disordered diffraction patterns shown in fig. 5a, and indexed as Cu(100)—(2×2)—O. After an exposure of 300 L the spots in the ($\frac{1}{2}$, 0) positions become weak and at 1000 L the diffraction mesh is clearly ($\sqrt{2}\times\sqrt{2}$) R45°. The diffraction pattern after an exposure of 30 000 L at 10^{-5} Torr

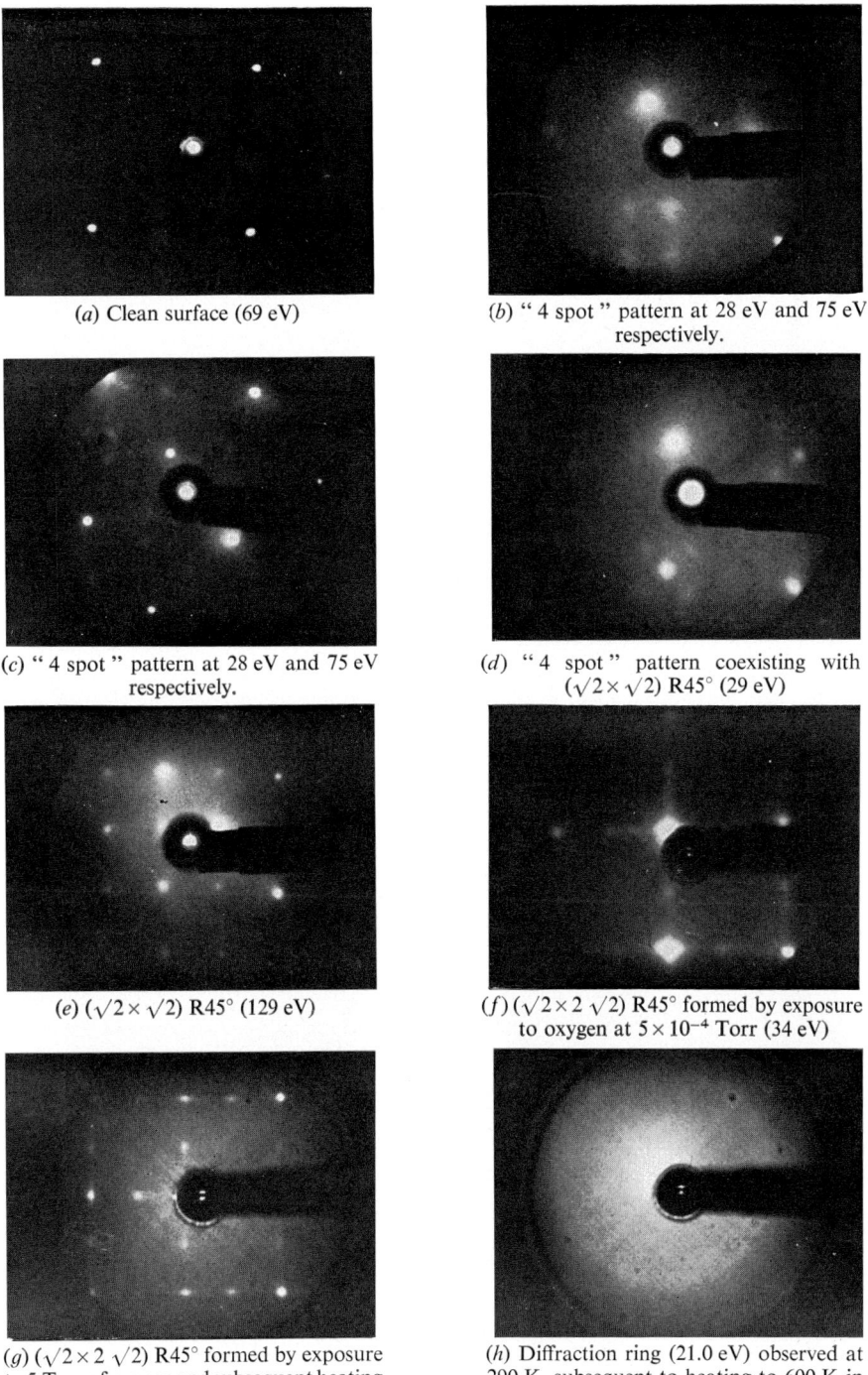

(a) Clean surface (69 eV)

(b) "4 spot" pattern at 28 eV and 75 eV respectively.

(c) "4 spot" pattern at 28 eV and 75 eV respectively.

(d) "4 spot" pattern coexisting with ($\sqrt{2} \times \sqrt{2}$) R45° (29 eV)

(e) ($\sqrt{2} \times \sqrt{2}$) R45° (129 eV)

(f) ($\sqrt{2} \times 2\sqrt{2}$) R45° formed by exposure to oxygen at 5×10^{-4} Torr (34 eV)

(g) ($\sqrt{2} \times 2\sqrt{2}$) R45° formed by exposure to 5 Torr of oxygen and subsequent heating to 520 K in U.H.V. (50 eV)

(h) Diffraction ring (21.0 eV) observed at 290 K, subsequent to heating to 600 K in 0.5 Torr of oxygen.

FIG. 2.—Diffraction patterns resulting from the exposure of Cu(100) to oxygen at and above 290 K (see table 2 for oxygen exposure data).

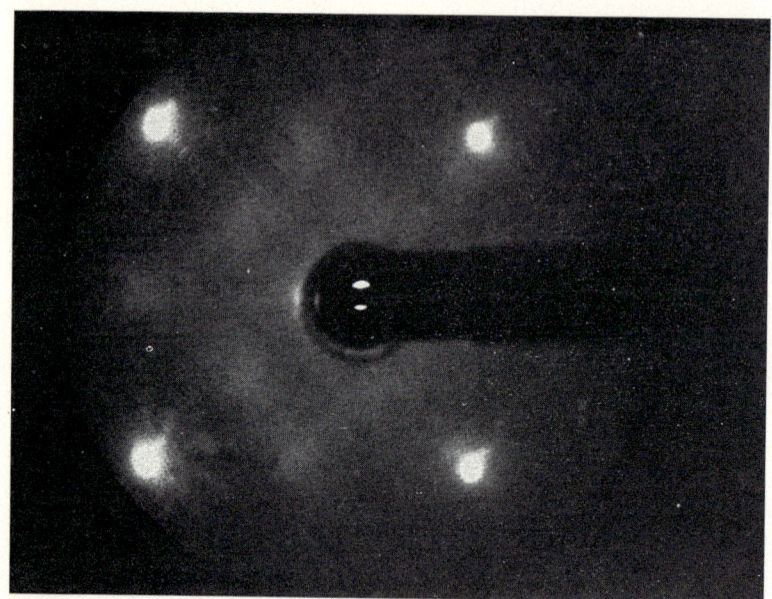

(a) Cu(100), (2×2) mesh (69 eV)

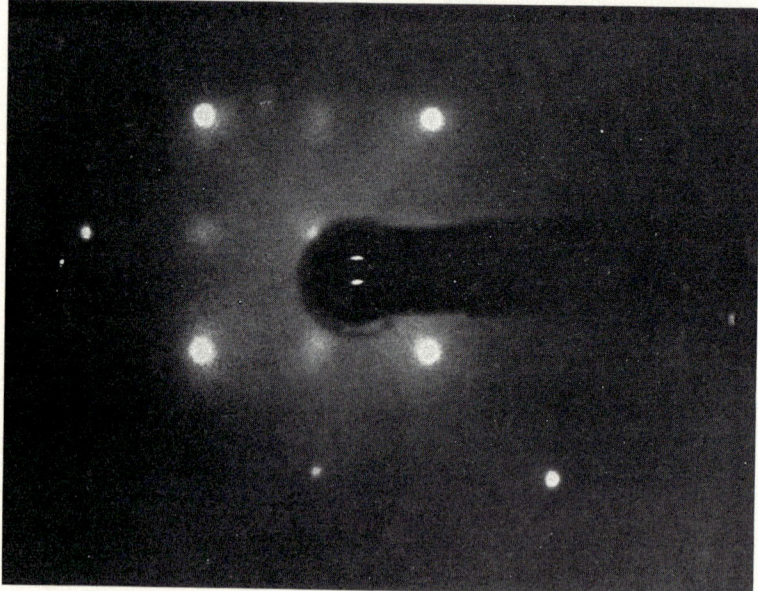

(b) Cu(100), ($\sqrt{2} \times \sqrt{2}$) R45° mesh (131 eV).

FIG. 5.—Diffraction patterns resulting from the exposure of the Cu(100) surface at 80 K to oxygen (see table 2 for oxygen exposure data).

is shown in fig. 5b. Oxygen 1s XPS peaks at the exposures mentioned above are shown in fig. 6b as a plot of the O(1s) peak height and peak area versus exposure. An important feature of the O(1s) profiles is the large FWHM value (>4 eV) observed at all coverages, which compares with a value of about 2.4 eV observed at 290 K.

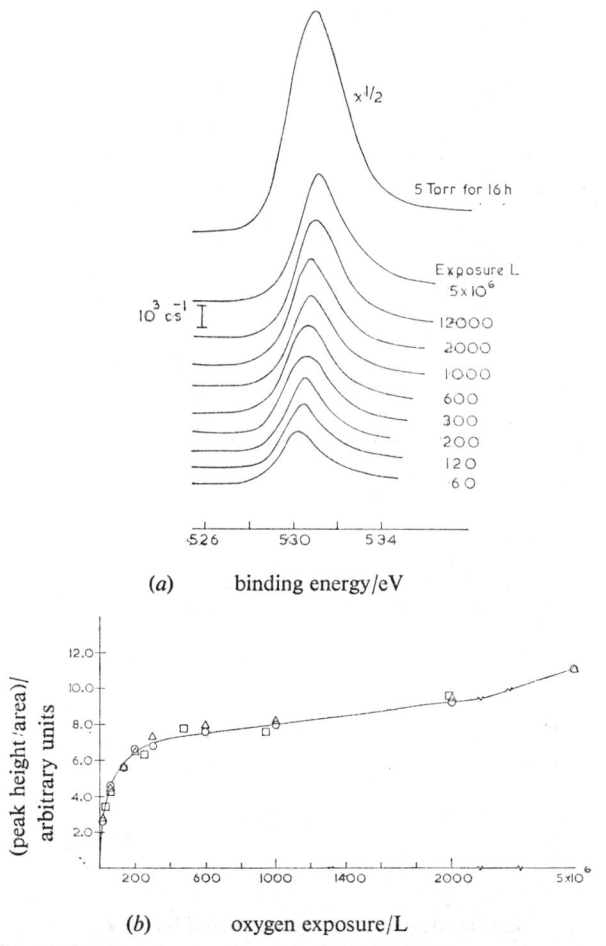

FIG. 3.—(a) Oxygen (1s) photoelectron peaks observed during exposure to oxygen at 290 K. Binding energy calibrated with respect to Cu $2p_{\frac{3}{2}}$ = 932.8 eV. The statistical noise level is < 50 c s^{-1}. (b) Oxygen (1s) peak height and peak area as a function of oxygen exposure (Langmuir), at 290 K. ○, peak height, △, peak area, □, KL_2L_2 Auger derivative peak to peak height determined using retarding-field AES in a separate vacuum system.

Warming *in vacuo* (~10^{-10} Torr) of the (2×2) mesh generated at 80 K resulted in its transformation between 170 and 220 K into the " 4 spot " pattern, (fig. 2b, 2c) also observed during the early stages of oxygen interaction at 290 K. The transformation was accelerated by exposure to the electron beam, and at 220 K was complete over all areas of the crystal, even those not hitherto exposed to the beam. We suggest the role of the electron beam is purely thermal. Adsorption onto a Cu(100) surface at 170 K resulted in the formation of the " 4 spot " mesh, there was no evidence for the (2×2) structure observed at 80 K.

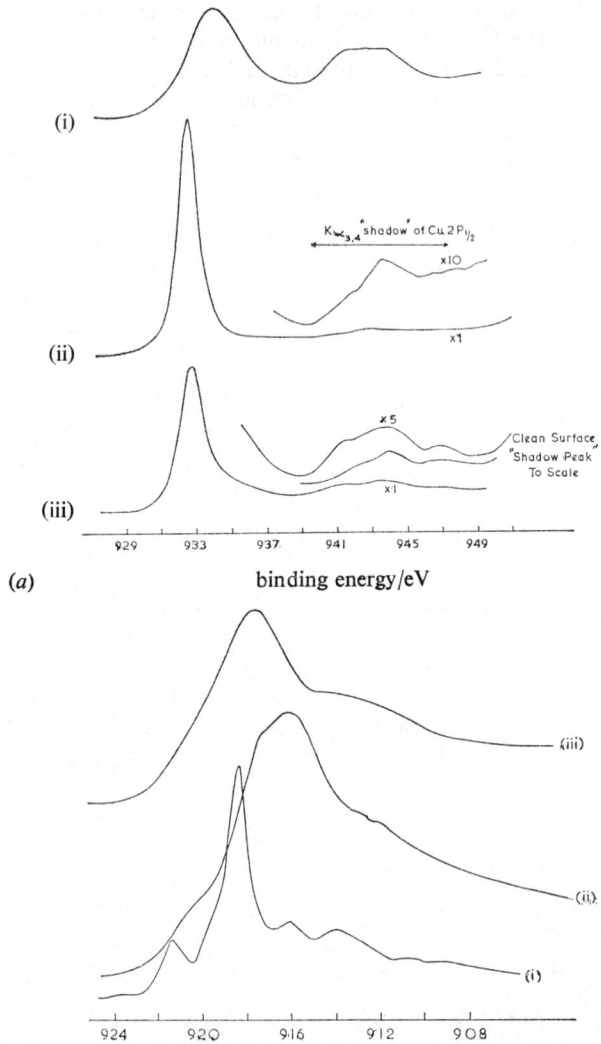

FIG. 4.—(a) The Cu $2p_{\frac{3}{2}}$ peak for: (i) Clean Cu(100) showing the $K\alpha_{3,4}$ "shadow" of the $2p_{\frac{1}{2}}$ peak. (ii) Cu(100)+5 Torr oxygen at 290 K, also showing the "shadow peak" to the correct scale. (iii) Copper (II) oxide. (B.D.H. Ltd.) (b) The $L_{23}M_{45}M_{45}$ Auger peak, curves as for fig. 4a.

The only change observed on warming the $(\sqrt{2}\times\sqrt{2})$ R45° mesh from 80 to 290 K was an increase in the ordering of the diffraction pattern. As shown in table 2 this was accompanied by significant oxygen desorption, the O/Cu ratio decreasing from 0.023 to 0.014, and a decrease in the FWHM value to 2.3 eV.

DISCUSSION

INTERPRETATION OF SURFACE STRUCTURES

(i) AT 80 K

At low temperature the structural picture is relatively straightforward; at low coverage a disordered (2×2) mesh is observed which transforms at higher oxygen exposure and coverage to the ($\sqrt{2} \times \sqrt{2}$) R45° mesh also seen at 290 K. The oxygen 1s photoelectron peak, although always broad, shifts gradually with increasing coverage from a value of 530.5 eV at low coverage to a value of about 532 eV after 300 L exposure, splitting into two peaks at 532 eV and 533.6 eV at much higher exposure. On warming the adlayer to 290 K the maximum O(1s): Cu (2p) ratio for the ($\sqrt{2} \times \sqrt{2}$) R45° mesh decreases from 0.033 to 0.017, the 533.6 eV peak disappears, and the FWHM values is almost halved. These observations taken with the known correlation between heat of adsorption and binding energy lead us to conclude that strongly and weakly adsorbed oxygen are present in approximately equal concentrations at 80 K. The weakly held oxygen forms throughout the exposure.

We suggest that the (2 × 2) and ($\sqrt{2} \times \sqrt{2}$) R45° structures are a consequence of oxygen adatoms strongly held at 4-fold coordination sites. Within these unit cells weakly held ($\lesssim 60$ kJ mol^{-1}) oxygen adatoms exist situated at the interstices of and above the strongly adsorbed oxygen.

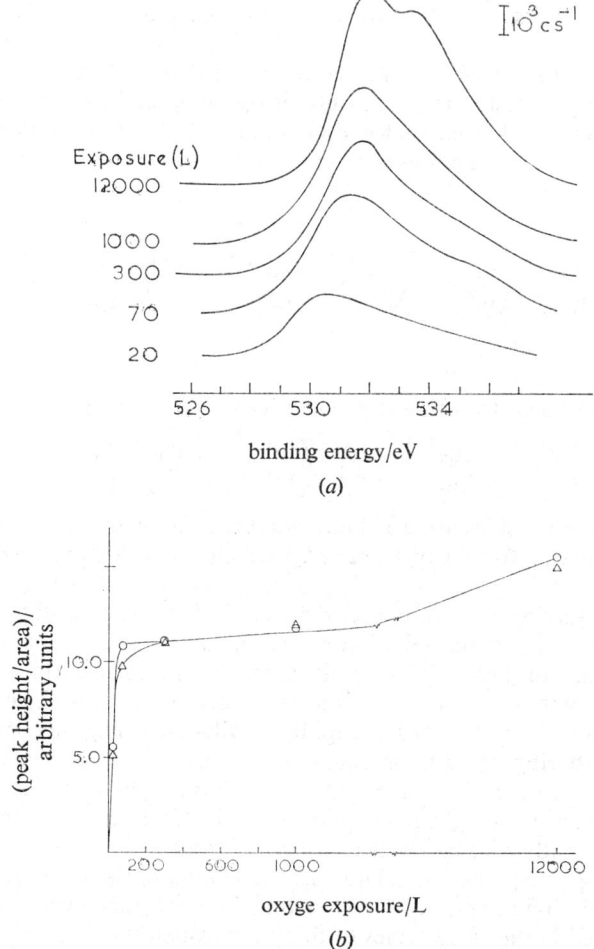

FIG. 6.—(a) Oxygen (1s) photoelectron peaks observed during exposure at 80 K to oxygen. (b) Oxygen (1s) photoelectron peak heights and areas observed during exposure to oxygen at 80 K. △, peak area; ○, peak height.

(ii) AT 290 K

The oblique, "4 spot" structure. This structure is not easy to interpret and we believe that two alternative explanations need to be considered. First, as discussed by Ellis,[19] it is possible that each group of four spots results from splitting of the $(\frac{1}{2},\frac{1}{2})$ spots of a $(\sqrt{2}\times\sqrt{2})$ R45° mesh, due to anti-phase domain interference. There are several reasons why this explanation is considered unlikely, although the fact that the "4 spot" mesh gives way directly to the $(\sqrt{2}\times\sqrt{2})$ R45° pattern is very much in its favour. The effect of out-of-phase domains on the size and definition of diffraction spots has been discussed in detail.[14] If the 4 spots in each group around the $(\frac{1}{2},\frac{1}{2})$ position result from interdomain interference we would expect that the size of the domains and, therefore, the extent of the splitting would change with coverage. Also, we might expect a change in domain size with adsorption temperature, since the rate of oxygen adatom diffusion will be temperature dependent. There is, however, no change in the position of the diffraction spots with either coverage or adsorption temperature, although as the coverage increases the "extra spots" are observed at progressively higher voltages. It is, therefore, necessary to discount this otherwise attractive explanation.

Following Lee and Farnsworth[1] we consider that this mesh can be analyzed as 4 domains of an oblique mesh composed only of oxygen adatoms, we have derived a virtually identical unit cell from that earlier reported. Using the matrix notation, the unit cells of the domains are defined as:

Reciprocal Space:

$$\begin{pmatrix} 0.5 & 0.354 \\ 0.5 & -0.646 \end{pmatrix} \begin{pmatrix} 0.5 & 0.646 \\ 0.5 & 0.354 \end{pmatrix} \begin{pmatrix} 0.646 & 0.5 \\ 0.354 & -0.5 \end{pmatrix} \begin{pmatrix} 0.354 & 0.5 \\ 0.646 & -0.5 \end{pmatrix}$$

Real Space:

(where the rows are taken to define the unit cell sides, as in reciprocal space):

$$\begin{pmatrix} 1.292 & -1.0 \\ 0.708 & 1.0 \end{pmatrix} \begin{pmatrix} 0.708 & -1.0 \\ 1.292 & 1.0 \end{pmatrix} \begin{pmatrix} 1.0 & -0.708 \\ 1.0 & 1.292 \end{pmatrix} \begin{pmatrix} 1.0 & -1.292 \\ 1.0 & 0.708 \end{pmatrix}$$

In real space the unit cell sides are 0.312 nm and 0.420 nm, with the angle being 87° 7' which may be compared to that of Lee et al.,[1] which was 0.308 nm × 0.426 nm, with $\theta = 86° 48'$.

Part of the diffraction pattern expected from all 4 domains is shown in fig. 7a, while fig. 7b shows one domain only, in real space. All of the 16 spots of the (1, 0) family are observed, 4 of those of (1,1) family coincide with clean surface features, and the remaining 8 are not observed. This is not considered a serious objection since the backscattering factor for oxygen falls off rapidly [20] with increasing energy and may also fall as the backscattering angle approaches 90°. The oxygen adatoms are (fig. 7b) located along the "troughs" of the Cu(100) surface. The degree of long range ordering at the surface is thought to be quite small; it should be remembered that the LEED technique is rather insensitive to surface disorder.

The "4 spot" pattern has not been observed in all studies of oxygen adsorption on Cu(100). Thus Simmons et al.[3] report that its appearance was irreproducible, which we also noted in the early stages of this study, when the first ordered structure to be observed after an exposure of about 600 L was the $(\sqrt{2}\times\sqrt{2})$ R45° mesh. Several factors may be important in this context including imperfections in surface preparation, crystallographic orientation and electron optics.

There appeared to be some evidence in the early stages of this study that, similar to the Ag(100)+O_2 system,[21] small quantities of gas phase carbon monoxide could influence the diffraction patterns observed. In many subsequent experiments, however, the oblique mesh was observed even when the working pressure of the system was 8×10^{-11} Torr and care was taken to minimize regurgitation and the exposure of the oxygen to hot filaments. Significantly, also, XPS indicated that the crystal was free of observable carbon before and after oxygen exposure.

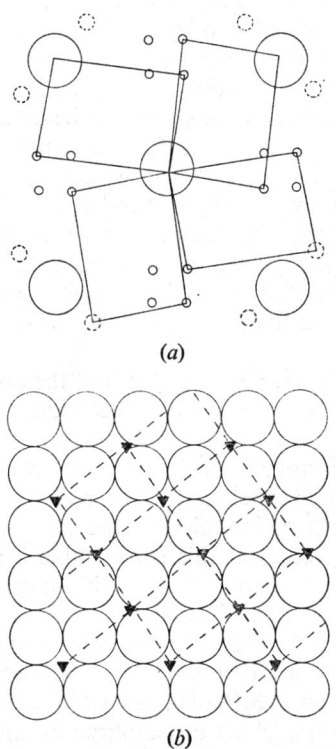

FIG. 7.—(a) Four domains of the " 4 spot " mesh in reciprocal space. Large circles, clean surface spots; small full circles, observed extra spots; dashed circles, absent spots. (b) One domain of the " 4 spot " structure in real space. Large circles, copper substrate atoms; triangles, oxygen adatoms. Note that the oxygen atoms always lie in the " troughs " of the Cu(100) structure.

The $(\sqrt{2} \times \sqrt{2})$ *R45° Mesh.* There is continuing interest in determining at which stage in metal-oxygen interaction movement of metal atoms commences. The LEED approach to this problem was initially intuitive and only within the last year have different theoretical groups agreed on the interpretation of the Ni(100)−($\sqrt{2} \times \sqrt{2}$) R45°−0 mesh, using dynamical calculations.[22, 23] McDonnell et al.[9] have carried out intensity analysis of the Cu(100)−($\sqrt{2} \times \sqrt{2}$) R45°−0 structure using the averaging scheme proposed by Lagally, Ngoc and Webb.[24] They conclude that this structure is reconstructed, with both copper and oxygen atoms in the surface layer. They also report some calculations by Duke and Lipari [25] on their data which also appear to support a reconstructed model. This must be viewed with some caution since Duke et al. simultaneously concluded that the same mesh on the (100) plane of nickel was reconstructed, a view which they have subsequently retracted in favour of a surface layer involving only chemisorbed oxygen.[26]

From the standpoint of photoelectron spectroscopy the situation is equivocal; thus we have argued that an oxygen 1s binding energy of about 530.3 eV observed at 290 K was, in general, characteristic of oxygen chemisorption [12a] and certainly not indicative of oxygen incorporation into the sub-surface region; other views have also been expressed.[12b]

We consider that the present combination of LEED with XPS allows us to state with some certainty that for the Cu(100)+O_2 system reconstruction occurs *only* on going from the ($\sqrt{2} \times \sqrt{2}$) R45° mesh to the ($\sqrt{2} \times 2\sqrt{2}$) R45° structure. The ($\sqrt{2} \times \sqrt{2}$) R45° mesh is generated by oxygen adsorption at 80 K and does not appear to change as the crystal is warmed to 290 K. Thus, if reconstruction occurs at 290 K it is necessary to argue that it must also occur at 80 K, which seems unlikely, particularly when work function data is considered (see below). Also, at 290 K no change occurs in the O(1s) binding energy when the LEED pattern transforms from the " 4 spot " structure to ($\sqrt{2} \times \sqrt{2}$) R45°. The " 4 spot " surface structure has been recognized to be composed only of oxygen,[1] and the XPS data indicate that, during this change in surface structure, the electronic environment does not suffer the significant change which might be expected to result from reconstruction. We, therefore, believe that the ($\sqrt{2} \times \sqrt{2}$) R45° structure involves strongly adsorbed oxygen adatoms at positions of maximum (fourfold) coordination, sitting on top of the copper (100) unconstructed surface.

The ($\sqrt{2} \times 2\sqrt{2}$) R45° Mesh. We suggest that the appearance of this mesh indicates the onset of surface reconstruction and incorporation of oxygen into the metal lattice, for the following reasons. This structure is not formed during adsorption at 80 K where reconstruction is not expected; it can, however, be regenerated by heating *in vacuo* the (100) surface after exposure to 5 Torr of oxygen at 290 K. The relative size of the oxygen and copper photoelectron peaks shows that several monolayer equivalents of oxygen are present at this stage (table 2); reconstruction and incorporation have, therefore, taken place. Only at this stage is a *fully developed* ($\sqrt{2} \times 2\sqrt{2}$) R45° mesh observed (fig. 2g). The pattern seen at lower pressure exposure at 290 K (fig. 2f) shows noticeable disorder, in accord with the known pressure dependence of copper oxidation. Further confirmation that oxidation is occurring is provided by the shift in oxygen 1s B.E. of 0.3 eV which occurs on the formation of this structure, and by the very low sticking probability into this structure which would be expected to be associated with reconstruction.

When the Cu(100)−($\sqrt{2} \times 2\sqrt{2}$) R45° mesh is present on the surface the copper 2p photoelectron peaks show no satellites. There is now much evidence to show that such satellites are characteristic of copper in the +2 oxidation state,[27] the electron configuration of the copper atoms, therefore, maintains the closed 3d shell at this stage of oxidation. Further, we consider that the unit mesh may be related to copper (I) oxide. The (110) plane of this cubic structure is shown in fig. 8, together with the ($\sqrt{2} \times 2\sqrt{2}$) R45° unit cell. The two cells are quite similar in dimension and almost identical in area. This suggestion also explains the decrease from 4 fold to twofold symmetry which the formation of ($\sqrt{2} \times 2\sqrt{2}$) R45° mesh involves.

STABILITY OF SURFACE STRUCTURE

An oxygen molecule impinging onto a Cu(100) surface will either be adsorbed as an adatom or be desorbed; there are clearly two aspects—the rate of adsorption and the surface configuration adopted. The sticking probability at 80 K is clearly greater than at 290 K (cf. fig. 3a with fig. 6b). Using the knee in the (peak height, coverage) relationships gives *average* values over the initial monolayer of about 2×10^{-2} (80 K) and 5×10^{-3} (290 K). This relationship is in keeping with the participation of a weakly-held precursor state, the concentration of which is greater at the lower temperature.

In view of the geometry of the close packed Cu(100) surface we might, therefore, (in the simplest possible case) envisage a 1×1 structure in the LEED pattern. However, the potential may be influenced by direct or indirect interactions between adatoms so that the equilibrium configuration at any coverage will be a direct consequence of these interactions. It is important to recall [28] that with the two dimensional Ising

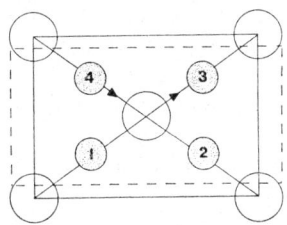

FIG. 8.—The copper (I) oxide (110) unit cell. Open circles, oxygen atoms; filled circles, copper atoms. All oxygen atoms and copper atoms 1, 2 are in plane; atom 3 is above and atom 4 below the plane. The dashed rectangle is the Cu(100) ($\sqrt{2} \times 2\sqrt{2}$) R45°—O unit cell drawn to the same scale,

model, equivalent to an adsorbed layer with $\theta = 0.5$ on a square lattice forming a ($\sqrt{2} \times \sqrt{2}$) R45° structure under the influence of repulsive interactions between nearest neighbours, disorder sets in when the interaction energy is less than $1.76\,kT$. The relevant interaction energies are, therefore, surprisingly small. In the present system an oxygen adatom formed on dissociation will dissipate some of its excess kinetic energy in diffusion. It comes to rest at 80 K at potential minima (sites of four fold co-ordination) leading to the observed (2×2) and ($\sqrt{2} \times \sqrt{2}$) R45° structures.

The " 4 spot " is the most elusive of the structures we have observed, it appears not to form at 80 K. However, it does form on warming to 180 K presumably due to the larger thermal energy available. But its existence is only transient since it is " forced " at essentially constant θ to take up the ($\sqrt{2} \times \sqrt{2}$) R45° structure. Both of these structures have (assuming a single domain) the same maximum theoretical coverage of one oxygen atom per two Cu surface atoms and we must assume that the " 4 spot " domain structure has associated with it repulsive forces of the inter-domain type, which are relieved by transformation to the ($\sqrt{2} \times \sqrt{2}$) R45° structure. Such forces although speculative have been discussed briefly by May.[29] The most stable structure observed is the ($\sqrt{2} \times 2\sqrt{2}$) R45°; clearly its stability is to be associated with surface compound (Cu_2O) formation.

OXYGEN INTERACTION AT PRESSURES GREATER THAN 10^{-3} Torr

For the reaction $2\,CuO(s) = \tfrac{1}{2}O_2(g) + Cu_2O(s)$ the standard free energy [30] at 298 K, $\Delta G°_{298}$, is -107.5 kJ mol^{-1} and $\Delta H°_{298} = -144$ kJ mol^{-1}. This gives an equilibrium oxygen pressure of about 10^{-34} Torr at 298 K, so that CuO is favoured over Cu_2O. At 600 K and 900 K the oxygen pressures are 5×10^{-10} Torr and 10^{-1} Torr respectively. Thus increasing temperature favours the decomposition *in vacuo* of CuO to Cu_2O. We now examine our experimental results in the light of the thermodynamic data.

At 290 K exposure of Cu(100) to oxygen at 5 Torr leads to the formation of a surface layer composed of both Cu_2O and CuO, clearly Cu_2O is kinetically stabilised at this temperature. The Cu(II) ion is recognised by the Cu (2p) satellites which were absent at an oxygen pressure of 10^{-4} Torr. Such satellites are now accepted as reflecting the presence of Cu(II).[27] Furthermore, our Auger data (fig. 4a) confirm the presence of Cu_2O in that the $L_{23}M_{45}M_{45}$ peak shifted ~ 2 eV to lower kinetic energy with respect to the same peak in either Cu (metal) or Cu(II) oxide.[31] The oxygen uptake as indicated by an O (1s) : Cu (2p) ratio of 0.099 (table 2) clearly corresponds to several monolayer equivalents. Further support for this is provided by the observed 40 % decrease in the intensity of the Cu ($2p_{\frac{3}{2}}$) peak: this compares with about 15 % usually associated with the formation of a chemisorbed monolayer.

It is interesting to note that in contrast to our results Evans et al.[32] report satellites when evaporated copper films were exposed to molecular oxygen at 290 K at as low an oxygen pressure as 10^{-4} Torr. The polycrystalline films appear to be much more reactive to oxide (CuO) formation than does the single crystal. This is compatible with the "ease of formation" of oxide nuclei on copper reported by Mitchell and Lawless,[33] (100) being the most difficult, followed by (111) and (110); the easiest being the (311).

The comparatively thick layer of oxide formed at room temperature on Cu(100) shows no order in the diffraction pattern, but on heating to 520 K in U.H.V. the $(\sqrt{2} \times 2\sqrt{2})$ R45° mesh reappears, at which stage the O (1s) : Cu (2p) ratio is 0.059, i.e., approximately three times the value corresponding to the $(\sqrt{2} \times \sqrt{2})$ R45° structure at 290 K (5×10^{-6} Torr, 10^3 L). At this point the satellites have disappeared, clearly copper (II) oxide has decomposed as expected on thermodynamic grounds. A further possibility is the occurrence of the solid state reaction:

$$Cu + CuO = Cu_2O, \quad \Delta G°_{298} = -20 \text{ kJ mol}^{-1}.$$

The appearance at this stage of a structure simply related to copper metal suggests that the Cu_2O structure is distorted to achieve epitaxy with the underlying metal.

Heating in 0.5 Torr of oxygen to 520 K followed by cooling in oxygen caused the reappearance of copper (II) and the emergence of diffraction rings. These rings indicate that the surface is composed of randomly oriented small crystallites and it is believed that diffraction features of this type have not previously been reported in LEED studies of oxidation. The unit cell parameters are not easy to derive, since the rings are only observed at 2 low voltages and the inner potential is not accurately known. Assumption of a value of 10 eV yields lengths of 0.47 and 0.26 nm; the shorter value is very close to the metallic copper close packing distance 0.254 nm, although the existence of random patches of copper seems unlikely. Copper (II) oxide has a monoclinic unit cell, and the lack of a simple epitaxial relationship between the oxide and the metal probably explains the random growth observed. The derived distance of 0.47 nm is close to the b side of the monoclinic CuO cell, which is 0.466 nm.

TABLE 3.—CHANGE OF WORK FUNCTION (eV) OF Cu ON ADSORBING OXYGEN

$\Delta\phi_1$	$\Delta\phi_2$	$\Delta\phi_3$	$\Delta\phi_4$
0.55	0.20	0.40	0.80

WORK FUNCTION STUDIES OF THE $Cu + O_2$ SYSTEM

Although no work function data are available for the Cu(100) surface over the temperature range 80 to 290 K, some general comments pertinent to the present discussion may be made and based on published results [10] with polycrystalline films (table 3). In brief, chemisorption at 80 K is followed first by some desorption and then oxygen incorporation (reconstruction) at 290 K.

$\Delta\phi_1$ refers to oxygen adsorption at 80 K, and $\Delta\phi_2$ is the value after warming to 290 K; $\Delta\phi_3$ is the value observed at 290 K at 10^{-2} Torr and $\Delta\phi_4$ at 290 K and an oxygen pressure of 50 Torr. If we assume as a first approximation that $\Delta\phi$ is a function of the surface concentration of oxygen adatoms, then it is clear that substantial desorption of oxygen occurs on warming from 80 to 290 K. Some oxygen incorporation also occurs. Furthermore the " oxygen surface concentration " is very pressure dependent at 290 K. Reconstruction with the formation of Cu_2O was indicated. The analogy with the present data is very clear and adds further weight to our conclusions.

We wish to acknowledge the support of the Science Research Council; one of us (M. J. B.) is grateful for an I.C.I. Fellowship.

[1] R. N. Lee and H. E. Farnsworth, *Surface Sci.*, 1965, **3**, 461.
[2] G. Ertl, *Surface Sci.*, 1967, **6**, 208.
[3] G. W. Simmons, D. F. Mitchell and K. R. Lawless, *Surface Sci.*, 1967, **8**, 130.
[4] A. Oustry, L. Lafourcade and A. Escaut, *Compt. Rend. B*, 1972, **274**, 1402.
[5] A. Oustry, L. Lafourcade and A. Escaut, *Surface Sci.*, 1973, **40**, 545.
[6] A. Oustry, L. Lafourcade, A. Escaut, C. Butto and P. Larroque, *Compt. Rend. B*, 1974, **278**, 189.
[7] K. Okada, T. Matsushika, H. Tomita, S. Motoo and N. Takahashi, *Shinkû*, 1970, **13**, 371.
[8] L. McDonnell and D. P. Woodruff, *Surface Sci.*, 1974, **46**, 505.
[9] L. McDonnell, D. P. Woodruff and K. A. R. Mitchell, *Surface Sci.*, 1974, **45**, 1.
[10] C. M. Quinn and M. W. Roberts, *Trans. Faraday Soc.*, 1964, **60**, 899; *Nature*, 1963, **200**, 648.
[11] T. Delchar and F. C. Tompkins, *Proc. Roy. Soc. A*, 1967, **300**, 141.
[12](a) R. W. Joyner and M. W. Roberts, *Chem. Phys. Letters*, 1974, **28**, 246.
(b) see also several contributions to *Faraday Disc. Chem. Soc.*, 1974. **58**.
[13] K. Kishi and M. W. Roberts, *J.C.S. Faraday I*, 1975, **70**.
[14] C. S. McKee, D. L. Perry and M. W. Roberts, *Surface Sci.*, 1973, **39**, 176.
[15] R. W. Joyner and M. W. Roberts, *Chem. Phys. Letters*, 1974, **29**, 447.
[16] S. J. Atkinson, C. R. Brundle and M. W. Roberts, *Faraday Disc. Chem. Soc.*, 1974, **58**.
[17] J. S. Ahearn, J. P. Monaghan and J. W. Mitchell, *Rev. Sci. Instr.*, 1970, **48**, 1853.
[18] C. R. Brundle, M. W. Roberts, D. Latham and K. Yates, *J. Electron Spectr.*, 1974, **3**, 241.
[19] W. P. Ellis, in *Optical Transforms*, ed. H. S. Lipson, (Academic Press, London, 1972).
[20] M. Fink, M. R. Martin and G. A. Somorjai, *Surface Sci.*, 1972, **29**, 303.
[21] H. H. Engelhardt, A. M. Bradshaw and D. Menzel, *Surface Sci.*, 1973, **41**, 410.
[22] J. E. Demuth, D. W. Jepsen and P. M. Marcus, *Phys. Rev. Letters*, 1973, **31**, 540.
[23] J. B. Pendry, personal communication quoted by McDonnell et al. in ref (8).
[24] M. G. Lagally, T. C. Ngoc and M. B. Webb, *Phys. Rev. Letters*, 1971, **26**, 1557.
[25] C. B. Duke and N. O. Lipari, quoted by McDonnell et al. in ref (8).
[26] C. B. Duke, N. O. Lipari and G. E. Laramore, *Electron Fisc. Appl.*, 1974, **17**, 139.
[27] D. C. Frost, A. Ishitani and C. A. McDowell, *Mol. Phys.*, 1972, **24**, 861.
[28] T. D. Lee and C. N. Yang, *Phys. Rev.*, 1952, **87**, 410.
[29] J. W. May, *Proc. Roy. Soc. A*, 1972, **331**, 185, 195.
[30] *Selected Values of Chemical Thermodynamic Properties*, ed. F. D. Rossini, (National Bureau of Standards, Washington, D.C. 1952).
[31] J. E. Castle and D. Epler, *Proc. Roy. Soc. A*, 1974, **339**, 49.
[32] S. Evans, E. L. Evans, D. E. Parry, M. J. Tricker, M. J. Walters and J. M. Thomas, *Faraday Disc. Chem. Soc.* 1974, **58**.
[33] K. R. Lawless and D. F. Mitchell, *Colloque International du C.N.R.S.* No. 122. Paris, 1965.

Valence Band Studies of the Copper+Oxygen System

By Stephen Evans, David E. Parry* and John M. Thomas

Edward Davies Chemical Laboratories, University College of Wales,
Aberystwyth, Dyfed, SY23 1NE

Received 24th March, 1975

An experimental reinvestigation (UPS and XPS) of the chemisorption of molecular oxygen on copper is described. Ultra-violet (HeI and HeII) photoelectron spectra of the oxidised surfaces are reported: XPS intensity data were used to monitor the uptake of oxygen, enabling an apparent discrepancy between previously published studies to be resolved. The spectra suggest that surface reconstruction occurs at ordinary temperatures even for very small oxygen uptakes.

Semi-empirical calculations have been performed for bulk Cu and Cu_2O, and for the (111) face of copper before and after the addition of an O monolayer. The valence electron band structures obtained are compared with experiment. The infinite crystal extended Hückel calculation for Cu metal reproduces the d-band well but provides a poor description of the s-band, suggesting that an enlargement of the basis set, in this and similar metal cluster calculations, would prove advantageous. For the copper+oxygen systems the effects of charge transfer, including the presence of a charged crystal lattice, were taken into account and a self consistent charge distribution was obtained for bulk cuprous oxide.

Less than a decade ago almost all reliable experimental determinations of the density of states of solids were obtained using soft-X-ray spectroscopy (SXS), a technique perfected by Skinner[1] in 1940. Photoemission studies employing monochromatic sources up to the LiF cutoff (11.8 eV) proved invaluable[2] for the exploration of energy levels up to ~10 eV below the Fermi level; these also yielded important data relating to surface phases and the electronic consequences of chemisorption. It was not, however, until the advent of XPS and high-energy UPS that the deep-lying regions of valence bands could be routinely explored. Both techniques are surface-sensitive. XPS generally lacks the resolution of UPS and probes more deeply into a solid because of the greater escape depths of more energetic electrons; but UPS, even when He II (40.8 eV) radiation is employed, suffers the disadvantage that the final states of the photoemitted electrons may not have a free-electron-like energy distribution.[3] Nevertheless, it is clear that ultra-violet photoelectron spectroscopy, and, even more so, photoelectron spectroscopy using synchrotron radiation (where by judicious choice of photon energy one can combine the advantages of UPS and XPS[4]) do offer extra experimental information relating to the electronic band structures of surface phases only a few atoms thick.

Consequently, it is appropriate that attempts be made to match theoretically derived band structures of surfaces with those obtained from experiment. We have chosen the copper+oxygen system for detailed consideration because much experimental information is already available[5-8] for comparison with an exploratory theoretical study. However, a further experimental reinvestigation of the interaction of molecular oxygen and copper, with particular reference to the low exposure region, was prompted by the recent publication by Wagner and Spicer[8] of a low-energy UPS study in which they concluded that an exposure as low as 10 L (1 L = 10^{-6} Torr s) of oxygen could produce monolayer coverage: our previous data indicated that even with much more reactive microwave-excited oxygen ca. 10^3 L exposure was required before a monolayer equivalent of oxygen was taken up. This discrepancy has been satisfactorily resolved using XPS intensity data.

For heuristic purposes, we have focused our theoretical effort on the close-packed face of copper, demonstrating first that the band structure of bulk metallic copper may be satisfactorily simulated by three-layer infinite sheets, (111) oriented, of Cu atoms. For such a solid it is both conceptually meaningful and computationally manageable to investigate the electronic changes which accompany a given prescribed kind of oxygen chemisorption on the uppermost layer. As it seems more instructive at this stage to examine the consequences of relatively large-scale changes of structure, rather than of small changes in interatomic distances, the band structure of bulk cuprous oxide was also calculated, since this might be expected to be similar to that for a grossly reconstructed surface. While much further work remains to be done, the results already seem likely to throw new light on the nature of the redistribution of charge which accompanies the chemisorption process.

EXPERIMENTAL

All measurements were made using an AEI ES 200A electron spectrometer (equipped with both Mg $K\alpha$ and u.v.(HeI and HeII) photon sources) as previously described.[6]

Freshly cleaned copper surfaces were exposed to molecular oxygen at temperatures in the 25–75°C range.* The progress of the oxidation was followed by XPS (Cu $2p$, O $1s$, C $1s$) and UPS. As reported previously,[5,6] the products obtained had closely similar UPS to those produced by oxidation with microwave-excited oxygen. The kinetic behaviour was, however, very different. Initially, oxygen uptake at pressures of ca. 10^{-7} Torr was very rapid, reaching a " plateau " value with ~ 10 L exposure, as did the work function of the surface, now 0.3–0.4 eV higher than that of bare copper metal. The O $1s$: Cu $2p_{\frac{3}{2}}$ XPS peak area ratio at this exposure was $\sim 1:50$ (uncorrected for the approximately linear [9] increase of analyser sensitivity with KE). Exposures of 100 L (at 10^{-6} Torr) increased this ratio only to 1:30, and very much larger exposures were needed to reproduce the " monolayer uptake " u.v. PE spectra of ref. (5) and (6) (fig. 1d), obtained in seconds with microwave-excited oxygen at 10^{-4} Torr (O $1s$: Cu $2p_{\frac{3}{2}}$ area ratio 1:6). The u.v. PE spectra obtained at lower exposures were all qualitatively similar, the effect of even 2 L exposure being clearly visible (fig. 1a). Spectra characteristic of Cu^{II} were only obtained with extremely large oxygen exposures (sometimes over 10^{12} L).†

Cuprous oxide (BDH) was pressed on to a copper grid and heated to 200°C to reduce surface Cu^{II} species [10] before the X-PE spectra were recorded. After heating, " shake-up " satellite structure could not be detected in the Cu $2p$ spectrum, thus confirming [5,6] that Cu^{II} species were indeed absent. The ubiquitous C $1s$ signal, however, although diminished in intensity by the heat treatment, could not be completely removed. The C $2s$ region in the valence band (fig. 4a) is therefore indicated in the figure, as the weak structure here is unlikely to be due to Cu_2O.

THEORETICAL

Valence electron band structures were calculated for crystals of copper metal and cuprous oxide, and for the (111) face of crystalline copper with and without a monolayer of oxygen atoms on the surface, using a semi-empirical, linear combination of Bloch orbitals, method. The model for the (111) face consisted of the three uppermost two-dimensionally infinite layers of Cu atoms. The oxygen monolayer was positioned

---

* These temperatures were produced by heat radiated from the X-ray source.

† The actual exposures required for these slow processes following the initial fast chemisorption were very variable, having changed during an investigative period of 15 months by several orders of magnitude, for apparently identical experimental conditions. This suggests that factors so far unidentified also affect the oxidation rate: no contamination peaks were detected in the X-PE spectra.

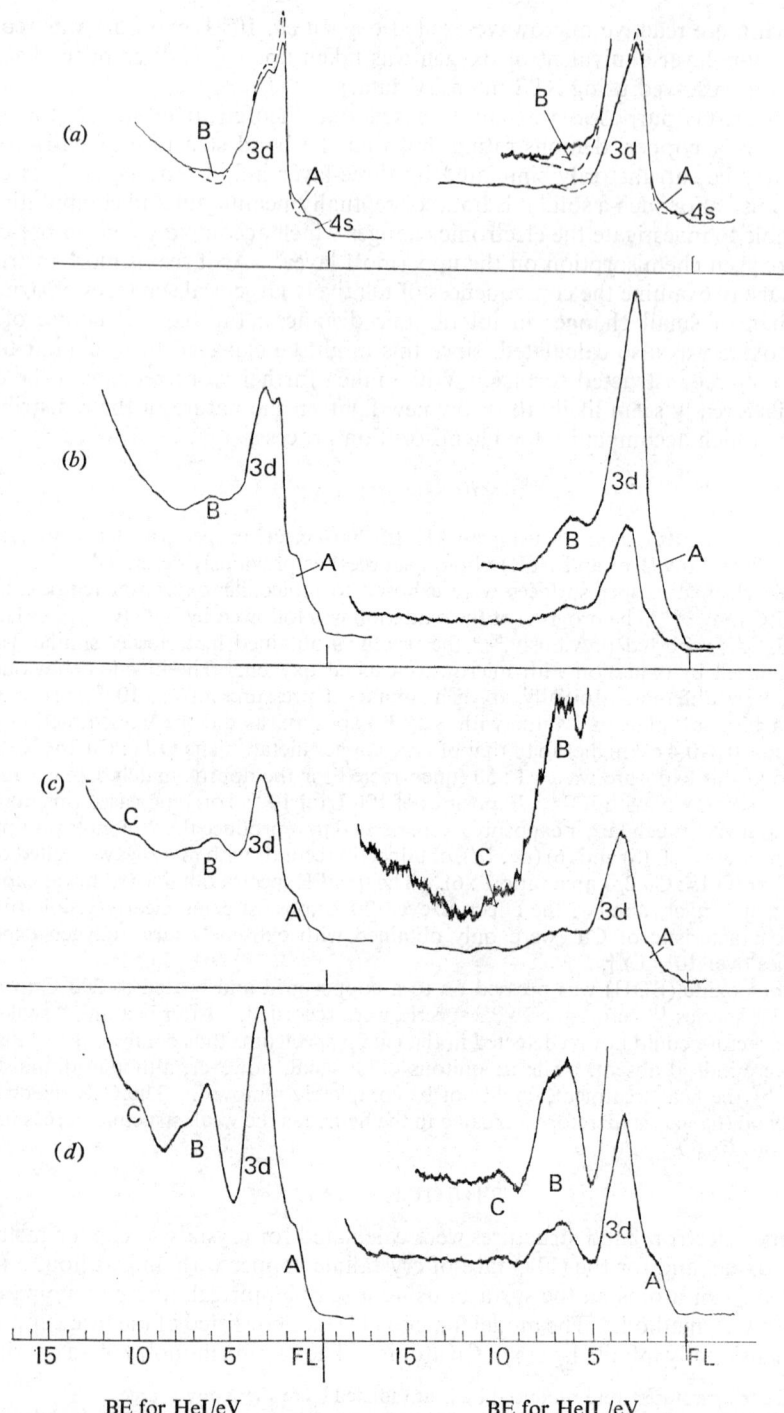

FIG. 1.—Ultra-violet photoelectron spectra of polycrystalline copper: (a) clean (dashed line) and after exposure to $\sim$2 L molecular oxygen (full line); (b) after $\sim 10^4$ L exposure to molecular oxygen [from fig. 5(a) of ref. (6)]; (c) after $\sim 10^5$ L exposure to molecular oxygen [note the emergence of the high BE band C]; (d) after exposure to $\sim 10^3$ L microwave-excited oxygen [monolayer equivalent uptake: from fig. 5(c) of ref. (6)].

so that each oxygen atom was 2.0 Å [11] directly above each copper atom in the top layer (layer 1) of this model surface.

The linear combination of Bloch orbitals method and the manner in which matrix elements between Bloch orbitals may be expressed in terms of matrix elements between atomic orbitals are well known.[12,13] Approximations to the latter may, therefore, be of the same form as in standard molecular orbital methods.

For bulk copper metal, and the clean (111) surface, extended Hückel theory [14] was employed, with the Cusachs prescription [15,16] for the off diagonal matrix elements of the one electron Hamiltonian $\mathscr{H}$ between orbitals a and b.

$$H_{ab} = -S_{ab}(2-|S_{ab}|)(I_a+I_b)/2.$$

S_{ab} is the overlap of a and b (calculated for inter-atomic distances of up to 6 Å throughout this work) and I_a is the valence state ionisation potential (VSIP) of a. The Cusachs prescription avoids the complication of an arbitrary parameter that arises with the more commonly used Wolfsberg-Helmholz and Ballhausen-Gray prescriptions (infinite crystal calculations have been performed using these for diamond [13] and lead metal [17] respectively) and is probably superior to them.[16]

The $4s$ and $3d$ VSIP's for copper were obtained by calculating average energies [18] for configurations of Cu°, Cu⁺, and Cu²⁺ using published [19] tables of Slater-Condon parameters. Differences between these average energies, together with spectral ground state IP's,[20] yield the energies required for the various valence state ionisation processes. The following linear expressions are fits to these calculated VSIP's and were used to adjust the VSIP's in the matrix elements of $\mathscr{H}$ as functions of atomic charge and configuration

$$I_{4s} = 5.7 + 9.85q + 1.9n_s$$

$$I_{3d} = 70.7 + 9.85q - 6.0n_d,$$

n_s and n_d being the number of $4s$ and $3d$ electrons on a Cu atom with charge q (charges are expressed in units of $|e|$ throughout). Similar expressions for the oxygen $2s$ and $2p$ VSIP's were taken from the literature.[21]

A minimal basis set of occupied valence atomic orbitals (those discussed above) was used. Except for the Cu $3d$ orbitals, each valence orbital was represented by the

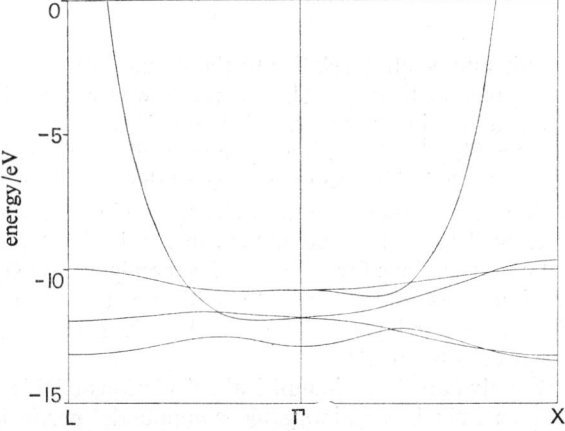

FIG. 2.—Calculated valence electron band structure for copper metal. The s-band extends to an energy of 70 eV.

appropriate single Slater type orbital (STO) [22] with Clementi [23] exponent. For each $3d$ orbital a double-zeta function [24] was used. It has been emphasized elsewhere [25] that a single STO does not reproduce the outer part of the $3d$ orbital accurately. Indeed when one was used, with Clementi [23] exponent, in the calculation for copper metal the d-band was then predicted, as expected, to be far too narrow.

Assuming each Cu atom to be in a $d^{10}s$ configuration in the metal, the valence electron band structures were calculated for bulk copper metal (fig. 2) and the model surface, and the density-of-states obtained by interpolation (fig. 3). The band structure for the bulk metal may be compared with, e.g., the well known calculations by Burdick [28] and by Segall.[29] The good agreement between results for the bulk metal and the model surface suggested that the latter would represent the metal surface

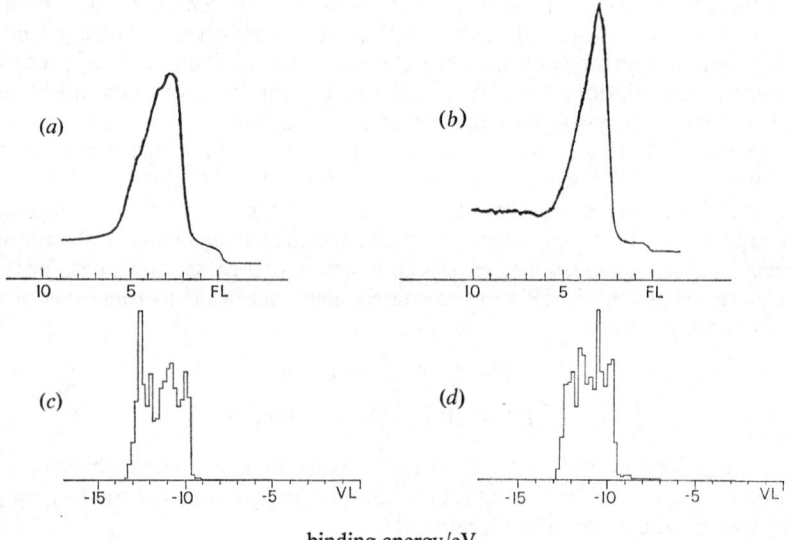

Fig. 3.—Valence bands for copper metal: (a) XPS experiment, monochromatised AlKα, resolution 0.55 eV, from ref. (27); (b) UPS experiment, HeII, resolution 0.2 eV, from ref. (6); (c) DOS from infinite crystal extended Hückel calculation; (d) DOS from 3-layer Cu(111) calculation.

well. The d-band width, and position relative to the vacuum zero, compare well with experiment (fig. 3) and previous theory. The s-band, however, is predicted to extend to very high energies, resulting in a very weak density-of-states and an unrealistically high position for the Fermi level. This discrepancy is easily attributed to the use of a minimal basis set. The $4s$ orbital has to reproduce all the non-d nearly-free-electron valence and conduction band structure and is clearly inadequate for this purpose. Consequently there seemed to be little advantage in continuing the calculations to obtain self consistent configurations for copper. The configuration suggested by the non-iterative calculations was $\sim d^{9.9} s^{1.1}$. In addition there was slight charge transfer (~ -0.001) to the outer layers of the model surface, in agreement with the trends noted for cluster calculations.[11]

To our knowledge, only two [30] semi-empirical calculations of this type have been previously performed on bulk binary inorganic compounds, partly because of uncertainty concerning the effects of charge transfer between atoms, which are not taken into account in extended Hückel theory. The method is generalised here to account systematically for the effects of charge transfer, including those of a charged lattice. The diagonal matrix element of $\mathscr{H}$ is written

$$H_{aa} = -I_a + W_a.$$

The simplest approximation for the calculation of the lattice term W_a is the point charge model, i.e., if a is on atom A, then the contribution to W_a of atom C, with charge q_C, distant R_{AC} from A, is $-q_C e^2/R_{AC}$. The summation over all atoms is easy for molecules, but care is required for crystals. For both two-dimensional [31] and three-dimensional [32] Bravais lattices expressions for W_a linear in the charges on the atoms in the crystal unit cell have been derived. To allow for the extended nature of the s and p valence orbitals the contribution to W_a for these orbitals from nearest neighbour atoms was taken to be of Mataga-Nishimoto,[33] rather than Coulombic, form.

It has been shown [34] that the correct expression for the off-diagonal matrix elements, when lattice terms are included, is

$$H_{ab} = S_{ab}\{-(2-|S_{ab}|)(I_a+I_b)/2+(W_a+W_b)/2\}.$$

Calculation to self consistency thus requires an initial calculation of all the overlap integrals and all parameters required for calculating the I_a and W_a as functions of charge and configuration. Gross atomic charges and orbital populations were used here (obtained by Mulliken population analysis [35]) since W_a depends linearly on gross atomic charges.[32] Performing the population analysis is not a trivial matter for an infinite crystal, since in principle a careful average over the Brillouin zone is required.

Self consistent charges and configurations calculated for bulk cuprous oxide were

$$q(\text{Cu}) = -\tfrac{1}{2}q(\text{O}) = 0.43$$

with copper and oxygen in configurations $d^{9.75} s^{0.82}$ and $s^{1.88} p^{4.98}$ respectively. The Brillouin zone was sampled at only five randomly chosen points to obtain these values. It was found that different random samples of five points did not yield very different charges and configurations, suggesting that the above results are precise to, say, $\pm 5\%$. This encouraging feature is mainly due to the equality of configurations calculated at symmetry-related points in the Brillouin zone. Such symmetry does not apply in general to the calculation of individual orbital populations and density matrices; for this a sample of symmetric sets of points in the zone must be used.

The calculated density-of-states is shown in fig. 4 and appears to be in good agreement with experiment. Structure on the low binding energy side of the d-band is mainly O $2p$. On the high binding energy side the structure to a large extent is Cu $4s$ in character. In an augmented plane wave calculation [39] Cu $4s$ levels were predicted to lie just below the band gap. Optical data however indicate [40] that it is the oxygen levels that lie higher in the valence band. The size of the band gap predicted, ~ 0.05 eV, is very small compared to the experimental value of ~ 2.2 eV.[40] In contrast, a test calculation, with neutral atom configurations $d^{10}s$ and $s^2 p^4$ assumed, predicted a gap of ~ 6 eV. It was found that the calculated band gap varied approximately linearly with the degree of charge transfer assumed, the experimental band gap being reproduced with $q(\text{Cu}) = -\tfrac{1}{2}q(\text{O}) \simeq 0.28$, which suggests that the self consistent charges calculated may be overestimates.

Attempts to perform a similar self consistent calculation for an oxygen monolayer on the (111) face of copper met with difficulty. The addition of the monolayer to the 3-layer model surface changes the position of the Fermi level in the energy band structure and a large number of points in the surface Brillouin zone must be sampled if this position is to be determined precisely. Such a complication is an artefact of the model system, for a monolayer on a bulk crystal (i.e., a quasi infinite number of layers of metal atoms) would not shift the Fermi level. Nevertheless, if the calculated position of the Fermi level of the " clean " model surface had been realistic it would probably have been worthwhile to carry the calculations to self consistency. Since

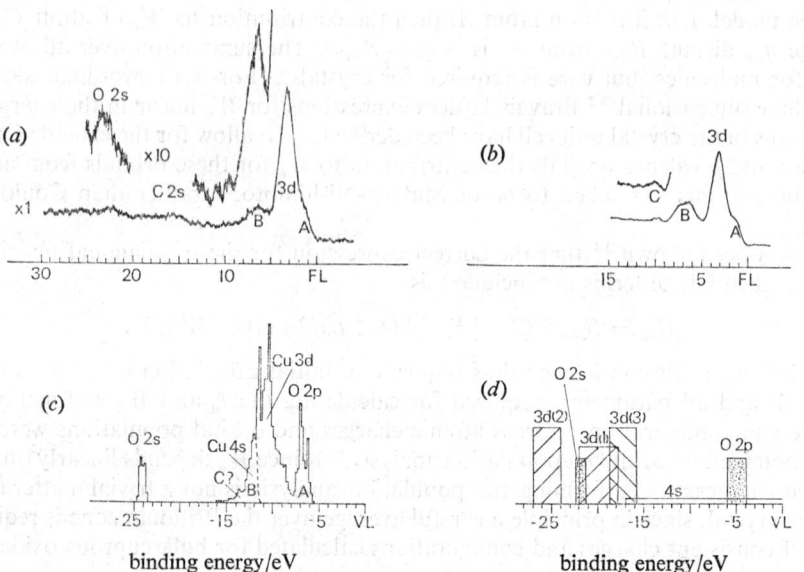

FIG. 4.—Valence bands for the copper-oxygen system: (a) XPS (MgKα) valence band of bulk cuprous oxide [this spectrum is essentially in agreement with those reported previously [6, 36-38]]; (b) HeII valence band of copper surface after oxygen uptake approximately equivalent to a monolayer (from fig. 1d); (c) calculated DOS for bulk cuprous oxide indicating predominant A.O. contributions to each region; (d) schematic diagram of calculated DOS for Cu(111)+O model system with estimated charges (see text).

the results for bulk Cu_2O had indicated the presence of significant charge transfer a few trial calculations were, however, made for the surface system and trends in the electron distribution noted. A representative set of charges and configurations was then used to calculate the band structure of the system, viz:

oxygen	Cu layer 1	Cu layer 2	Cu layer 3
$q = -1.0$	$q = +0.8$	$q = +0.25$	$q = -0.05$
$s^2 p^5$	$d^{10} s^{0.2}$	$d^{10} s^{0.75}$	$d^{10} s^{1.05}$

The band structure calculated is shown schematically in fig. 4d. It differs considerably from that for bulk Cu_2O, especially in the splitting of the d-band. The high binding energy d-band consists of orbitals in layer 2, the potential energy of the d electrons in this layer being lowered considerably by the large positive charge on the neighbouring layer 1. Also the d electrons on layer 3 move, to a lesser degree, to higher energy, broadening the original d-band. In a bulk crystal (cf. Cu_2O) an atom surrounded by positively charged neighbours itself possesses a complementary negative charge (and *vice versa*) so that there need be no large shifts in binding energies. It must be emphasized that this calculation merely gives the band structure for an *assumed* charge distribution; it is quite possible that the self consistent charge distribution could be substantially different.

DISCUSSION

Our experimental results agree well with those recently reported by Wagner and Spicer [8]: however, these workers assumed, in the absence of XPS intensity data, that the cessation of the work function change at ~ 10 L exposure to molecular oxygen occurred at monolayer uptake. Comparison of our XPS intensity data at completion of this initial fast uptake with that reported previously [6] for the " plateau " value found during oxidation using microwave-excited oxygen (monolayer equivalent) suggests that the oxygen uptake at this point is in fact only ca. 0.1 monolayer. Even allowing for the uncertainties inevitably present in the parameters used to estimate coverage (i.e., photoionisation cross-sections, mean free paths, etc.[9]) we think the XPS data show clearly that the uptake must be well under a monolayer equivalent.

We had noted previously [6] that the spectra after monolayer uptake were remarkably similar to those of bulk Cu_2O: the lack of marked qualitative change in the He II spectra between 2 L oxygen and monolayer uptake suggests further that the initial surface reconstruction is rapid at ordinary temperatures even for very low oxygen exposures. Such a conclusion is not incompatible with the LEED experiments reported by Simmons et al.[7] in which ordered patterns appeared at exposures between our 0.1 and 1 monolayer points: small nuclei of a Cu^I oxide with coordination similar to bulk copper(I) oxide may grow randomly at first, only gradually reorganising into a well-structured layer of oxide.

The X-PE valence band spectrum of Cu_2O (fig. 4) clearly shows bands corresponding to A and B in the UV-PE spectra (fig. 1), but the presence of band C is difficult to establish. Previous spectra [6, 37] show some evidence for it, though the close proximity of a C $2s$ contaminant band in both our spectra and, apparently,* in all but one [37] of those previously reported [6, 36-38] make an unequivocal identification difficult.

The large difference in oxidation rate between ground-state molecular and microwave-excited oxygen (which contains oxygen atoms) after the first rapid uptake is complete suggests that the rate of this surface reconstruction is not the rate-determining step: with molecular oxygen as oxidant the overall rate is apparently limited by the low probability of dissociative chemisorption on the partially-oxidised surface. This behaviour contrasts with that found for lead,[41] where the rate of production of PbO was found to be independent of the nature of the oxidant, and determined by the rate of the slow reorganisation following the initial oxygen uptake.

While it is gratifying to note the good agreement between theory and experiment for bulk Cu_2O, it would, however, be premature to take the calculated band structure for an oxygen monolayer on copper as additional evidence for extensive surface reconstruction. Further calculations, incorporating improvements discussed below, are needed for this and other metal+adatom systems before definite predictions can be made.

The first step should be to increase the size of the basis set in the calculation for copper metal by including Cu $4p$ orbitals. A better description of the higher conduction bands should result and bring about a reduction in width of the s-band (cf. ref. (28) and (29)). It is anticipated that such changes in the calculation for bulk Cu_2O will have little effect on the occupied valence bands, owing to the presence of a band gap (the size of which could well change, however). Minimal valence basis set calculations as described here appear to offer an economical means of predicting the valence band structure of a variety of compounds, e.g., graphite monofluoride.[42]

* Hüfner and Wertheim have attributed a band in this region to shake-up [36] and plasmon loss [37] processes, but the simpler explanation advanced above appears more probable.

Although no adjustable parameter enters the calculation explicitly, the present simple, but arbitrary, method [33] of accounting for the lattice potential due to neighbouring atoms is open to modification. A suitable calibration procedure may be the comparison of calculated band gaps with experiment.

It is interesting to compare the results for metals with those from cluster calculations, which if carried out in the extended Hückel approximation will be subject to the s-band deficiencies discussed above. A feature of cluster calculations is that they tend to underestimate the d-band width, a 9-atom nickel cluster [11] predicting 1.8 eV while experiment indicates ~ 3 eV for the bulk metal. Preliminary calculations by us, using the approximations and input data of Anderson and Hoffmann,[11] verify their results for the 9 atom cluster and yield, as they surmised, a width ~ 3 eV for the bulk metal. This does appear to suggest, in agreement with the reasoning of Heine and co-workers,[43] that band widths predicted by many cluster calculations may be too small. However, cluster models do appear to be superior in the respect that they offer an economical means, as part of a quantitative investigation of relaxation effects, of calculating the energies of localised holes in the final states [44] of the photoemission process. Relaxation corrections from cluster calculations could then be applied to the results of infinite crystal calculations.

We thank the S.R.C. for support and Mr. R. G. Pritchard for experimental assistance.

[1] H. W. B. Skinner, *Phil. Trans. A*, 1940, **239**, 95.
[2] W. E. Spicer, *J. Physique*, 1973, **34**, Suppl. C6, p.16–19.
[3] G. K. Wertheim, Proc. *NATO Advanced Study Institute on Electronic States of Inorganic Compounds: New Experimental Techniques*, Oxford, Sept. 8–18th 1974 (in press).
[4] D. E. Eastman, *Proceedings of International Conference on Vacuum UV Radiation Physics*, Hamburg, July 22–26, 1974.
[5] S. Evans, E. L. Evans, D. E. Parry, M. J. Tricker, M. J. Walters and J. M. Thomas, *Faraday Disc. Chem. Soc.*, 1975, **58**, 97.
[6] S. Evans, *J.C.S. Faraday II*, 1975, **71** 1044.
[7] G. W. Simmons, D. F. Mitchell and K. R. Lawless, *Surface Sci.*, 1967, **8**, 130.
[8] L. F. Wagner and W. E. Spicer, *Surface Sci.*, 1974, **46**, 301.
[9] P. Cadman, S. Evans, J. D. Scott and J. M. Thomas, *J.C.S. Faraday II*, 1975, **71**, 1777.
[10] T. Novakov and R. Prins, in *Electron Spectroscopy*, ed. D. A. Shirley, (North Holland, Amsterdam, 1972), p.821: also published as *Solid State Comm.*, 1971, **9**, 1975.
[11] A. B. Anderson and R. Hoffman, *J. Chem. Phys.*, 1974, **61**, 4545.
[12] J. C. Slater, *Quantum Theory of Molecules and Solids*, Vol. 2, (McGraw Hill, New York, 1965).
[13] R. P. Messmer, *Chem. Phys. Letters*, 1971, **11**, 589.
[14] R. Hoffman, *J. Chem. Phys.*, 1963, **39**, 1397.
[15] L. C. Cusachs, *J. Chem. Phys.*, 1965, **43**, S157.
[16] D. G. Carroll and S. P. McGlynn, *J. Chem. Phys.*, 1966. **45**, 3827.
[17] A. Breeze, *Solid State Comm.*, 1974, **14**, 395.
[18] J. C. Slater, *Quantum Theory of Atomic Structure*, (McGraw Hill, New York, 1960).
[19] E. Tondello, G. de Michelis, L. Oleari and L. di Sipio, *Coord. Chem. Rev.*, 1967, **2**, 65.
[20] C. E. Moore, *Nat. Bur. Stand. Circular No. 467*, 1952, vol. 2.
[21] L. C. Cusachs and J. W. Reynolds, *J. Chem. Phys.*, 1965, **43**, S160.
[22] J. C. Slater, *Phys. Rev.*, 1930, **36**, 57.
[23] E. Clementi and D. L. Raimondi, *J. Chem. Phys.*, 1963, **38**, 2686.
[24] J. W. Richardson, W. C. Nieuwpoort, R. R. Powell and W. F. Edgell, *J. Chem. Phys.*, 1962, **36**, 1057.
[25] D. A. Brown and N. J. Fitzpatrick, *J. Chem. Soc. A*, 1966, 941.
[26] G. Burns, *J. Chem. Phys.*, 1964, **41**, 1521.
[27] S. Hüfner, G. K. Wertheim and J. H. Wernick, *Phys. Rev. B*, 1973, **8**, 4511.
[28] G. A. Burdick, *Phys. Rev.*, 1963, **129**, 138.
[29] B. Segall, *Phys. Rev.*, 1962, **125**, 109.
[30] A. Breeze and P. G. Perkins, *J.C.S. Faraday II*, 1973, **69**, 1237: *Solid State Comm.*, 1973, **13**, 1031.

[31] D. E. Parry, *Surface Sci.*, 1975, **49,** 433.
[32] D. E. Parry, *J.C.S. Faraday II*, 1975, **71,** 337.
[33] N. Mataga and K. Nishimoto, *Z. phys. Chem.*, 1957, **13,** 140.
[34] P. W. Smith, R. Stoessiger and A. G. Wedd, *Theoret. chim. Acta*, 1968, **11,** 81.
[35] R. S. Mulliken, *J. Chem. Phys.*, 1955, **23,** 1833.
[36] G. K. Wertheim and S. Hüfner, *Phys. Rev. Letters*, 1972, **28,** 1028.
[37] S. Hüfner and G. K. Wertheim, *Phys. Rev. B*, 1973, **8,** 4857.
[38] R. F. Roberts, *J. Electr. Spectr.*, 1974, **4,** 273.
[39] J. P. Dahl and A. C. Switendick, *J. Phys. Chem. Solids*, 1966, **22,** 31.
[40] S. Brahms, J. P. Dahl and S. Nikitine, *J. Physique*, 1967, **28,** (suppl.) C3-33.
[41] S. Evans and J. M. Thomas, *J.C.S. Faraday II*, 1975, **71,** 313.
[42] D. E. Parry, J. M. Thomas, B. Bach and E. L. Evans, *Chem. Phys. Letters*, 1974, **29,** 128.
[43] R. Haydock, V. Heine, M. J. Kelly and J. B. Pendry, *Phys. Rev. Letters*, 1972, **29,** 868.
[44] L. Ley, F. R. McFeely, S. P. Kowalczyk, J. G. Jenkin and D. A. Shirley, *Phys. Rev. B*, 1975, **11,** 600.

Note added in proof:

Further calculations for Cu metal using Burns',[26] rather than Clementi's,[23] exponents for the $4s$ STO indicate that the more compact Burns STO provides a much better description of the s-band. Inclusion of the virtual $4p$ orbitals in the basis set effects no significant improvement.

Chemisorption of Oxygen on Iridium

By G. Brodén and T. N. Rhodin*

School of Applied and Engineering Physics,
Cornell University, Ithaca, N.Y. 14853

Received 14th March, 1975

A photoemission and LEED–Auger study has been made to investigate the effect of oxygen adsorption on the electronic and atomic structure of a reconstructed Ir(100) surface. The clean surface is found to have a $5d$-band width of 8.0 eV. Origin of an intrinsic surface state and other possible effects of the reconstruction are discussed. Upon adsorption of O_2 an oxygen $2p$ peak emerges 11.1 eV below the vacuum level. The influence of the electronegativity difference between oxygen and the substrate metal on the photoemission is also considered.

INTRODUCTION

Platinum group metals, of which iridium is a member, are widely used as catalysts for reactions involving hydrocarbons.[1] To gain a more fundamental understanding of catalysis requires a better understanding of the chemisorptive process. Two essential concepts in chemisorptive bonding are covalency and charge transfer. The latter is related to the electronegativity difference between the elements involved in the bonding.[2] Oxygen is one of the most electronegative elements in the periodic table and chemisorption of oxygen on iridium therefore provides a good starting point to study the effects of charge transfer in chemisorption.

Iridium is a f.c.c. transition metal with a valence electron configuration of $(5d)^7 (6s)^2$. A (100) surface of a f.c.c. metal would normally exhibit a (1×1) LEED pattern. However, previous LEED investigations[3] indicate that the LEED pattern of the clean surface is (5×1). This is thought to be due to a relaxation of the uppermost atomic layer hence forming a close-packed-type layer on top of the quadratic atomic arrangement of the (100) surface. The reconstruction of the surface may induce changes in the electron distribution close to the surface and give rise to surface states. Such surface states have been predicted and observed for reconstructed semiconductor surfaces.[4,5]

Relatively little work has been done on the microscopic properties of well-characterized iridium surfaces. Apart from LEED work,[3,6] field emission energy distributions[7] have been obtained for the Ir(100) as well as other surfaces of iridium.

EXPERIMENTAL

A schematic drawing of our experimental set-up is shown in fig. 1. By means of a mechanical manipulator, our sample can be moved between two levels. On the lower level, photoemission spectroscopy can be done and on the upper level, LEED–Auger spectroscopy. The light-source is a cold-cathode helium or neon discharge lamp giving monochromatic radiation at 16.8, 21.2 and 40.8 eV. It has features which make it similar to lamps constructed by Starbuck[8] and Rowe.[9] The discharge lamp can be sealed off from the UHV chamber with a straight-through valve and the incident angle of the radiation can be varied by rotating the sample. The present results were obtained with an angle of incidence of the radiation of 40° to the surface normal. The photoemitted electrons are energy analyzed with a Physical Electronics double-stage cylindrical mirror analyzer. Our $\hbar\omega = 21.2$ eV spectra were measured at a pass energy of 20 eV (nominal energy resolution 0.3 eV) giving 50,000 counts/s on the d-band peak of iridium. Similarly, at $\hbar\omega = 40.8$ eV the pass energy was 30 eV (energy resolution 0.4 eV) giving an intensity on the d-band of 600 counts/s. Inherent with the above-mentioned energy analyzer is a transmission function which tends to over-emphasize the contributions from low-energy electrons. The analyzer accepts electrons on an annulus around the analyzer axis thus giving an electron energy distribution which is fairly well integrated over steric angles. The LEED–Auger spectroscopy is done using a 4-grid LEED system as shown in fig. 1. The base pressure in the UHV-system is 2×10^{-10} Torr and the pressure rises to 1×10^{-9} Torr of helium when the light source is operating.

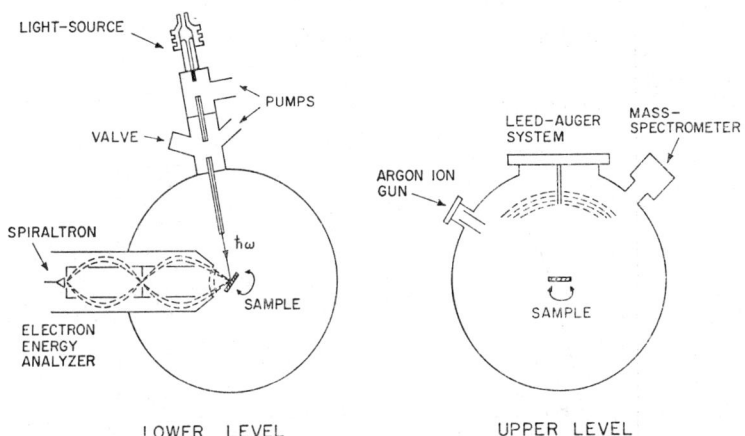

FIG. 1.—Schematic diagram of experimental set-up. The sample can be translated between a lower level where photoemission spectroscopy is done and an upper level used for LEED–Auger spectroscopy.

The main impurity on our uncleaned sample was carbon. Most of the carbon was sputtered away by argon bombardment. To remove residual carbon, it was necessary to follow up with repeated heat treatments in oxygen (10 min at 1400-1500 K and 5×10^{-8} Torr O_2). The clean surface had a (5×1) LEED pattern and showed no impurities as judged by the relative peak heights in the Auger electron spectra, i.e.,

(C(272 eV)/Ir(171 eV) $\sim$ O(510 eV)/Ir(171 eV) < 0.04).

It was found that if the clean surface was exposed to the Auger beam for about 5 min, development of a carbon peak at 272 eV initiated. This effect was dependent only on the exposure time to the Auger beam and did not depend on the total time elapsed since the cleaning of the surface. Great care has to be exercised in using Auger spectroscopy on clean iridium. In fig. 2a a schematic LEED pattern of the clean Ir(100) surface is shown. The pattern is made up of two domains, (1×5) and (5×1), of about equal intensity.

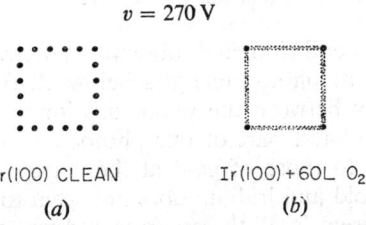

FIG. 2.—(a) LEED pattern of clean Ir(100) surface. The pattern is made up of two domains, (5×1) and (1×5), of about equal intensity. (b) LEED pattern of iridium surface exposed to 60 L of oxygen at 300 K. Traces of a (5×1) pattern could be discerned even at saturation coverage of oxygen.

RESULTS

In fig. 3, three photoemission curves corresponding to different photon energies for clean Ir(100) are shown. The energy scale is the traditional one where zero energy is assigned to electrons excited from the Fermi level. Most of the electrons in the lowest-energy humps are due to inelastic scattering in the sample. The structure within 10 eV from the Fermi edge is due to emission from the valence band. The structure changes considerably with photon energy. This is expected because in the bulk model of photoemission wave-vector conservation in the optical transitions is known to cause such shifts in the structure.[10] Some agreement between a direct

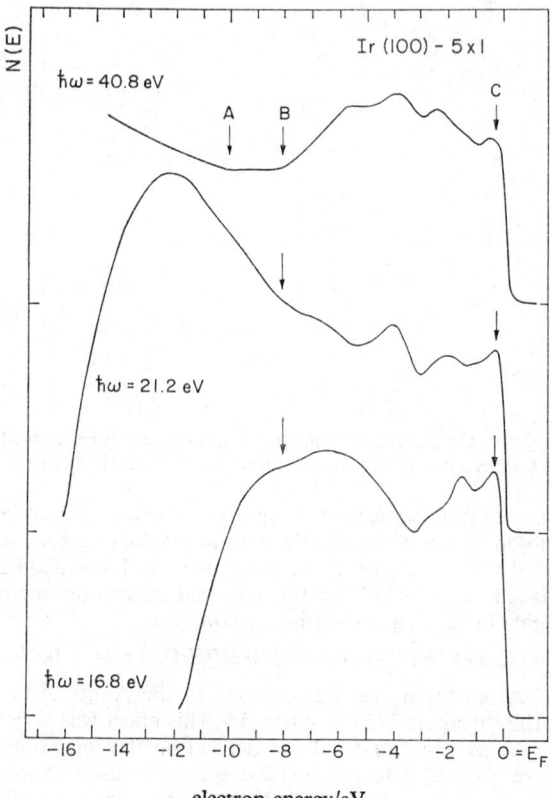

FIG. 3.—Photoemission curves of a clean Ir(100) surface having a (5×1) LEED pattern. Work function $\phi = 5.1$ eV.

transition model of photoemission and observed spectra has been obtained for polycrystalline iridium [11] at photon energies below 12 eV. This latter experiment also indicated a similarity between the photoemission spectra of the metals in the 5d-series. Indeed, the 5d-band part of our photoemission curves at $\hbar w = 16.8$ eV and 21.2 eV look similar to those of gold at the corresponding photon energies.[10] This similarity between gold and iridium does not seem to apply at $\hbar w = 40.8$ eV.[12] Structure B, (fig. 3) is present in all three curves and we assign it to transitions from the bottom of the 5d-band giving an occupied d-band width of 8.0 eV. This value is in good agreement with a recent XPS-measurement [13] on Ir. From band structure calculations [13, 14] one can deduce a d-band width of 8.4 eV. Finally, an estimate of the work function for clean Ir(100) of 5.1 eV can be obtained from the width of the curves (fig. 3). This compares with a thermionically determined value [15] of 5.3 eV.

On adsorption of oxygen the LEED pattern changes into a somewhat disordered (1×1) pattern (fig. 2b). Some traces of the (5×1) structure may still be discerned even at saturation coverage of oxygen. This result does not agree with an earlier LEED study by Grant [6] who, on adsorption of oxygen, obtained a LEED pattern tending towards (2×1). Photoemission spectra for oxygen adsorption are shown in fig. 4. In the lower part of fig. 4 the difference curve for 120 L of O_2 (240 s at 5×10^{-7} Torr O_2) adsorbed at room temperature is shown. This exposure corresponds to a saturated coverage of oxygen. The corresponding work function shift is $\Delta \phi = +0.6$ eV. In the difference curve the structure labelled D, is ascribed to emission

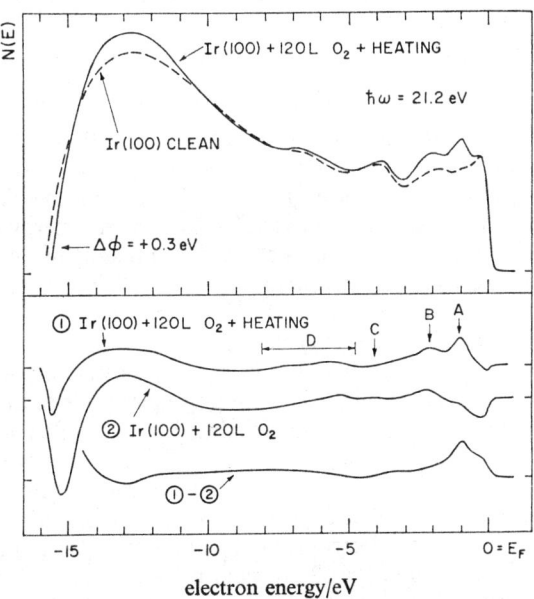

FIG. 4.—*Upper part:* Solid line shows the photoemission curve of a Ir(100) surface exposed to 120 L (240 s at 5×10^{-7} Torr) of O_2 and subsequently heated to 900 K for 1 min. Dashed line shows the clean Ir(100) spectrum. *Lower part:* ① Shows the difference between the two curves in the upper part of the figure. ② Shows the difference curve between a Ir(100) surface exposed to 120 L O_2 at 300 K and the clean Ir(100) surface. ①-② Shows the difference between the curves ① and ②. This curve emphasizes the changes brought about by the heat treatment of the oxygen-covered surface.

from the 2p-states of oxygen. Its peak is 5.4 eV below the Fermi level, i.e., 11.1 eV below the vacuum level. The width of the structure is about 3.6 eV. This energy position of the 2p-states agrees relatively well with those obtained for oxygen adsorption on other metals. For Ni(111) [16] the position is 11.4 eV and for W(110) [17] 12.5 eV. There are also some changes induced in the upper part of the valence band. These structures are labelled A, B and C in fig. 4. After heating the sample to 900 K for 1 min, a photoemission spectrum as shown in the upper part of fig. 4 was obtained. The corresponding difference curve shows a strong increase of the structure labelled A and the disappearence of emission at C. The structure B remains unaffected. These changes can be better presented by taking the difference between the difference curve obtained after heating and the difference curve obtained at room temperature.

The resulting curve is the lowest one in fig. 4. The LEED pattern of this heated surface still corresponded to a (1 × 1) structure which was somewhat better ordered than the one shown in fig. 2b. On further heating at higher temperatures the oxygen gradually desorbed. After heating at 1300 K for 1 min, it was completely desorbed.

The magnitude of the oxygen $2p$ bump is very small compared to the valence band emission. This is in contrast to the results for oxygen on W(110) [17] and Ni(111).[16] One reason may be that the photoemission yield is larger for iridium than for tungsten and nickel. Another reason could be a lower saturation coverage for oxygen on iridium.

A problem encountered when exposing the iridium surface to oxygen was co-adsorption with CO. Evidently the Ir(100) surface was co-adsorbing CO to some degree. This tendency was reduced by having a liquid-nitrogen-cooled cold finger in the vacuum chamber. A similar problem seems to have been observed in the adsorption of oxygen on Ag(110).[18]

DISCUSSION

The conclusion can be drawn from fig. 4 that oxygen is dissociatively adsorbed on Ir(100). Judging from photoemission results for gaseous O_2, associative adsorption of oxygen would probably give at least three peaks in a photoemission spectrum at $\hbar\omega = 21.2$ eV.[19] This is clearly not the case. The binding energy of the $2p$ levels in atomic oxygen is 13.6 eV.[20] Allowing for a relaxation shift of about 3.0 eV (as for CO on Ir(100)) one obtains a binding energy of 10.6 eV for oxygen on Ir(100). This compares with an observed value of 11.1 eV. The discrepancy can very well be accounted for by chemical bonding effects.

One may note (fig. 4) that the $2p$ bump is asymmetric and skewed towards higher energies, i.e., the peak is situated on the low binding energy side of the structure. This is also the case with the $2p$-bumps obtained for the adsorption of oxygen on other metals like nickel [16] and tungsten.[17] It is proposed that this asymmetry is associated with the electronegativity difference between oxygen and the substrate metal. For example, when the electronegativity difference is large as for strontium and barium, the peak is symmetric.[2, 21, 22] A smaller electronegativity difference as tabulated [2] for Ir, Ni and W is associated with asymmetric $2p$ oxygen peaks. One may also note that the maximum work function change on adsorption of oxygen on Ir(100), +0.6 eV, is in keeping with a negative charge transfer to the oxygen atom consistent with its large electronegativity.[2]

As previously indicated, reconstruction of the Ir(100) surface can result in changes of the electronic distribution close to the surface which could in turn possibly be associated with surface states.[4, 5] This feature of the electronic structure would be very sensitive to small amounts of chemisorption since this effectively tends to destroy the reconstructed surface as demonstrated, for example, in fig. 2. We have observed that the peak C (fig. 3) to be the most chemisorption-sensitive structure in the clean photoemission spectrum. Its occurrence may thus be associated with the reconstructed surface. Exposure to CO removes this peak completely whereas room temperatures adsorption of O_2 and H_2 leaves a remnant of the peak. In contrast, the LEED pattern for oxygen adsorption is not a well-ordered (1 × 1) pattern but shows streaks between the integral order diffraction spots and even traces of a (5 × 1) pattern. Therefore electronic structure associated with the reconstructed surface may not be completely removed.

As shown in fig. 4, adsorption of oxygen induces changes in the valence band region of iridium. Apart from changes caused by the removal of reconstruction one may also expect changes due to the chemisorptive bonding itself. We have plotted (fig. 5) the difference curves obtained for a full coverage adsorption of O_2, H_2 and CO at room temperature. All curves have three common features of spectral structure denoted A, B and C. These vary somewhat in magnitude and perhaps also in energy but the pattern is still remarkably similar for all. This implies that the changes in the valence band region of Ir(100) may depend more on the geometry of the bonding site than on the adatom as such. As shown in fig. 2b, the LEED pattern on adsorption of O_2 (also H_2) is disordered. This disorder would give rise to a large number of different types of bonding sites. Heating the sample after the adsorption of oxygen can be interpreted as an annealing process giving rise to a more ordered surface. The ensuing reduction in the number of different types of bonding sites could explain the growth on heating of peak A and the disappearance of emission at C as shown (fig. 4).

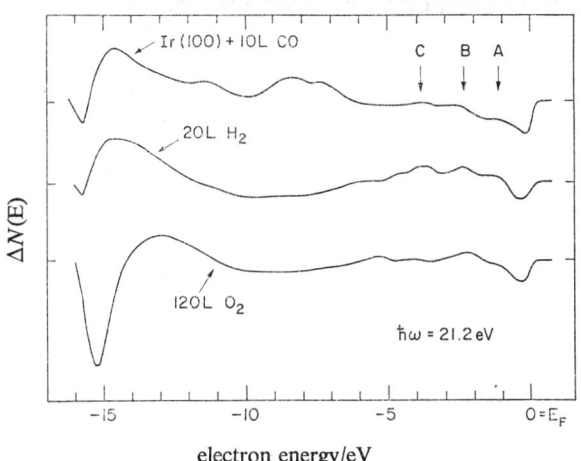

FIG. 5.—Difference-curves obtained for room temperature adsorption of CO, H_2 and O_2. The exposures correspond to saturation coverage.

Additional studies are in progress to evaluate the validity of the preceding suggestions, particularly with reference to the interpretation of peak shape and the possible occurrence of an intrinsic surface state on clean Ir(100)—(1 × 5).

Research Support from the National Science Foundation Grant DMR 71-017 69 702 and from the Advanced Research Projects Agency through the Cornell Materials Science Center is gratefully acknowledged. Preparation of iridium single crystals by B. F. Addis, Cornell MSC-facility is appreciated as well as technical assistance from C. Brucker and J. Starbuck.

[1] G. C. Bond, *Catalysis by Metals* (Academic Press, London, 1962), chap. 19, p. 441.
[2] L. Pauling, *The Nature of the Chemical Bond* (Cornell University Press, Ithaca, N.Y., 1960), 3rd ed., p. 93.
[3] A. Ignatiev, A. V. Jones and T. N. Rhodin, *Surface Sci.*, 1972, **30**, 573.
[4] J. A. Appelbaum and D. R. Hamann, *Phys. Rev. Letters*, 1973, **31**, 106.
[5] J. E. Rowe and H. Ibach, *Phys. Rev. Letters*, 1973, **31**, 102.
[6] J. T. Grant, *Surface Sci.*, 1969, **18**, 228.
[7] N. J. Dionne, Ph.D. Thesis (Cornell University, Ithaca, N.Y., 1975).
  N. J. Dionne and T. N. Rhodin, *Phys. Rev. Letters*, 1974, **32**, 1311.
[8] J. Starbuck, M.S. Thesis (Cornell University, Ithaca, N.Y., 1974).

[9] J. E. Rowe, S. B. Christman and E. E. Chaban, *Rev. Sci. Instr.*, 1973, **44,** 1675.
[10] D. E. Eastman and J. K. Cashion, *Phys. Rev. Letters*, 1970, **24,** 310.
[11] M. Traum and N. V. Smith, *Phys. Rev. B*, 1974, **9,** 1353.
[12] J. Freeouf, M. Erbudak and D. E. Eastman, *Solid State Comm.*, 1973, **13,** 771.
[13] N. V. Smith, G. K. Wertheim, S. Hufner and M. Traum, *Phys. Rev. B*, 1974, **10,** 3197.
[14] N. V. Smith, *Phys. Rev. B*, 1974, **9,** 1365.
[15] O. Weinreich, *Phys. Rev.*, 1951, **82,** 573.
[16] D. E. Eastman and J. E. Demuth, *Proc. 2nd Int. Conf. Solid Surfaces*, 1974; *Japan J. Appl. Phys.*, Suppl. 2, part 2, 1974, p. 827.
[17] J. M. Baker and D. E. Eastman, *J. Vac. Sci. Tech.*, 1973, **10,** 223.
[18] H. A. Engelhardt, A. M. Bradshaw and D. Menzel, *Surface Sci.*, 1973, **40,** 410.
[19] D. W. Turner, C. Baker, A. D. Baker and C. R. Brundle, *Molecular Photoelectron Spectroscopy* (Wiley-Interscience, London, 1970), p. 52.
[20] *Atomic Energy Levels*, circ. Nat. Bur. Stand. 467, ed., C. Moore (Washington D.C., 1949), vol. I, p. 45.
[21] K. A. Kress and G. J. Lapeyre, *Phys. Rev. Letters*, 1972, **28,** 1639.
[22] C. R. Helms and W. E. Spicer, *Phys. Rev. Letters*, 1972, **28,** 565.

Chemisorption of Organic Molecules on Single Crystal Surfaces of Transition Metals

Adlayer Stoichiometry, Arrangement and Electronic States

By T. A. Clarke, I. D. Gay, B. Law and R. Mason*

School of Molecular Sciences, University of Sussex

Received 10th March, 1975

XPS and UPS of alkenes and haloalkenes, chemisorbed on a clean Pt(100) surface, are reported. The dissociative chemisorption of vinyl fluoride and vinyl chloride at low coverages is modified when the clean surface is pre-treated with carbon monoxide. Arguments are presented that the activation of carbon–hydrogen bonds takes place preferentially at four-fold bridging sites on the surface. The differing patterns of sorption of ethylene on the Pt(100) and Ni(111) surfaces are discussed and some general comments provided on the importance of intra-adlayer interactions in determining some surface reactions.

INTRODUCTION

Electron diffraction and photoelectron spectroscopy offer the prospect of defining the surface crystallography and electronic states of clean metal surfaces and their modification following sorption of inorganic or organic ligands. The chemisorption of carbon monoxide and alkenes by surfaces of transition metals poses important questions in surface chemistry and in studies of the mechanisms of heterogeneous catalysis. This paper sets out to summarise results, referring largely to the organometallic chemistry of a single crystal surface of platinum, which will relate to the following questions: (i) Is chemisorption of a given ligand an associative, dissociative or combination of events? (ii) Is there a changing pattern of reactivity towards alkenes at surface sites having different local symmetries? (iii) Are the present experimental and theoretical methods adequate to provide an assignment of electronic states of the " surface molecule " and is there a correlation with the energy levels of simple discrete molecules, such as the metal carbonyls and metal–alkene complexes?

EXPERIMENTAL

All experiments were carried out in our combination ultra-high vacuum (base pressure *ca.* 5×10^{-11} Torr) instrument. No use of the Auger spectroscopic facilities was made, other than for assessing the state of the clean surface, in view of the generally observed electron beam-induced dissociation of organometallic surface species. The metal-ligand surfaces are spectroscopically stable to X-rays (chromatic Al K-radiation) and to He(I) and He(II) resonance radiation. In both the XPS and UPS studies, photons are incident at an angle of 45° to the sample normal and the usual kinetic energy analysis of the emitted photoelectrons was supplemented by computer averaging techniques; typically, the spectra illustrated below refer to accumulations over 0.5 h. Work function changes on chemisorption were measured by the zero kinetic energy method of Evans.[1] LEED of the surfaces was carried out by simply translating and rotating the crystal within the apparatus.

Cleaning and polishing of the platinum (100) and (111) surfaces was initially with diamond paste and with γ- and α-alumina. Mounting on the goniometer head was via platinum clips and surface carbon was removed by heating the sample at 700 °C in oxygen (10^{-6} Torr);

surface oxygen was removed by ion bombardment and careful annealment provided surfaces with LEED patterns representative of the clean surfaces.[2, 3] Chemisorption of gases was effected by exposing a steady-state flux corresponding to pressures of up to 5×10^{-7} Torr. One disadvantage of our present equipment is that the gases are not led directly to the surface; monitoring of ligand coverage was, therefore, by measurement of integrated C 1s emission intensities relative to saturation values, rather than through an acceptance of nominal exposure conditions.

RESULTS

The stoichiometry of an organometallic surface can be rather precisely estimated by XPS, given that the photoemission from associatively chemisorbed ligands should closely relate to that from the molecules in the gas phase, both in respect to the core binding energies and, more importantly, the ratio of the emission intensities from two or more chemically non-equivalent atoms;[4-6] moreover, the core emission intensity

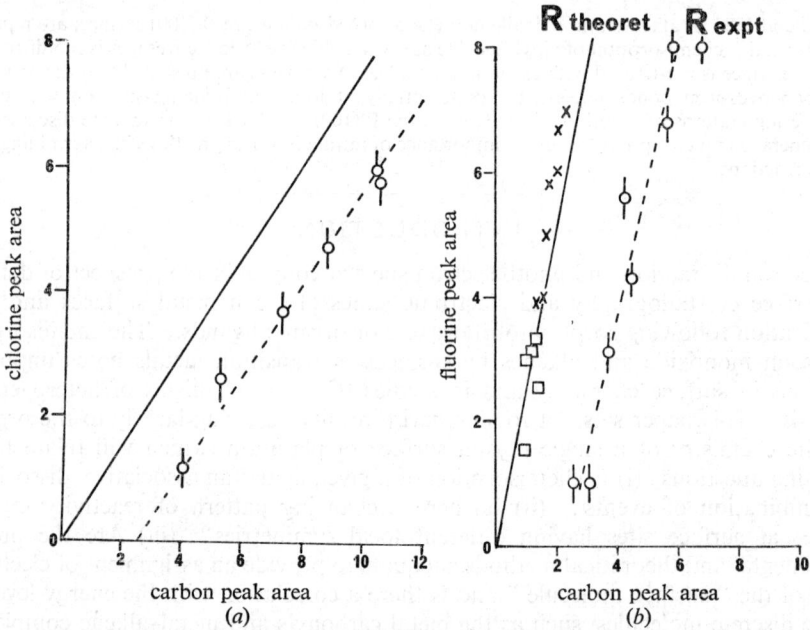

FIG. 1.—(a) XPS of the Pt(100)-vinyl chloride surface. The solid line refers to the ratio, $\mathscr{R}$(Cl/C) (0.78) of the Cl 2p and C 1s emission intensities from the gas phase XPS spectrum. (b) XPS of the Pt(100)-vinyl fluoride surface (open circles). The solid line, $\mathscr{R}$ theoret. (2.27), corresponds both to the ratio of ionisation cross-sections and the relative intensities of the F 1s and C 1s emission intensities in the gas-phase spectrum. The remaining experimental data refer to $\mathscr{R}$(F/C) (vinyl fluoride) following preadsorption of 0.3 monolayer carbon monoxide (×) and of 0.6 monolayer carbon monoxide (□).

ratios of atoms such as chlorine and fluorine with that from carbon in, say, undissociated haloalkenes is well-represented by the appropriate ratio of ionisation cross-sections. In fig. 1, we illustrate heteroatom–carbon emission intensity ratios for chemisorption of vinyl chloride and vinyl fluoride on a clean Pt(100) surface and corresponding data for the platinum surface pretreated with varying coverages of carbon monoxide.

He(I) ($\hbar\omega = 21.2$ eV) and He(II) ($\hbar\omega = 40.8$ eV) spectroscopy of Pt(100)–C_2H_4 and Pt(100)–C_3H_6 surfaces are shown in fig. 2 and 3. Energies are relative to the

Fermi level. The "spokes" are at energies, E_L, calculated from the gas phase ligand ionisation potentials, I_L, by the relation

$$E_L = I_L - E_F - \Delta\phi + R.$$

The ligand ionisation potentials, I_L, are taken as 10.7, 12.8, 14.8, 16.0 and 19.1 eV (ethylene);[7] 9.9, 12.1, 13.1, 14.2, 16.0 and 18.1 eV (propene);[8] and 11.4, 16.8 and 18.7 (acetylene).[9] E_F is 5.5 eV and work function changes are -0.8 eV (C_2H_4), -1.1 eV (C_3H_6) and -1.1 eV (C_2H_2).[2] R, a relaxation shift,[10] is assumed constant for all levels at -2.5 V. This latter parameter may be responsible for uncertainties

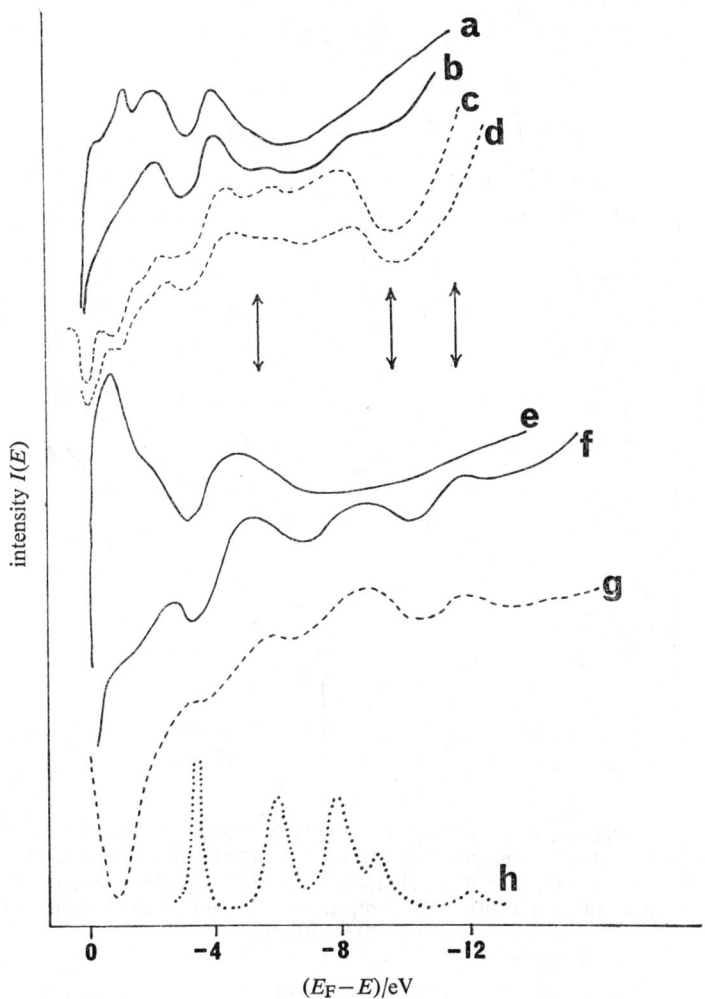

FIG. 2.—The UPS of the Pt(100)–C_2H_4 surface: (a) photoemission spectra for clean Pt(100) surface (He(I) radiation); (b) the He(I) emission spectrum for Pt(100)+100 L (C_2H_4) surface; (c) the difference emission spectrum (b)–(a); (d) the difference spectrum (He(I) radiation) between emission from a clean Pt(100) surface and the Pt(100)+5 L (C_2H_4) surface; (e) the He(II) photoemission spectrum of the clean surface; (f) the He(II) spectrum for the Pt(100)+100 L (C_2H_4) surface; (g) the difference spectrum (f)–(e) and, (h) schematic representation of the gas-phase emission spectrum of ethylene. The arrowed solid lines refer to peaks in a "shifted" acetylene gas-phase spectrum.

attaching to the assignment of surface electronic states. There is no *prima facie* reason why it should be identical for all levels, independent of their symmetries and energies—indeed, the reverse is true: but the discussion set out below does not seem, in contrast to the uncertainty attaching to the M–CO surface,[11] to depend critically on the empirically chosen value(s) of R, given that there are more than two significant peaks in the emission spectra which help to remove ambiguities of assignment.

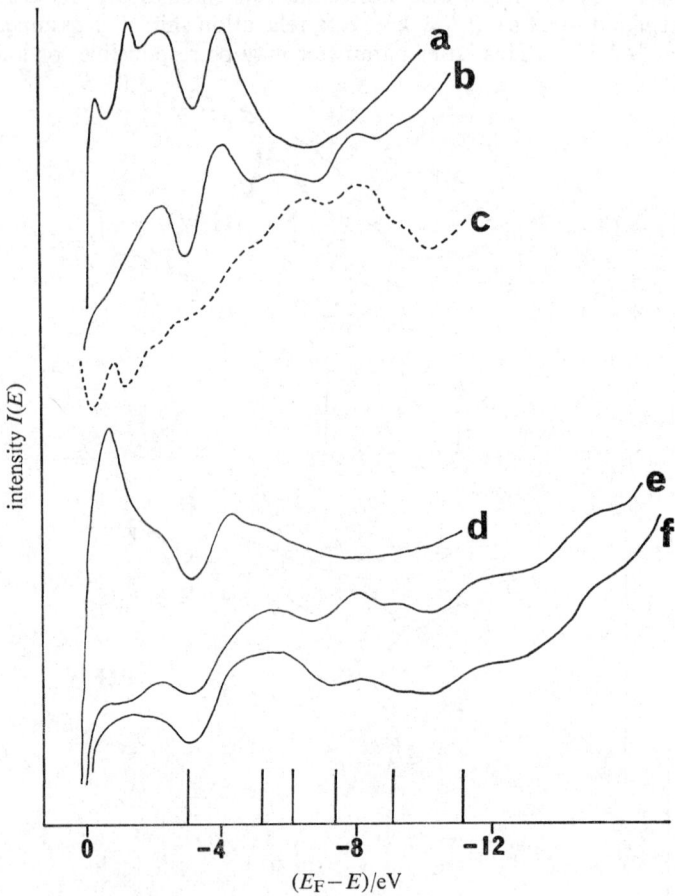

FIG. 3.—UPS of the Pt(100)-propene surface: (*a*) the photoemission spectrum for the clean Pt(100) surface ($\hbar\omega = 21.2$ eV); (*b*) the emission from the Pt(100)–100 L propene surface (He(I)); (*c*) the difference emission spectrum (*b*)–(*a*); (*d*) the He(II) excited emission spectrum from the clean surface; (*e*) the He(II) spectrum of the Pt(100)–100 L propene surface, and (*f*) the He(II) spectrum of the Pt(100)–20 L surface.

DISCUSSION

The results contained in fig. 1 can be summarised as follows.

(i) Both vinyl fluoride and vinyl chloride are dissociatively chemisorbed at low coverages of the Pt(100) surface, given that the observed ratio of the emission intensities, $\mathscr{R}(X/C)$, is much less than that observed in their gas phase spectra or, in the case of vinyl fluoride, than that which can be predicted from the relative carbon and fluorine ionisation cross-sections. The initial dissociative pattern is followed by an associative régime (after approximately one-quarter monolayer coverage) for the data then correspond, within experimental error, with those expected theoretically. We have noted [6] earlier that all the di- and perchloroalkenes are associatively chemisorbed; 1, 1- and 1-2 difluoroethene are dissociatively and associatively sorbed respectively and remaining perfluoroalkenes are associatively bound to the surface. Our rationale [4-6] of these results was in terms of intermolecular HX elimination from the surface, following the metallation of a carbon–hydrogen bond and hydride attack on adjacent carbon–halogen bonds. The changing pattern of sorption of vinyl halides as a function of coverage leads to the obvious suggestion that only certain sites on the metal surface have the geometrical and/or electronic properties necessary for initiating the surface oxidative addition reactions.

(ii) Preadsorption of fractional monolayers of carbon monoxide provides for the associative chemisorption of vinyl fluoride, at least for *total* ligand coverages of up to three-quarters or so (it is re-emphasised that coverage refers to a fraction of the saturation C $1s$ emission intensity which, in the mixed ligand surface, consists of two resolved peaks centred at 284.5 eV (C_2H_3F) and 287.1 eV (CO) relative to the Pt $4f_{\frac{7}{2}}$ emission at 71.1 eV). At high carbon monoxide–vinyl fluoride and low (*ca.* 0.3) carbon monoxide–vinyl fluoride coverage, there is, again, some dissociation of the alkene in that $\mathcal{R}(F/C)$ is greater than the theoretical value (2.27) and one may infer that the carbon monoxide is now precluding metal–carbon bonded species, generated from the alkene, from the surface. This result needs extension to vinyl chloride and other haloalkenes before it can be discussed quantitatively, so that we content ourselves here with a discussion of the significance of the changing pattern of dissociative sorption of vinyl fluoride at low coverages of the clean surface and the associative pattern observed for relatively low alkene coverage of the Pt(100)–0.3 CO surface.

The choice of carbon monoxide as a "blocking" ligand was based first on a recognition of its very small electronic perturbation of the Pt(100) surface ($\Delta\phi = +0.2$ V); one could thus be clear that the electrophilicity of the metal atoms at the surface was not significantly affected by the preadsorption of carbonyl ligands. Secondly, there is some evidence, admittedly often of an indirect and inconclusive kind, that carbon monoxide binds preferentially, at low coverage, to the four-fold bridging sites of the Ni(100) and Pt(100) surfaces. Tracy [12] argued for four-fold coordination of carbon monoxide, in the $C(2 \times 2)$ structure observed for sorption of 0.5 monolayer CO on Ni(100), from the behaviour of the substrate diffraction beams on disordering of the $C(2 \times 2)$ pattern. Doyen and Ertl [13] have provided a theoretical energy profile for chemisorption of CO on Ni(100): the minimum is at the four-fold bridging site and is some 40 kJ mole^{-1} less than that for terminal bonding. Unfortunately, these results and predictions cannot be carried over without reservations to the crystallography of the Pt(100)–CO surface. No detailed theoretical calculations of potential energy profiles are available. The energy surface [13] for the Pd(100)–CO surface shows a preference for bridge bonding but now there is little difference between the two-fold and four-fold sites and the LEED pattern at $\theta = \frac{1}{2}$ corresponds to a $C(4 \times 2)$–45° structure, rather than the $C(2 \times 2)$ symmetry for Ni(100)–CO. For a regularly packed adlayer this could arise from two-fold site binding at low coverage but the situation is again different for CO sorption on the Pt(100) surface. We have simulated the displaced (4×2) arrangement for the Pt(100)–CO surface by the use of molecular dynamics methods based on similar potential functions to those which mimicked the coverage dependences of the surface crystallography of the Ni(100)–CO surface.[14] LEED of the Pt(100)–CO surface has been interpreted by us [3] as indicating a mixture of four-fold bridge and terminal carbonyl bonding, although it is obvious that this cannot be accepted as proven until complete LEED intensity calculations have been completed. Our working hypothesis at present is that it is the four-fold bridging sites which are responsible for initiating the surface reactions and, fortunately, this can be tested directly: a comparison of the dissociative–associative patterns for adsorption on the Pt(111) surface and the Ni(100) and Ni(111) surfaces and their dependence on preadsorption of fractional monolayers of carbon monoxide should be reasonably unequivocal given the better understanding of the crystallography of the nickel–carbon monoxide surfaces. These data will be reported at the Discussion.

We now turn to the low energy photoelectron spectroscopy of metal–alkene and other metal–ligand surfaces. The feasibility of facile intermolecular dehydrogenation of ethylene at the Pt(100) surface is indicated by the HX elimination reactions of the haloalkenes but one might also predict that the dehydrogenation would not be complete. This would then be in contrast to Demuth and Eastman's results [10] for ethylene chemisorbed on Ni(111) where it seems that ethylene is associatively sorbed at 100 K but is totally dehydrogenated at 230 K. Ethylene and propene have essentially identical van der Waals envelopes to those of vinyl fluoride and vinyl chloride respectively. The spectroscopy of the Pt(100)–C_2H_4 and Pt(100)–C_3H_6 at varying coverages and at *ca*. 290 K (fig. 2 and 3), and also of trifluoropropene (*vide infra*) is consistent with the major surface species being the alkenes rather than the alkynes. The table summarises our data on chemisorbed alkenes.

TABLE 1.—A SUMMARY OF IONISATION ENERGIES FOR Pt(100)–ALKENE SURFACES

ionisation energies (eV). E_{obs} (He(I)) and E'_{obs} (He(II)) are relative to the Fermi level and calculated values $E_L = I_L - E_F - \Delta\phi + R$

ligand								
$H_2C=CH_2$	E_{obs}	?	5.8		8.5			
	E'_{obs}	3.0	5.5		8.6		11.5	
				(v. broad)				
	E_L	3.5	5.6	7.6	8.8		11.9	($\Delta\phi = -0.8$ V)
$H_3C-CH=CH_2$	E_{obs}	2.8	5.0	6.4	8.2	9.3		
	E'_{obs}	2.6	5.7		8.0	9.3	11.0	15.2
			(v. broad)					
	E_L	3.0	5.2	6.2	7.3	9.1	11.2	($\Delta\phi = -1.1$ V)
$F_3C-CH=CH_2$	E_{obs}				7.8	9.5		
	E'_{obs}				8.0	9.7	13.0	
	E_L	unavailable						
H—C≡C—H	E_L		5.5			9.8	11.8	($\Delta\phi = -1.1$ V)
$H_2C=C=CH_2$	E'_{obs}			6.4	8.6		11.6	
	E_L	3.6			8.6		10.9	($\Delta\phi = -1.4$ V)
				(v. broad)				
$H_2C=C=CH_2$ (cumulene)	E'_{obs}	2.8(?)		6.6		9.0	12.0	
	E_L	2.7	4.9	5.8	7.0	8.9		($\Delta\phi = -1.5$ V)

The difference photoemission spectra of the metal–ethylene surface (fig. 2, c and h) are seen, so far as the bands lying at energies greater than 4 eV or so above the Fermi level are concerned, to be those which could be expected from perturbed ligand states. As the tabulated data also indicate, the four peaks in the surface emission spectrum centred at approximately 3.0, 5.5, 8.6 and 11.5 eV below the Fermi level are assigned to ligand levels whose unperturbed ionisation energies are, 10.7 12.8, 14.8 and 16.0, and 19.1 eV respectively. The additional structure in the Pt(100)-propene spectra (fig. 3) reinforce the ethylene assignments, and together with the results for trifluoropropene, are consistent with a predominantly associative binding of the alkenes. Fig. 2(c) shows the difference spectrum (He(I) radiation) for low ethylene coverages of the Pt(100) surface. This was obtained in order to determine whether the predominant ligand species was then acetylene. There is a peak in the emission spectrum which could correspond to ionisation from a molecular orbital built up largely from an acetylene π-level but the total spectrum is better assigned in terms of shifted ethylene states. We can seemingly rule out the stoichiometric dehydrogenation of ethylene at the Pt(100) surface but, remembering the low emission intensities of the second and third bands of acetylene, we would not be justified in excluding the possibility of partial dehydrogenation, equivalent to the dissociative sorption of the haloalkenes.

This contrast between the reactions of ethylene at the Ni(111) and Pt(100) surfaces deserves some comments, even if they are presently preliminary to obtaining complete data for the Pt(111) and the Ni(100) surfaces. There is, first, a clear difference to be recognised between the reactions of haloalkenes on the Pt(100) and Pt(111) surfaces [5, 6] in that the more close-packed surface provides for almost complete dissociation of the vinyl halides. The dependence of the reactions on substrate geometry was related to the increased facility of HX elimination from the (111) surface, given the additional short non-bonded interactions between adjacent molecules positioned on a $p(2 \times 2)$ substrate mesh. These adlayer interactions would be enhanced further on the Ni(111) surface and geometrical considerations alone may be sufficient to explain the differences between the reactions on the Pt(100) and Ni(111) surfaces. The essential point is the relative activation energies for metallation of the carbon–hydrogen bonds and for hydride transfer between adjacent molecules. The latter will reflect the average intermolecular contacts within the adlayer and will, therefore, follow the order Ni(111) < Ni(100) < Pt(111) < Pt(100). It may be much more difficult to delineate the relative electrophilicities of the nickel and platinum surface atoms which will be of importance in determining rates of initial carbon–hydrogen bond fission but experimental data on mixed ligand surfaces should be useful.

The nature of the bonding of associatively chemisorbed alkenes to surfaces of nickel, palladium and platinum has been discussed extensively. The photoemission spectra of fig. 2 and 3 allow two further comments.

The decrease, on chemisorption, in the pseudo-density of states of the metal valence band is similar to that which has been described for the Pt(100)–CO surface.[11] It is only the doublet component of the valence band centred at 2 eV or so below the Fermi level which is significantly modified and we have suggested [15] that this reflects " back donation " of the metal d electrons, having considerable t_{2g} character, to ligand antibonding states. In the case of the alkene–metal bond, net charge transfer is from the ligand to the metal and the " forward " donation process must be into the metal s band or suitable hybrid states. The difference spectra appear to show additional emission intensity in the second doublet component of the valence band but this must be treated with caution since some emission at that energy could be expected from levels which are predominantly ligand in character. Simple perturbation theory argues for much greater stabilisation, on sorption, of the highest filled ligand π-level compared with the ligand σ-states and the weakness of our economical assumption—the relaxation shift R is a constant for all levels—is that one is in danger of throwing away the baby with the bath water! We have provided substantial evidence that the 5σ level of carbon monoxide is stabilised by 2 eV or so on sorption at the Pt(100) surface, whereas the ligand 1π and 4σ levels are essentially non-bonding.[11] The emission spectra of the Pt(100)-propene surface (in particular, fig. 3f) suggest that the highest filled ligand state is stabilised also by $ca.$ 1.5–2 eV on bond formation; the emission curves 2(c), 2(d), 2(f) and 2(g) could accommodate a similar interpretation for ethylene if the peak centred at $ca.$ 5.8 eV or so below the Fermi level were regarded as a composite one, similar to the peak centred at $ca.$ 8 eV below the Fermi-level in the emission spectrum from the Pt(100)–CO surface. A close parallel exists between the photoelectron spectra of metal–carbon monoxide surfaces and simple metal carbonyls [11] and a systematic survey of metal–alkene complexes would help to remove the present ambiguities in the assignment of states for metal-alkene surface molecules.

We are grateful to the Science Research Council for support and to Simon Fraser University for granting sabbatical leave to I. D. G.

[1] S. Evans, *Chem. Phys. Letters*, 1973, **23**, 134.
[2] A. E. Morgan and G. A. Somorjai, *J. Chem. Phys.*, 1969, **51**, 3309.
[3] T. A. Clarke, R. Mason and M. Tescari, *Proc. Roy. Soc. A*, 1972, **331**, 321.
[4] T. A. Clarke, I. D. Gay and R. Mason, *Chem. Comm.*, 1974, 331.
[5] T. A. Clarke, I. D. Gay and R. Mason, *Chem. Phys. Letters*, 1974, **27**, 562.
[6] T. A. Clarke, I. D. Gay and R. Mason, *Proc. Battelle Symp., Physical Methods in Heterogeneous Catalysis*, 1975, in press.
[7] D. W. Turner, C. Baker, A. D. Baker and C. R. Brundle, *Molecular Photoelectron Spectroscopy*, (Wiley-Interscience, London and New York, 1970), p. 179.
[8] M. Horn and J. N. Murrell, *J. Organometal. Chem.*, 1974, **70**, 51.
[9] D. W. Turner, C. Baker, A. D. Baker and C. R. Brundle, *Molecular Photoelectron Spectroscopy* (Wiley, 1970), p. 190.
[10] J. E. Demuth and D. E. Eastman, *Phys. Rev. Letters*, 1974, **32**, 1123.
[11] T. A. Clarke, I. D. Gay, B. Law and R. Mason, *Chem. Phys. Letters*, 1975, **31**, 29.
[12] J. C. Tracy, *J. Chem. Phys.*, 1972, **56**, 2736.
[13] G. Doyen and G. Ertl, *Surface Sci.*, 1974, **43**, 197.
[14] T. A. Clarke, I. D. Gay and R. Mason, *Surface Sci.*, in press.
[15] T. A. Clarke, I. D. Gay and R. Mason, *Chem. Phys. Letters*, 1974, **27**, 172.

Adsorption of Methanol, Formaldehyde and Ammonia on W(100) Studied by Ultra-Violet Photoelectron Spectroscopy

By William F. Egelhoff, John W. Linnett and David L. Perry*

Department of Physical Chemistry, University of Cambridge,
Lensfield Road, Cambridge CB2 1EP

Received 25th March, 1975

The adsorption of methanol, formaldehyde and ammonia on W(100) at room temperature has been studied by ultra-violet photoelectron spectroscopy. It is shown how UPS may be used as a fingerprint technique to monitor the chemical nature of the ad-layer during the progress of a decomposition reaction on the surface. Both methanol and formaldehyde, at saturation coverages, adsorb as molecular complexes with spectra which are dependent upon the coverage of the tightly bound β-CO layer formed during the early stages of the adsorption. The photoelectron spectra of these complexes of methanol and formaldehyde, and also of ammonia, adsorbed on surfaces of different composition are discussed in relation to the gas phase spectra and the available free valence of the substrate.

The application of ultra-violet photoelectron spectroscopy (UPS) to the study of adsorption on metal surfaces has expanded rapidly in recent years, since the sensitivity of the technique to adsorbed monolayers was first demonstrated.[1, 2] The use of the results obtained, to gain a complete understanding of the electron energy levels of the adsorbate-surface complex, awaits a more satisfactory theoretical model for both chemisorption and for photoeffects in metals and in adsorbed layers. However, UPS has led to a number of useful insights into chemisorption. For example, when combined with thermal desorption and work function changes or with LEED experiments, it has been possible to correlate the photoelectron spectra with adsorbate structure and to evaluate the reality of different states identified by thermal desorption as chemically distinct binding states at the actual temperature of adsorption.[3-6]

In certain cases, where the interaction between adsorbate and surface leads to discrete features in the photoelectron spectrum, analysis has been attempted by comparing the new surface energy levels with those of the gas phase species. However, the general difficulties of obtaining a complete interpretation are illustrated by the substantial disagreement over orbital identification in the well-studied system of CO adsorbed on Ni.[7] In this case, adsorption is molecular and the observed energy levels might be expected to be simply related to the gas phase CO spectrum. Interpretation of spectra where the adsorption is strong and some degree of dissociation is presumed, is much more difficult as the spectra tend often to lack sharp features.

It was this type of difficulty of interpretation which led us to examine the adsorption of some larger molecules on metal surfaces. These molecules might be expected to adsorb at low temperatures as species resembling those occurring in the gas phase and, therefore, enable some orbital identification to be attempted. We report in this paper the results of a study of the adsorption of methanol, formaldehyde and ammonia on W(100) at room temperature. In addition to UPS, thermal desorption was also studied and work function changes were measured. The results obtained enable some assessment to be made of the bonding of these species to the surface and the effect on this of the underlying surface structure.

EXPERIMENTAL

All experiments were carried out in a photoelectron spectrometer operating at a base pressure of 2×10^{-10} Torr of active gases, using HeI (21.2 eV) radiation (1 Torr = 1.33×10^2 Pa.) Electrons photoemitted from the W(100) surface within a cone of apex angle 4°, were energy analysed by a 127° cylindrical analyser with a measured resolution of 1 %. The axis of this cone, the direction of propagation of the light beam and the axis of rotation of the crystal were mutually perpendicular (see fig. 1). On account of the marked angular dependence of UPS spectra from single crystal surfaces, spectra were obtained at $\theta = \sim 13°$, i.e., at near-glancing photon incidence and near normal electron emission. Electron intensities were typically 25 000 counts/s enabling spectra to be obtained in a single scan of approximately 60 s with a signal/noise ratio of 100 : 1.

The W(100) crystal, a disc of diameter 8 mm and thickness 0.45 mm, was aligned, cut and mechanically polished by the usual techniques. It was supported by 0.2 mm tungsten wires attached to 2 mm diameter tungsten rods. A tungsten filament, tightly wound in a flat coil and positioned behind the crystal, enabled the crystal to be heated in two ways. For cleaning purposes, the crystal was heated by electron bombardment to 2300 K; in subsequent thermal desorption experiments, radiative heating by the filament gave a constant heating rate of $6° \text{s}^{-1}$ up to 750 K. The crystal temperature was measured by means of a W–W/Re thermocouple. The crystal was cleaned by many hours of flashing to 2000 K in oxygen at 10^{-6} Torr to remove carbon and afterwards by flashing to 2300 K in vacuo to remove oxide. This produced a crystal with a reproducible clean-surface spectrum in each experiment. Periodic oxygen treatment was required to remove small quantities of carbon which gradually accumulated at the surface. The gas phase composition was monitored by a quadrupole mass spectrometer. In all experiments the crystal was allowed to cool for up to five minutes after a high temperature flash, to a temperature of < 310 K, before the surface was exposed to the adsorbing gas. Owing to strong adsorption of methanol and formaldehyde on the walls of the apparatus, quoted exposures are only approximate and may be in error by up to a factor of two.

RESULTS AND DISCUSSION

METHANOL AND FORMALDEHYDE ADSORPTION

The adsorption and decomposition of formaldehyde on W(100) has been studied previously in flash desorption and work function experiments.[8] No analogous study for methanol has been reported, but the kinetics of decomposition on different metal surfaces has been the subject of several investigations.[9, 10] In this study we have applied the combined techniques of flash desorption, UPS and work function measurements to the investigation of both methanol and formaldehyde adsorption on W (100). A preliminary account of some early results in this study has been published previously.[11] The flash desorption results will be published in detail elsewhere.[12]

The HeI photoelectron spectrum of a W(100) surface exposed to CH_3OH is shown in fig. 1. The clean-surface spectrum is consistent with that obtained and analysed in the light of band structure calculations by Feuerbacher and Christenson [13] and will not be discussed further here. A prominent feature of the clean W(100) spectrum is the peak at 0.5 eV below the Fermi level, designated a surface state.[14] The intensity of this peak is extremely sensitive to the presence of low coverages of adsorbed species and provides, therefore, a sensitive cross-check on surface cleanliness. The interaction of CH_3OH with W(100) passes through two stages which may be clearly distinguished in fig. 1. The first stage, which occurs at exposures of up to approximately 1 Langmuir (1 Langmuir, L = 1×10^{-6} Torr s), produces changes mainly in the valence band region of the spectrum (fig. 1(b)), whereas at higher exposures a two peaked spectrum is observed at lower energies (fig. 1(c) and (d)).

At low exposures the behaviour of methanol resembles very closely that of formaldehyde reported previously.[8] That is, both molecules adsorb dissociatively on the clean surface at 310 K. The evidence for this is that the photoelectron spectra, the work function changes, and the flash desorption curves can be reproduced exactly by small, carefully proportioned exposures to H_2 and CO. Fig. 2 shows the spectra

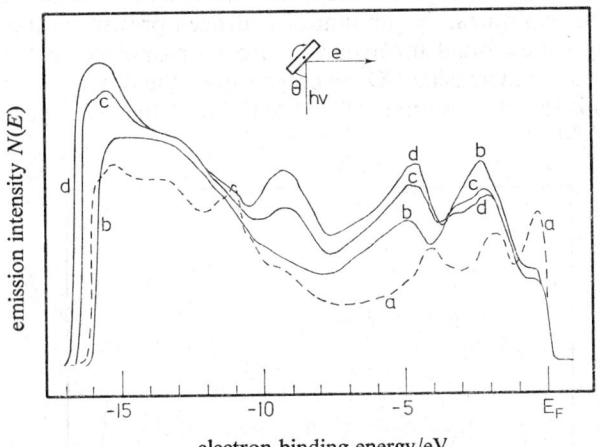

FIG. 1.—The HeI photoelectron spectrum of a W(100) surface exposed to CH_3OH at room temperature ($\theta = \sim 13°$): (a) clean surface; (b) 1.0 L CH_3OH; (c) 2.0 L CH_3OH; (d) 3.0 L CH_3OH. (1 L = 1 × 10^{-6} Torr s.)

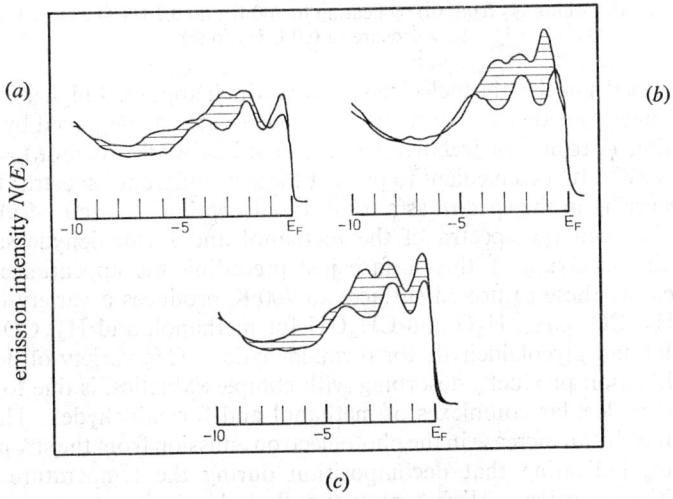

FIG. 2.—The HeI photoelectron spectrum of a clean W(100) surface exposed to (a) 0.4 L CO; (b) 0.5 L H_2; (c) 0.2 L CH_3OH (shaded areas indicate an increase in intensity following adsorption).

of low exposures of CO, H_2 and CH_3OH. It may easily be seen that the sum of the changes in intensity induced by CO and H_2 is equivalent to the CH_3OH spectrum. In fact the CH_3OH spectrum may be synthesised experimentally as follows. Fig. 3(a) is the spectrum after exposure to ~ 0.3 L CH_3OH. Spectrum 3(b) was taken following desorption of H_2 by heating to 700 K. No CO desorbed during this heating and spectrum 3(b) is characteristic of β-CO.[3] Fig. 3(c) and (d) shows the crystal spectrum after exposure to (c) 0.2 L CO followed by (d) 0.9 L of H_2. Spectra (a) and (d) are identical. These and similar experiments with formaldehyde confirm the conclusions drawn from flash desorption experiments [8, 12, 15] that CH_3OH and HCHO dissociate on the W(100) surface to adsorbed CO and hydrogen and that, as in CO and hydrogen co-adsorption, there is no evidence for the formation of a (CO-H) molecular complex.

With increasing exposures, the amount of hydrogen present on the surface in the same two binding states found for a pure hydrogen monolayer begins to decrease. This is expected, as the increasing CO coverage causes the desorption of hydrogen.[15] The result of this is that at exposures of ~ 1 L CH_3OH the UPS spectrum (fig. 1(b)) resembles that of β-CO.

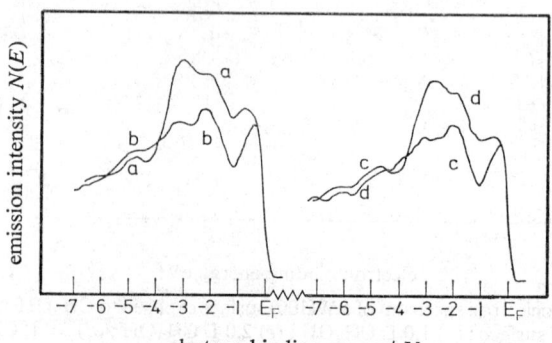

FIG. 3.—Simulation of a low exposure methanol spectrum. (a) 0.3 L CH_3OH exposed to clean surface; (b) after desorbing H_2 from (a) by heating to 700 K; (c) 0.2 L CO exposed to clean surface; (d) exposure of 0.9 L H_2 to (c).

The new peaks in the photoelectron spectra which appear at higher exposures cannot be attributed to either CO or hydrogen. These are accompanied by a decrease in work function ϕ from $\Delta\phi$ (relative to clean surface work function) $= +0.4$ eV to $\Delta\phi = -0.8$ eV. It is convenient to present these as difference spectra; that is, as the changes occurring in the spectra as a result of adsorption. Spectra 4(a) and 4(b) are differences between the spectra of the methanol and formaldehyde saturated surfaces and the spectrum of this surface just preceding the appearance of the new peaks. Heating these saturated surfaces to 700 K produces a variety of desorption products; H_2, CO, CH_4, H_2O and CH_3OH for methanol, and H_2, CO, CH_4, H_2O, CO_2, HCHO and glycolaldehyde for formaldehyde. This variety of decomposition and recombination products, desorbing with complex kinetics, is due to the presence of adsorbed molecular complexes of methanol and formaldehyde. The desorption is accompanied by an increase in the photoelectron emission from the strongly adsorbed CO ad-layer, indicating that decomposition during the temperature flash causes increased CO adsorption. High temperature flash desorption experiments show that, at exposures of methanol just preceding the appearance of the molecular complex in the photoelectron spectrum, the CO coverage is less than 0.5 of a monolayer. Full CO coverage is achieved only after several cycles of desorbing the complex in flashes to 700 K and readsorbing methanol. For both methanol and formaldehyde, some of the complex ad-layer dissociates during the temperature flash, and fills the gaps in the incomplete CO layer to give eventually a complete β-CO layer which is identical in work function, photoelectron spectrum and desorption states with a (1×1) β-CO surface produced by exposure to CO.[3]

The photoelectron spectra of the molecular complexes adsorbed on the complete β-CO surface are not the same as those of the molecular complexes produced by saturation of the clean surface. This is illustrated in the difference graphs fig. 4 (c) and (d) which are the difference between the spectra after methanol and formaldehyde saturation of the 1 × 1 β-CO structure and the spectrum of the 1 × 1 β-CO structure. The molecular complexes on the 1 × 1 β-CO structure show less evidence

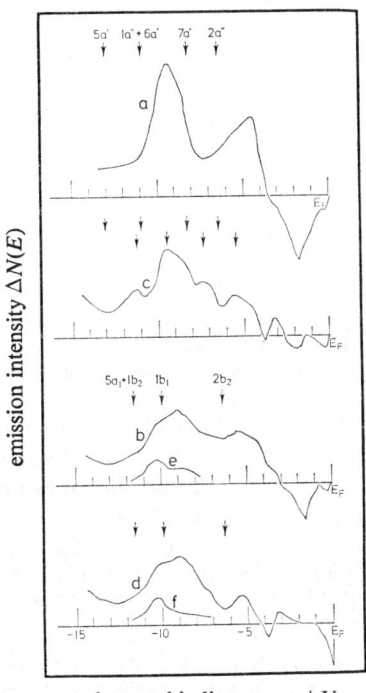

FIG. 4.—The difference spectra for saturation coverages of methanol and formaldehyde adsorbed on a W(100) surface. The peak positions of each band in the gas phase spectrum are indicated. (a) CH_3OH on a clean surface; (b) HCHO on a clean surface; (c) CH_3OH on a β-CO covered surface; (d) HCHO on a β-CO covered surface; (e) residual spectrum after heating (b) to 500 K; (f) residual spectrum after heating (d) to 500 K.

of decomposition in their flash desorption products than do the molecular complexes produced by saturation of the clean surface. The adsorbing species is a larger fraction of the desorption products. This is illustrated in part by fig. 5, in which the flash desorption spectra of methanol and formaldehyde from the adsorbed layers are presented. This indicates that the complete 1 × 1 β-CO layer is less chemically active towards the adsorbing molecules than the incomplete β-CO layer. Reduction of reactivity would be expected if the adsorption of CO reduces the free valence of the surface.

The difference between the photoelectron spectra of the two types of formaldehyde complex is much less marked than for methanol. The arrows in fig. 4 indicate the positions of peaks in the gas phase photoelectron spectrum of methanol, labelled according to the symmetry of the ground state orbital which is ionised.[16, 17] The zero of the energy scale (the equivalent of the vacuum level in gas phase spectroscopy) was fixed at the value of the clean surface work function (4.5 eV) above the Fermi level. Although this is a convenient experimental fixed point, it is an arbitrary reference level, since the contribution of the surface dipole cannot be determined experimentally.

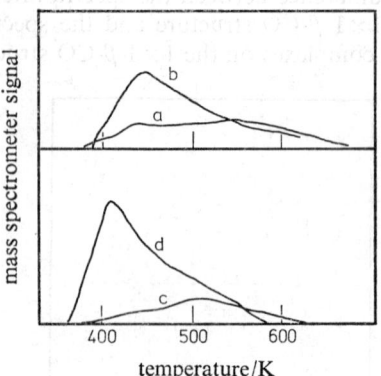

FIG. 5.—(a and b) Mass 30 thermal desorption curves for formaldehyde adsorbed on a clean (a) and on a CO saturated (b) surface; (c and d) Mass 32 thermal desorption curves for methanol adsorbed on a clean (c) and a CO saturated (d) surface.

In fig. 4 it is clear that the spectrum of the molecular complex of methanol adsorbed on the saturated β-CO surface is much more like that of gaseous methanol than is the complex on the more reactive surface. The third band, as in the gas phase spectrum, is twice as broad as the others. The peak at -3 eV is probably not associated directly with the adsorbed methanol complex. A similar peak appears in fig. 4(d) for the formaldehyde molecular complex, and seems more likely to be due to a redistribution of the valence electrons of the underlying surface. The maxima in the methanol spectrum, which are marked by arrows, are shifted, relative to the gas phase spectrum, towards the Fermi level. This may be a symptom of the arbitrary positioning of the vacuum level, but a shift of this type would be expected to result from an increase in the electronic reorganisation energy. This is caused by increased screening of the final state positive ion by the sea of electrons at the surface. These effects are well established.[18] The upwards shift is not equal for all orbitals, the non-bonding $2a''$ and $7a'$ orbitals are shifted approximately 0.5 eV less towards the Fermi level than the bonding $6a'$, $1a''$ and $5a'$ orbitals. The contribution of the ground state binding energies to these shifts cannot be determined without a knowledge of the relevant relaxation energies. However, if we suppose that the relaxation energy is the same for each orbital, approximately 1.5 eV, then we may conclude that the non-bonding orbitals are reduced, relative to the bonding electrons, by about 0.5 eV in the adsorbed state. This suggests that methanol is adsorbed, as one might guess intuitively, via the lone pair orbitals. The work function change of $\Delta\phi = -0.7$ eV, due to adsorption of this complex, is consistent with the transfer of electrons from the complex to the surface.

In the case of formaldehyde, the spectra of the two adsorbed complexes are not very different. The peaks of the gas phase spectrum have been included in fig. 4, but a one-for-one correspondence cannot be made. This is particularly true since the shoulders at -10 eV in fig. 4(b) and (d) may be due to adsorbed CO_2. Carbon dioxide is a thermal desorption product of both the clean surface and the 1×1 β-CO structure after saturation with formaldehyde. Since CO_2 has a higher desorption temperature than all other products except β-CO, it is easy to stop the heating (at

500 K) just before CO_2 desorbs and compare this spectrum to the spectrum when CO_2 has been desorbed. The resulting difference graphs, fig. 4(e) and (f), have a peak at the position where a shoulder was observed for the saturated surface. It is not clear whether this shoulder is due to CO_2 present on the room temperature saturated surface.

Although an unambiguous identification of the peaks in the formaldehyde spectrum with molecular orbitals is not apparent, it is worth noting that the adsorbed molecular complexes of formaldehyde produce peaks in the same position as those of the methanol decomposition intermediate. Although the different desorption products of these species indicate that they are not identical, the similarity of the photoelectron spectra may well indicate that the methanol decomposition intermediate has only two hydrogen atoms bonded to the CO core. This is reasonable because, roughly speaking the addition of two hydrogen atoms to formaldehyde to make methanol populates an additional molecular orbital. Thus a simpler spectrum may indicate fewer hydrogen atoms attached to the CO core. In any case, it appears that extensive decomposition of methanol is associated with vacancies in the 5-coordinate adsorption sites. A W(100) surface with a high concentration of these vacant sites would be expected to be more reactive than a surface on which all such sites were occupied. This is simply a result of greater free valence of the tungsten atoms around an unoccupied site.

ANGULAR DEPENDENCE OF THE METHANOL SPECTRA

A further point of interest in the methanol work concerns the angular dependence of the spectra of the molecular complexes. The molecular complexes of methanol and formaldehyde would be expected to consist of molecular orbitals localized around the cluster of atoms forming each complex. It would be expected that such molecular orbitals would be observed at all angles of photon incidence because an electron in such a molecular orbital experiences a potential parallel as well as perpendicular to the surface.[19] Indeed, just such behaviour is found. The photoelectron spectra of methanol saturated on the 1×1 β-CO structure were observed over a range of angles between 10° and 80°. At all angles the same peaks seen in fig. 4(c) appeared and vanished as methanol was adsorbed and desorbed. There was no large fall-off in the intensity near normal photon incidence as has been observed in the study of hydrogen adsorbed on W(100), due to the delocalisation of the adsorbate orbitals. The actual intensity of the peaks due to the adsorbed complex followed approximately the intensity variation of the clean surface in having a maximum around $\theta = 35°$, falling off at larger and smaller angles.

ADSORPTION OF METHANOL ON CARBIDE AND OXIDE SURFACES

The similarity of the photoelectron spectra of carbided W(100) surfaces and their equivalent CO covered surfaces has been reported previously.[3] It was of interest, therefore, to adsorb methanol on the $c(2 \times 2)$ and (1×1) carbide layers. The spectra are shown in fig. 6(b) and (c) respectively and the spectrum for a $c(2 \times 2)$ CO substrate, fig. 6(a), is shown for comparison. The similarities between the spectra of fig. 4 and fig. 6 are obvious; for both $c(2 \times 2)$ surfaces a two peaked spectrum and for both (1×1) surfaces a four peaked spectrum is seen. Again the $c(2 \times 2)$ surface is much more active for decomposition of CH_3OH in thermal desorption experiments.

Methanol was also adsorbed on a well characterised oxide surface. This oxide, giving a (2×1) LEED pattern, contains roughly a monolayer of oxygen atoms.[4] The difference graph of the spectrum of the methanol saturated surface oxide minus the spectrum of the surface oxide is shown in fig. 6(d). In this case the molecular complex shows peaks characteristic of the previously studied complexes which showed little decomposition. However, the peaks are not in the same positions as in previous cases; they are all shifted to increased binding energy. This probably indicates that the electronic reorganisation energy associated with ionisation is less than in the previous cases. This would be consistent with the intuitive feeling that the electrons in a surface oxide should be less polarisable than those in a surface carbide.

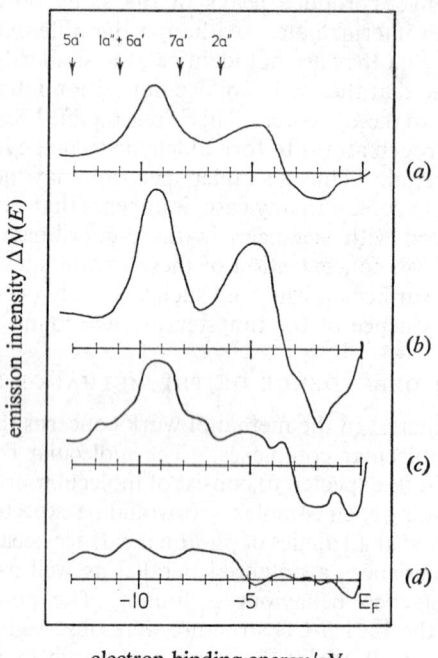

FIG. 6.—Difference spectra for methanol saturated on different surfaces: (a) the $c(2 \times 2)$ CO surface; (b) the $c(2 \times 2)$ carbide surface; (c) the (1×1) carbide surface; (d) the (2×1) oxide surface.

ADSORPTION OF AMMONIA

The catalytic decomposition of ammonia on W(100) has been studied by flash desorption and reported in several recent papers.[20-22] It was established that at low coverages dissociation into adsorbed hydrogen and nitrogen occurred. The photoelectron spectra obtained from exposure of the clean surface to increasing amounts of ammonia initially shows peaks clearly attributable to adsorbed hydrogen and nitrogen. As the exposure exceeds 0.5 L the adsorbed hydrogen and nitrogen peaks continue to increase, but new peaks appear at lower energy which are attributed to a molecular complex of ammonia. Clearly, the general behaviour is quite similar to that of methanol and formaldehyde. However, a problem is encountered in separating that part of the spectrum which is due to the molecular complex. This is because, unlike the methanol and formaldehyde cases, the decomposition of ammonia into adsorbed hydrogen and nitrogen continues to occur after the adsorption of the molecular complex. Thus the difference spectrum of fig. 7(a) (the spectrum following exposure of the clean surface to 2.5 L ammonia minus the spectrum after heating this to 800 K) includes peaks due to adsorbed hydrogen. The only thermal desorption product of this surface, up to 800 K, is hydrogen, indicating that the molecular complex is decomposing, leaving nitrogen atoms behind to fill gaps in the 1×1 structure. Such behaviour has been observed previously.[21]

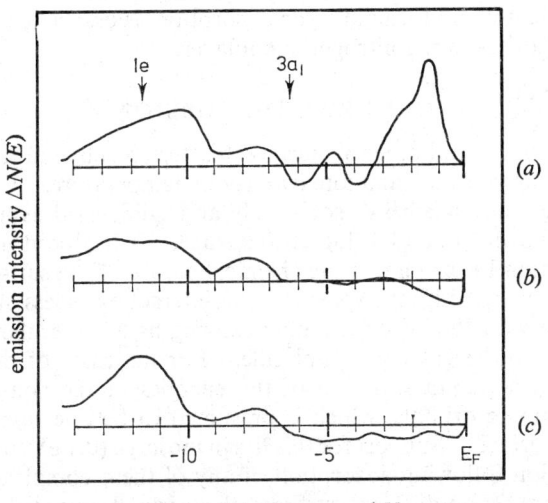

FIG. 7.—The difference spectra for ammonia adsorbed on (a) the clean W(100) surface; (b) the (1×1) nitrogen surface; (c) the (1×1) CO surface. The arrows indicate peak positions in the gas phase spectrum.

To simplify the discussion of the spectra, the adsorption of ammonia on the 1×1 β-CO structure will be discussed first. Fig. 7(c) shows the difference graph of the spectrum of the 1×1 β-CO structure exposed to 2 L of ammonia at 320 K minus the spectrum of the 1×1 β-CO structure. The desorption products of the ammonia saturated 1×1 β-CO structure (heated to 700 K) was almost entirely ammonia in a broad peak between 450 K and 550 K. By analogy with the behaviour of methanol, a spectrum closely resembling gas phase ammonia would be expected. Indeed, this is the case as seen by comparing fig. 7(c) with the arrows indicating the gas phase vertical ionisation potentials.[17] An interesting feature of this comparison is that for the adsorbed complex the peaks appear closer together than in the gas phase. Two possible explanations are readily apparent. First, the reorganisation energy associated with the ionisation of the $3a_1$ molecular orbital could be ~ 1 eV less than that of the $1e$ molecular orbitals. Second, it could be that bonding is primarily by partial donation of the ammonia lone-pair electrons ($3a_1$) into the vacant orbitals of the metal. This partial donation would lower the energy level of the lone pair orbital so that the $3a_1$–$1e$ separation would be smaller. Support for this partial donation comes from the decrease in the work function of 1.0 eV which accompanied this adsorption of ammonia, suggesting a transfer of electrons from ammonia to the surface. After desorption of ammonia the spectrum reverted to that of β-CO.

Fig. 7(b) shows a difference spectrum for ammonia adsorbed on the (1×1) nitrogen surface. The nitrogen surface was itself generated by several cycles of ammonia adsorption followed by heating to 700 K. The peaks of spectrum (b) are broader than those of spectrum (c); this could indicate a distribution of binding states. However, the two bands are in the same positions as spectrum (c). Returning to the spectrum of ammonia adsorbed on a clean W(100) surface, fig. 7(a), discussion is complicated by the interference of hydrogen peaks. However, two peaks appear again in the spectrum in approximately the same position. The peak shape is different from that of the two cases previously discussed, but this might be expected in view of the instability of the complex with respect to thermal decomposition. This increased reactivity, as in the case of methanol and formaldehyde adsorption, is associated with the presence of vacant four-fold sites in the nitrogen monolayer.

GENERAL DISCUSSION

The W(100) surface (and other W surfaces) is very reactive in its interaction with methanol, formaldehyde and ammonia at room temperature. Following low exposures, these molecules adsorb dissociatively and UPS, used as a fingerprint technique, has been shown to be of value at this stage. It has been used to follow the progress of a reaction by examining the chemical nature of the surface at the temperature of adsorption. At higher exposures, the interaction is less strong; adsorption is less dissociative with the adsorbate now existing as a molecular complex having some resemblance to the gas phase molecule. For the cases of methanol and ammonia, there is an apparent lowering of the energies of the non-bonding orbitals, relative to the bonding orbitals, which is presumed to be due to donation of these electrons into the surface. Support for this lies in the large (0.7 eV to 1.0 eV) reduction of the work function following adsorption of any of these complexes. In the study of methanol adsorption on different surfaces, there is a clear correlation between the spectrum of this complex and the coverage and hence reactivity of the underlying adlayer. In general, the chemical behaviour of the complexes described is analogous to the α state of CO on W(100) which adsorbs at room temperature in an undissociated state on a surface on which is already adsorbed the more strongly held β state of CO.

The financial assistance of the Paul Instrument Fund and the Science Research Council is gratefully acknowledged.

[1] W. T. Bordass and J. W. Linnett, *Nature*, 1969, **222**, 660.
[2] D. E. Eastman and J. K. Cashion, *Phys. Rev. Letters*, 1971, **27**, 1520.
[3] W. F. Egelhoff, J. W. Linnett and D. L. Perry, *Faraday Disc. Chem. Soc.*, 1974, **58**.
[4] A. M. Bradshaw, D. Menzel and M. Stenkilberg, *Faraday Disc. Chem. Soc.*, 1974, **58**.
[5] J. M. Baker and D. E. Eastman, *J. Vac. Sci. Tech.*, 1973, **10**, 223.
[6] B. Feuerbacher and M. R. Adriaens, *Surface Sci.*, 1974, **45**, 553.
[7] Discussion remarks. *Faraday Disc. Chem. Soc.*, 1974, **58**.
[8] J. T. Yates, T. E. Madey and M. J. Dresser, *J. Catalysis*, 1973, **30**, 260.
[9] J. G. Hardy and M. W. Roberts, *Chem. Comm.*, 1971, 494.
[10] D. W. McKee, *Trans. Faraday Soc.*, 1968, **64**, 2200.
[11] W. F. Egelhoff, D. L. Perry and J. W. Linnett, *J. Electr. Spectr.*, 1974, **5**, 339.
[12] W. F. Egelhoff, J. W. Linnett and D. L. Perry, to be published.
[13] B. Feuerbacher and N. E. Christenson, *Phys. Rev. B*, 1974, **10**, 2373.
[14] B. J. Waclawski and E. W. Plummer, *Phys. Rev. Letters*, 1972, **29**, 783.
[15] J. T. Yates and T. E. Madey, *J. Chem. Phys.*, 1971, **54**, 4969.
[16] J. H. D. Eland, *Photoelectron Spectroscopy* (Butterworth, London, 1974).
[17] D. W. Turner *et al.*, *Molecular Photoelectron Spectroscopy* (John Wiley, New York, 1970).
[18] J. E. Demuth and D. E. Eastman, *Phys. Rev. Letters*, 1974, **32**, 1123.
[19] J. W. Linnett, D. L. Perry and W. F. Egelhoff, *Chem. Phys. Letters*, 1975, **36**, 331.
[20] P. J. Estrup and J. Anderson, *J. Chem. Phys.*, 1968, **49**, 523.
[21] K. Matsushita and R. S. Hansen, *J. Chem. Phys.*, 1969, **51**, 472.
[22] P. T. Dawson and Y. K. Peng, *J. Chem. Phys.*, 1970, **52**, 1014; 1971, **54**, 950.

GENERAL DISCUSSION

Mr. K. Y. Yu (*Stanford*) said: To complement the extensive UPS data on the adsorption of many small molecules on Ni, I shall present some additional data on the chemisorption of CO, C_2H_4 and C_2H_2 on two neighbouring transition metals, Cu and Fe. In doing so, we focus on one of the problems of extracting bonding information from the UPS spectra of chemisorbed gases: i.e., to what extent will a large difference in the heats of adsorption of the same gas adsorbed on two metals influence their respective spectra. We start by comparing the spectra of chemisorbed CO on Ni and Cu (fig. 1). It has been very well established that the binding energy of CO on Ni is roughly a factor of two higher than on Cu (30 kcal/mol against 15 kcal/mol)[1] and that the electronic factor, rather than the structural factor accounts for

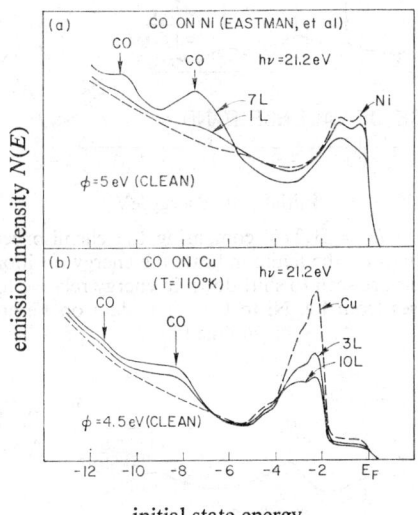

FIG. 1.—UPS spectra at $hv = 21.2$ eV comparing the chemisorption of CO on Ni and Cu. If one takes into account the difference in work function (4.5 eV for Cu, 5.0 eV for Ni), the ionization energies of the two observed resonant levels are also the same. The data were taken on polycrystalline films.

most of the difference. Yet we found that the two chemisorbed CO spectra are similar, in terms of the relative intensities and energy positions of the two observed broad resonance levels. The next figures (fig. 2 and 3) show the chemisorption of C_2H_4 on Fe, Ni and Cu respectively. For the chemisorbed hydrocarbons on Ni, the shift of the π levels relative to the σ levels (referenced to the gas phase spectra) was interpreted by Demuth and Eastman as a bonding shift (due to π-d interaction).[2] Using the magnitude of this π level shift as an input parameter they have also calculated the chemisorption energies, from the chemisorption model of Grimley (Mulliken's theory of Donor Acceptor Complex).[2] Fig. 2 and 3 show that the π level shift increases as one goes from Ni to Cu, while the reported heat of adsorption

[1] J. C. Tracy, *J. Chem. Phys.*, 1972, **56**, 2736.
[2] J. E. Demuth and D. E. Eastman, *Phys. Rev. Letters* 1974, **32**, 1123.

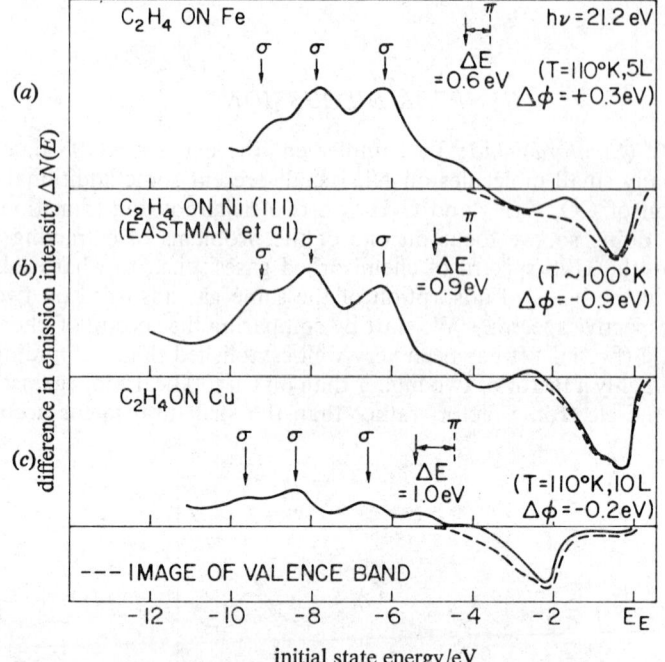

FIG. 2.—Difference spectra at $h\nu = 21.2$ eV comparing the chemisorbed C_2H_4 spectra of Fe, Ni and Cu. The σ and π labellings on the figure indicate the energy positions of the respective orbitals in the gas phase. The π levels are seen to shift down in energy relative to the σ orbitals. Note that the π level shift (ΔE) increases from Fe, Ni to Cu. The data on Fe and Cu were taken on polycrystalline films.

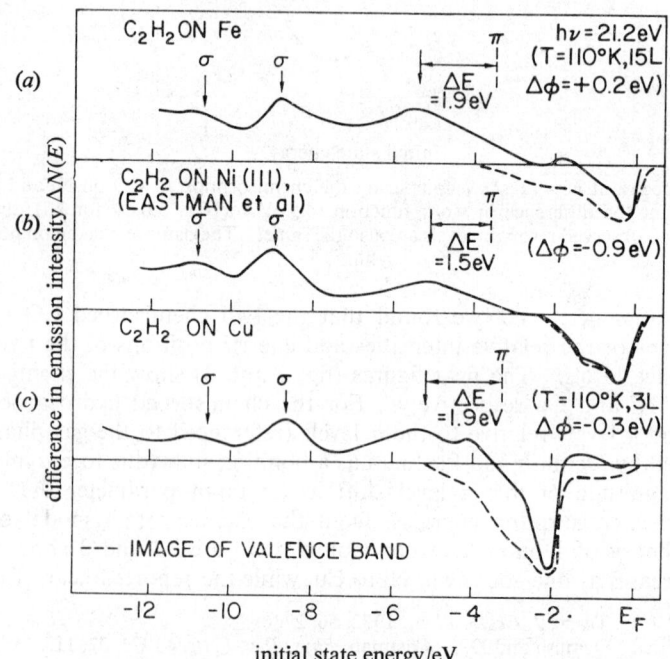

FIG. 3.—Difference spectra at $h\nu = 21.2$ eV comparing the chemisorbed C_2H_2 spectra of Fe, Ni and Cu. The data on Fe and Cu were taken on polycrystalline films.

(measured on polycrystalline films) of C_2H_2 and C_2H_4 on Ni is a factor of 3 higher than on Cu.[1] If we follow Demuth and Eastman's procedure to calculate the heats of adsorption for Fe and Cu, we find that the calculated energies do not correlate with the measured heat of adsorption either. We conclude that any interpretation of the UPS data based on gas level shifts alone is not sufficient to understand the energetics of chemisorption. The substrate contribution to the bonding and many electron effects have to be taken into account in any realistic modelling of chemisorption. In addition, we question the validity of assigning the π level shift as entirely due to bonding effect. The measured π level shift may, in fact, be a combination of relaxation and bonding effects (as suggested by Brundle in the text.)

Dr. C. R. Brundle (*IBM*) (*partly communicated*): It seems to me that the CO/Cu He data presented by Yu *are* different from those of CO/Ni, as would be expected from the arguments given by these authors, since in the Cu case the observed adsorbate peaks come to higher B.E. The difference is more obvious in the HeII spectrum

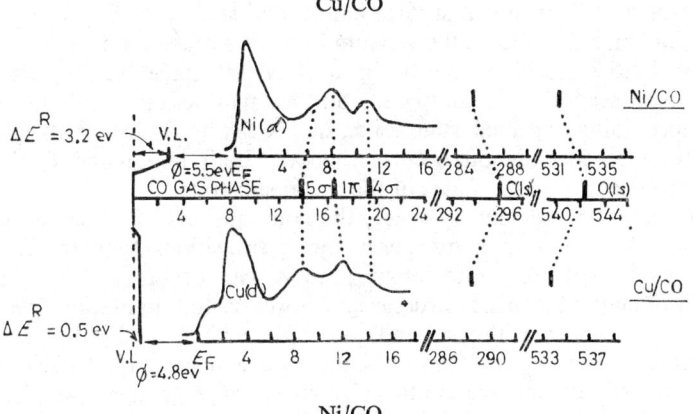

FIG. 1.—HeII UPS of Cu/CO and Ni/CO (films) taken at 80 K. XPS data are indicated schematically.

(fig. 1) where a third peak in the UPS becomes obvious. In addition the XPS results show that C(1s) and O(1s) are at higher B.E. than for Ni. I have proposed the assignment [2] shown in the diagram, which does indicate a difference in bonding in the two cases (much weaker for Cu, therefore, less relative movement of 5σ and lower ΔE^R), though an alternative assignment for the same UPS data has also been proposed.[3]

Dr. C. M. Quinn (*Birmingham*) said: Brundle and Carley have produced a detailed survey of available photoelectron spectra for adsorbed species on metals. I am concerned, however, by the emphasis placed on gas-phase and surface-phase comparisons, because it seems to me that such emphasis obscures the essential differences in the photoelectric process for the gas and solid phases and may lead to error. The comparisons being made are all the more disturbing if " gaseous values are lined up with the adsorbed state spectra simply by sliding the scales until a best fit is obtained with all or some of the orbitals " and then any " excessive " changes in orbital

[1] B. M. W. Trapnell, *Chemisorption* (Butterworth, London, 1955), p. 140.
[2] C. R. Brundle, *J. Electr. Spectr.*, 1975, **7**, 484.
[3] H. Conrad, G. Ertl, J. Küppers and E. E. Latta, *Solid State Comm.*, 1975, **17**, 613.

energies resulting from this procedure be accounted for in terms of " the large difference in final-state relaxation energies $\Delta\varepsilon^R$ in the two cases ".

For the adsorbed state it is necessary to consider two extreme possibilities in the photoelectric process,[1-3] volume and the surface photoelectric emission. In both cases the final state wave functions, in which photo-excited electrons appear in the vacuum, are the incoming wave functions of multiple scattering and LEED theory. The degree of damping of these final states within the metal substrate depends on whether they resonate with propagating Bloch states in the higher conduction bands of the material. The initial states for the adsorbed species may be of two kinds. The outermost valence orbitals are very likely to interact with the metal substrate conduction band and form virtual levels, indistinguishable parts of conduction band wave-functions, except that in the energy region of a virtual level the conduction band wave functions of the metal extend through the adsorbed species at the surface. For the inner orbitals of the adsorbed species, substantial energy differences probably prevent their mixing with substrate wave function, and isolated surface states will occur in the metal band gaps. Outer core orbitals on the adsorbed species may however, form two dimensional surface bands, and such an effect is very likely also to be found in band gaps above the vacuum level, as surface state resonances. Most of these possibilities are illustrated in fig. 6 of Willis's paper at this Discussion for one-dimensional sections in real space. Surface photoelectric emission occurs for highly damped initial or final states, i.e., states falling in the emitter band-gaps. Volume photoelectric emission occurs whenever both initial and final states are essentially propagating Bloch states of the metal substrate.

In wave-vector or momentum space, the consequences of these effects can also be examined. For a given light energy direct transitions to within a reciprocal lattice vector, the volume photoelectric process, can occur in only relatively few instances throughout the band structure. However, this limitation is not present when a final state occurs within a band gap of the emitter, and indeed it seems that surface photoemission may be the dominant process in u.-v. photoelectron spectroscopy since photoelectrons are emitted for all energies greater than the threshold energy even though direct band-to-band transitions may not be possible.

The principal consequences of these remarks are the following. First, if direct transitions (the volume photoelectric process) occur, highly angular and photon energy-dependent features should dominate the photoelectron spectra of solids whether adsorbed species are present or not. This follows because the parallel components of initial state wave-vectors are conserved to within reciprocal lattice vectors in the emission, and the normal components too are good quantum numbers. For adsorbed species creating virtual levels, simple extensions of the solid wave functions through the adsorbed layer, this situation should not be altered substantially except that some change in intensity may result for direct transitions in the energy region of the virtual level. In this region more wave function is present within an " escape depth " and the transition moment will be somewhat increased. This is quite different, however, to the gas-phase situation, and any direct transition observed need not correspond to a feature in the corresponding gas-phase spectrum.

Secondly, for surface photoelectric processes the momentum broadening required normal to the surface will cause an " angular broadening " in the photoemission, and spectral features due to this type of photoemission should be observable with analyser settings to the surface normal over the range of angles reflecting the broadening in

[1] C. Caroli, D. Lederer-Rozenblatt, B. Roulet and D. Saint-James, *Phys. Rev. B.*, 1973, **8**, 4552.
[2] P. J. Feibelman and D. E. Eastman, *Phys. Rev. B.*, 1974, **10**, 4932.
[3] B. Feuerbacher and N. Egede Christensen, *Phys. Rev. B.*, 1974, **10**, 2349.

the normal component of momentum. Such effects are now being observed with angular-resolving equipment.[1] Again the variations in spectra in angular-resolved form emphasise the dangers involved in making assignments for fixed angle spectra of adsorbed species by comparison with gas-phase results.

Dr. C. R. Brundle (*IBM*) (*communicated*): Comparison between free molecule and adsorbate induced spectra are not made in the way ascribed to us by Quinn, who has juxtaposed two unrelated sentences from our paper. Comparisons are made, *where appropriate*, on the basis of eqn (1):

$$E_g - E_{ads} = \phi + \Delta E^r + \Delta E^b \tag{1}$$

with all the qualifications discussed in the paper.

If the equation is to be applied and have any meaning it does imply that, for the adsorption case concerned, the valence and core levels of the free molecule are still recognizable in the adsorbed situation and that any relative changes in orbital IP's are ascribable to differences in $(\Delta E^r + \Delta E^b)$ for different orbitals.

It is clear that for the cases we have termed in our paper "non-interacting surface species"—i.e., condensation or weak physical adsorption (*a*) that all the gas phase orbitals *are* recognizable in the adsorbed situation and (*b*) that their relative I.P.'s do not change—i.e., $\Delta E^r + \Delta E^b$ is constant for all orbitals, and in all probability $\Delta E^b \simeq 0$. In this situation then (5 of the 8 molecules in our paper, for adsorption at 80 K) Quinn's strictures concerning the the possible strong angular and photon energy dependent modifying effects are unfounded. A successful comparison to gas phase values *is* made; there are no strong angular effects which result in peak shifts, excessive broadening, or splitting (condensed UPS spectra of some 20 molecules have appeared in the literature so far, mainly due to myself and co-workers, Eastman and Demuth, and Spicer and co-workers. The latter two groups use a retarding field analyser which collects over all angles, whereas I used a deflection analyser collecting over a small solid angle at a fixed orientation. Adsorbate orbital I.P.'s measured in these different angular arrangements nevertheless agree between groups. In addition Perry has mentioned at this meeting that an angular study of α-CO on W revealed no strong angular effects); and there are no photon dependent effects (I.P.'s and peak widths recorded by HeI and HeII are the same though relative intensities change, of course, owing to cross-section effects). It should also be remembered that our paper deals with a polycrystalline substrate, so that effects due to surface region periodicity might be partly averaged out anyway.

For *chemisorbed molecular* species treated according to eqn (1) (only chemisorbed CO and NO have their valence levels analysed in this way in this paper) more caution is needed since though all orbitals are detected, relative changes in I.P.'s are observed which we and many others attempt to ascribe to the strong bonding interactions between one or some of the adsorbate MO's and the substrate. Quinn's argument is that this is dangerous because the distinction between the gas-phase and solid-state photoemission processes (manifested in angular and *hv* dependent structure) may be responsible for the energy changes we are ascribing to chemical bonding. The experimental data available so far indicate that, for those systems studied and in the spectrometer geometries used, such worries are groundless. Both photon energy variations (synchrotron radiation,[2] HeI, HeII [3] and angular variations (polar angle

[1] D. R. Lloyd, C. M. Quinn and N. V. Richardson, *J. Phys. C.*, 1975, **8**, L1.
[2] T. Gustafsson, E. W. Plummer, D. E. Eastman and J. L. Freeouf, *Solid State Comm.*, 1975, **17**, 391.
[3] C. R. Brundle and A. F. Carley, paper at this Discussion.

only)[1] indicate that for CO chemisorption on a number of metals only peak relative intensities change, and they do so in a way appropriate to the free molecule itself. It should be noted, however, that the deflection analyser used by myself and Menzel et al.[1] collects over a largish azimuthal angle which may average out sharp azimuthal angular structure.

For the case of *dissociated* chemisorption Quinn's cautions *are* probably well-founded. For chemical reasons I would expect this to be the case when atoms become strongly bound to the substrate (the reason for dissociation) rather than to each other (molecular adsorption), so that a band structure description of the bonding becomes appropriate (resulting in the " virtual levels, indistinguishable parts of conduction band wave-functions " referred to by Quinn). For this reason we have not tried to interpret valence level spectra of the dissociated species in terms of any electronic structure effects, but have merely used the characteristic UPS and XPS features as a means of identifying dissociation products.

In conclusion, a complete range of interactions between adsorbate and substrate is to be expected, resulting in examples of slightly modified free molecule electronic structure through to full periodic solid-state band structure. This was acknowledged several years ago.[2] In analyzing experimental data one needs to be aware of this range and the probable consequences in terms of the photoemission process. I believe we and most others working in this field are well aware of these factors. There can however be no doubt that $h\nu$ and angular dependent studies are going to enhance the value of photoemission adsorption studies in the near future.

Prof. R. Mason (*University of Sussex*) said: Quinn's comments on the paper by Brundle and Carley raise the important question of what relation, if any, exists between unperturbed ligand electronic states and those of the surface molecule. The simplest bonding model would argue, as Quinn implies, for the essential integrity

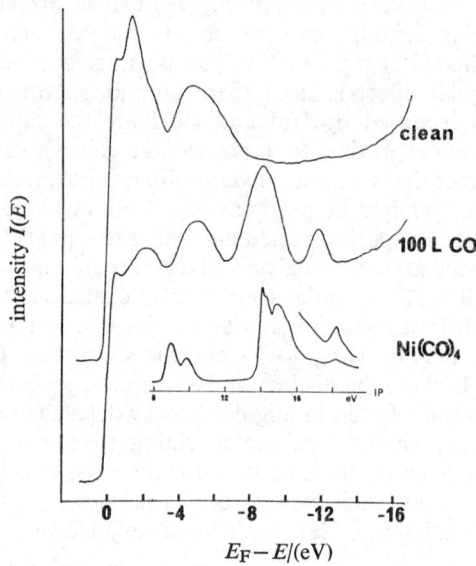

[1] D. Menzel and J. C. Fuggle, private communications.
[2] C. R. Brundle in *Defect and Surface Properties of Solids*, Vol. 1, ed. J. M. Thomas and M. W. Roberts, 1971, Specialist Periodical Report of the Chemical Society (London).

of only those deepest lying ligand states which will not be "mixed in", to any significant extent, with metal wave functions. Intuitively, a much more obvious starting point for a discussion of the photo-emission from a metal-ligand surface is the photoelectron spectrum of an appropriate organometallic molecule. The figure shows the emission spectra from a clean Pt(111) surface, from the Pt(111)–CO surface and from tetracarbonylnickel (O). A self-consistent molecular orbital calculation of the energy levels of Ni(CO)$_4$ allows one to be confident that the emission at ionisation emerges between 8 and 10 eV is from molecular orbitals which are almost completely metal "d" in character—note the splitting into two components of predominantly t_2 and e symmetry in a way which simple ligand field theory would predict. The "d" electron emission in Ni(CO)$_4$ has a FWHM of $ca.$ 1.5 eV compared with valence band widths in nickel and platinum of up to 6.5 eV but, as expected, the emission from the levels made up predominantly from ligand states (I.P. ~ 15 eV and 18 eV) in Ni(CO)$_4$ appear very similar to those from the surface molecules. A fairly straightforward survey of the photoelectron spectra of binuclear—and polynuclear—carbonyls would be worthwhile; that should provide a more definite understanding of the electronic states of bridged carbonyl ligands vis-à-vis terminal bonded groups as well as examining more closely the thesis that "cluster states" approximate those of the surface molecule.

Prof. C. S. Fadley (*Hawaii*) said: In discussing the results presented in this paper, Quinn has presented UPS data for single crystals which indicate pronounced changes in valence-band spectra as the angle of electron emission is varied with respect to the crystal axes, and has pointed out that the unambiguous interpretation of chemisorption phenomena may require angular-dependent studies of single crystals.

I would like to point out that such single-crystal effects have also been observed in XPS, both for core levels where the total peak intensity shows pronounced structure with angle,[1-5] and also for valence levels, where the relative intensities of various peaks are found to change with angle.[5]

The valence-level effects have first been observed very recently in our laboratory by Baird, Fadley and Wagner[5] for photoelectron emission from a Au(100) single

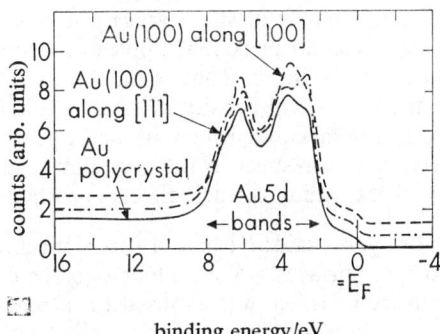

FIG. 1.—XPS valence-band spectra for gold obtained for a polycrystalline specimen and also for a specimen with (100) surface exposed and electron emission along either the [100] or [111] directions.

[1] K. Siegbahn, U. Gelius, H. Siegbahn and E. Olsen, *Phys. Letters*, 1970, **32A**, 221.
[2] C. S. Fadley and S. Å. L. Bergström, *Phys. Letters*, 1971, **35A**, 375.
[3] C. S. Fadley and S. Å. L. Bergström in *Electron Spectroscopy*, ed. D. A. Shirley, (North Holland, Amsterdam, 1972) p. 233.
[4] J. Brunner and H. Zogg, unpublished results.
[5] R. J. Baird, C. S. Fadley and L. F. Wagner, to be published.

crystal surface. Both core- and valence-level angular distributions were studied in more detail and with higher resolution than in the prior studies of Au(111) by Fadley and Bergström.[1, 2] The core-level angular distributions showed considerable fine structure and an overall symmetry characteristic of the (100) surface: a more detailed analysis of these data is now in progress. The gold valence-band spectrum, which is dominated by a well-known doublet primarily due to 5d-derived bands, was also observed to change markedly in both total intensity and fine structure with angle, and in particular was found to exhibit distinct changes in structure when viewed along different high symmetry directions. A partial summary of the valence-level results is presented in the figure, in which a spectrum obtained from a polycrystalline specimen deposited *in situ* is compared to spectra taken along the [100] and [111] directions. The polycrystalline spectrum agrees quantitatively with prior XPS investigations on similar specimens.[3, 4] Both of the single-crystal spectra deviate considerably from the polycrystalline result and also from one another. The spectrum obtained along [100] shows roughly the same fine structure within each component of the doublet as do the polycrystalline data, but the two components exhibit much different relative intensities in the two spectra: also, a slight change in the doublet separation occurs between [100] and polycrystal. The spectrum along [111] shows a much different relative intensity of the two components, with the maximum number of counts in each peak being essentially equal: also, the fine structure in the lower-binding-energy component is much different for [111] as compared to either [100] or polycrystal, and shows a pronounced dip at 3.5 eV that has not been observed previously. A more detailed account of this work is in preparation.[5]

These results suggest a strong dependence of XPS valence-band spectra on electron emission direction with respect to single crystal axes. It thus appears that angular-dependent XPS studies on single crystals may be very useful in providing the most complete characterization of either bulk- or surface-electronic structures. Also, valence-band XPS spectra obtained from single-crystal specimens may not be as directly related to density-of-states curves derived by integration over the entire Brillouin zone as has previously been assumed in some studies.[4, 6] For example, for cleaved silicon single crystals, Ley, Kowalczyk, and Shirley [6] have noted good overall agreement between the positions and relative internsities of peaks and shoulders predicted by a theoretical total density of states and those observed in experimental spectra. However, one experimental peak that appears to be associated with high density near the L_1 point in the Brillouin Zone is much more intense in the experimental spectrum than in the theoretical density of states: this lack of agreement in intensity could be due to preferential enhancement of the L_1 peak for electron emission along an unspecified direction with respect to the crystal axes at which these experiments were performed. Further study of such effects is certainly called for.

Prof. M. W. Roberts (*Bradford*) said: Kishi and I have recently completed a study of nitrogen adsorption on Fe surfaces. We neither observe chemisorbed nitrogen nor a nitride when a clean iron surface was exposed to nitrogen at low pressure at room temperature. We have however observed the " nitride " N(1s) peak at about 398 eV *under conditions when oxygen was also present* as an impurity in the nitrogen

[1] C. S. Fadley and S. Å. L. Bergström, *Phys. Letters*, 1971, **35A**, 375.
[2] C. S. Fadley and S. Å. L. Bergström in *Electron Spectroscopy*, ed. D. A. Shirley (North Holland, Amsterdam, 1972), p. 233.
[3] D. A. Shirley, *Phys. Rev.*, 1972, **B5**, 4709.
[4] S. Kowalczyk, L. Ley and D. A. Shirley, *Phys. Letters*, 1972, **41A**, 455.
[5] R. J. Baird, C. S. Fadley and L. F. Wagner, to be published.
[6] L. Ley, S. Kowalczyk and D. A. Shirley, *Phys. Rev. Letters*, 1972, **29**, 1088.

and indicated by the development of an O(1s) surface peak. If, therefore, ref. (24) of Brundle's paper refers to our unpublished work it has been incorrectly reported in the sense that our experiments were only done in the presence of traces of oxygen.

In view of the well known and frequently reported "activated chemisorption" of nitrogen by iron our results indicating the absence of chemisorbed nitrogen on clean polycrystalline Fe films at 290 K and 10^{-4} Torr N_2 are interesting. These and also the role of oxygen in possibly inducing nitrogen chemisorption, will be discussed elsewhere.

Dr. C. R. Brundle (*IBM*) (*communicated*): I am grateful for the correction concerning the communication from Dr. Kishi. Adsorption of N_2 on Fe to yield "nitride" has been claimed by XPS by other authors,[1] but the work was done at high pressure (200 Torr) and oxygen adsorption was also observed, in agreement with the comment of Roberts that oxygen is needed for the reaction. For Ni we have observed the same situation, i.e., the 397.8 eV N(1s) peak only appears for N_2 adsorption in situations where a stronger O(1s) peak also appears. This could be due to (*a*) initial interaction with trace oxygen followed by subsequent reaction with N_2; (*b*) co-adsorption reaction of N_2 and O_2; (*c*) the presence of gas phase NO or NO_2 impurities.

Dr. C. R. Brundle (*IBM*) said: In the light of the experimental results on Ni/NO presented in our paper, I would like to introduce some theoretical results of a colleague from IBM, Dr. Inder P. Batra, on this system. Using the SCF-X scattered wave method, Batra has calculated the energy levels of an Ni_5NO cluster where the NO

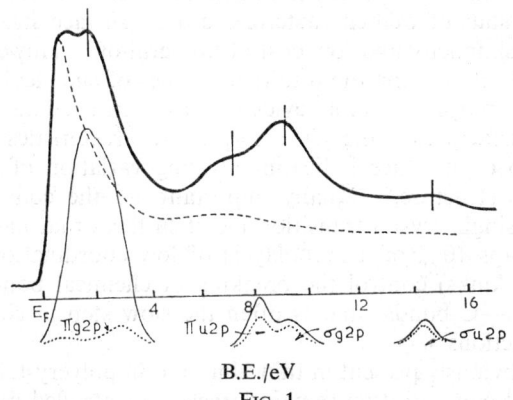

FIG. 1.

Lower—Theoretical spectrum (full trace) for an Ni_5CO cluster with d (see text) equal to 1.82 a.u. Dotted trace indicates NO orbital parentage.

Upper—HeII spectrum of Ni/NO at 80 K (full trace). Dashed trace shows a scaled clean Ni spectrum.

is in the fourfold site (N down) a distance, d, above the plane defined by 4 Ni atoms. The fifth Ni atom is in the plane below the other 4, directly below the NO molecule. By applying a broadening function to the calculated levels, the spectrum shown in the lower full curve of the figure is obtained. This can be compared with the experimental UPS data. The intensities merely represent the relative number of electrons, and no allowance is made for cross-section variation. Batra has also calculated the parentage

[1] F. Honda and K. Hirokawa, *J. Electr. Spectr.*, to be published.

of each cluster orbital and that portion which originates from the NO molecule is given by the dotted line in fig. 1. The levels at about 8, 9.5 and 14.5 eV are heavily NO in character, and in fact are not much altered from the original NO $(\Pi_u 2p)^4$, $(\sigma_g 2p)^2$, and $(\sigma_u 2p)^2$ molecular orbitals respectively—i.e., these orbitals are largely non-bonding in their interaction with the Ni cluster. All the interaction occurs with the $(\Pi_g 2p)^1$ anti-bonding electron which mixes heavily with the Ni d-electrons such that NO $\Pi_g 2p$ character spreads throughout the Ni d-band region (0–4 eV) as indicated in the figure.

The HeII spectrum for Ni/NO is reproduced above the calculated cluster spectrum. A normalized clean Ni spectrum is also shown so that one can more easily see where the NO-induced intensity appears. The conclusions in our paper that the 7–11 eV region represented both the $(\Pi_u 2p)^4$ and $(\sigma_g 2p)^2$ orbitals and the $(\Pi_2 2p)'$ orbital fell at around 1.8 eV are amply supported by the calculation. In addition our suggestion from the consideration of $E_g - E_{ads}$ values that the $\Pi_g 2p$ orbital was more involved in the bonding than the other orbitals is also borne out.

The gratification of agreement wears off somewhat, however, when one discovers that the only strong effect of altering d in the calculation is to alter the distribution of NO $\Pi_g 2p$ character over the 0–4 eV region. This means that in this case UPS is a rather insensitive tool because one must look for small spectral differences in the region which is more difficult to interpret in any attempt to establish a structure by comparison of experiment to theory.

Prof. G. A. Somorjai (*Berkeley*) said: My comment concerns the use of polycrystalline foils or evaporated thin films of transition metals for surface studies. There are many well-known experimental advantages of utilizing foils or thin films instead of single crystals of well-characterized atomic surface structure that include larger surface area, simplicity and low cost of preparation. However, great caution must be exercised in interpreting the results of studies of chemical interactions when these surfaces are employed. Ample evidence has been accumulating from studies on single crystal surfaces that the chemisorption characteristics change markedly from crystal face to crystal face indicating striking variation of chemical bonding with atomic surface structure. Equally important are the conclusions of surface reaction studies on single crystal transition metal surfaces that indicate that a small concentration (perhaps 10% of a monolayer) of low coordination number surface sites (atomic steps, kinks) control the breaking of chemical bonds (O—O, N—O, H—H, C—H and C—C bonds) that is often the slow step in chemisorption or in chemical surface reactions.

These sites are obviously present in thin films and in polycrystalline foils but their atomic structure and concentration remain largely uncontrolled during the preparation of these surfaces. Thus, large differences in chemical behaviour are expected from sample to sample with slight variation of sample preparation procedure. Also, the chemisorption characteristics (sticking probability, reaction product distribution, surface reaction rates) widely vary in studies in different laboratories when these surface (foils and thin films) are used.

I would like to suggest that for studies of the surface chemical bond the use of well characterized single crystal surfaces be preferred. At the very least the results of chemisorption on one single crystal surface should be reported along with the results of chemisorption using foils and/or films and the similarities and differences be pointed out. In the absence of control of the surface structure in these studies, dissimilarities in chemisorption characteristics reported in different laboratories cannot easily be avoided.

Dr. U. Landman (*Rochester*) said: I also would like to question the use of polycrystalline samples in the study of detailed adsorption mechanisms. The crystallographic anisotropy of adsorption phenomena is well known and documented. In particular, the adsorption of oxygen on the low-index planes of Ni exhibits marked dependence upon the single-crystal plane employed. For example, the sequence of oxygen structures as observed by LEED and their temperature dependence differ from one crystal plane to another.[1-4] Analysis of LEED intensity data [5,6] resulted in proposed structural models which emphasize crystallographic anisotropy. In addition, the oxygen KL_IL_{II} Auger line-shape on the (111) face of Ni behaves markedly different from those on the (100) and (110) faces, and the change in work-function is considerably larger on the (111) face.[7] In view of this wealth of information, conclusions and comparisons drawn on the basis of studies performed on polycrystalline samples are hazardous and rather ambiguous. Since the distribution of crystal planes in the above samples is unknown it seems to me advisable to perform the above experiments on well characterized single-crystal samples.

Dr. P. R. Norton and **Dr. R. L. Tapping** (*AECL*) said: The occurrence of an oxygen species on nickel surfaces, with an O1s binding energy of $\sim$531.4 eV appears to be related to surface heterogeneity. On unsintered evaporated nickel films this species may constitute 30-50% of the adsorbed oxygen, whereas on well annealed polycrystalline foils, not more than $\sim$10% of the oxygen is present in this state. The 531.4 eV species is not detectable after room temperature adsorption of O_2 on well annealed (111) nickel surfaces, but may be produced by adsorption at 77 or $\sim$400 K. It may also be produced by heating the room temperature grown oxide on (111) nickel to $\sim$400 K, but is not formed when the oxide is grown on (111) nickel at 1000 K. The origin of this peak remains obscure and further experiments on (100) and (110) nickel are being carried out in an attempt to determine its nature.

Dr. C. R. Brundle (*IBM*) (*communicated*) The interpretation of Ni 2p behaviour at 300 K in the low oxygen coverage adsorption region given by Norton and Tapping is very similar to that presented by us elsewhere.[8-10] A lack of effect on the Ni 2p peak is taken to indicate a co-operative donation of Ni itinerant electrons to O such that Ni metallic character is maintained, and no Ni^{2+} is formed. Additional evidence, not previously reported, is also given by Norton and Tapping, namely the lack of effect on the Ni/d band structure. At higher coverage NiO nucleates at the surface and a Ni^{2+} component is observed. Norton and Tapping do not make any comments concerning the significance of the O(1s) B.E. during this sequence, however. I have suggested previously[8-10] that the fact that the *ca.* 530 eV O(1s) peak is in essentially the same position during the initial adsorption and NiO nucleating stages indicates that it is as much " O^{2-} " in the former situation as in the latter, which may in turn imply that the oxygen represented by the *ca.* 530 eV peak is incorporated into the

[1] A. U. McRae, *Science*, 1963, **139**, 379.
[2] G. E. Becker and H. D. Hagstrum, *Surface Sci.*, 1972, **30**, 505.
[3] L. H. Germer, J. W. May and R. J. Szostak, *Surface Sci.*, 1967, **7**, 430.
[4] J. E. Demuth and T. N. Rhodin, *Surface Sci.*, 1974, **42**, 261.
[5] J. E. Demuth, D. W. Jepsen and P. M. Marcus, *Phys. Rev. Letters*, 1973, **31**, 540; 1974, **32**, 1182, *Solid State Comm.*, 1973, **13**, 1311.
[6] M. A. Van Hove and S. Y. Tong, *J. Vac. Sci. Technol.*, 1975, **12**, 230.
[7] J. E. Demuth and T. N. Rhodin, *Surface Sci.*, 1974, **45**, 249.
[8] C. R. Brundle and A. F. Carley, *Chem. Phys. Letters*, 1975, **33**, 41.
[9] C. R. Brundle, *Surface Sci.*, 1975, **48**, 99.
[10] C. R. Brundle, *Faraday Disc. Chem. Soc.*, 1974, **58**.

surface, even in the pre-nucleation stage. Since it is not expressed in the paper, I would welcome Norton's opinion as to the electronic nature and location of the oxygen during the pre-nucleation stage.

In the work we have published on the $Ni+O_2$ system we have observed a second high B.E. $O(1s)$ peak at ca. 531.4 eV even at 300 K (as have others [1]), though this peak becomes relatively more important at liquid nitrogen temperatures. Norton and Tapping do not mention this peak at 300 K, though they do at 90 K. It does, however, seem to be present in their data at 300 K (fig. 2 and 3) as an asymmetry in the $O(1s)$ line. Examination of this high B.E. shoulder under higher resolution with a monochromator [2] indicates considerable variation of position and resolution of the component with surface treatment. This is compatible with the presence of more than one peak in this region, a situation which could remove the dichotomy between assignment to " Ni_2O_3 " bulk defect structure at very high coverage [1] and chemisorbed O overlayer at low coverage.[3] The rapid growth of an $O(1s)$ peak in this region at 80 K, with saturation at relatively low coverage, accompanied by only restricted growth of the ca. 530 eV peak has been observed for Zn [4] as well as for Ni.[3] The facile explanation is that in both cases one is stabilizing true chemisorption (the 531.4 eV peak) with respect to oxidation (the ca. 530 eV peak). In the case of Zn one cannot, of course, propose that the higher B.E. peak is associated with any structure in which Zn is in a higher oxidation state than 2+.

Dr. P. R. Norton (*AECL*) (*communicated*): The answers to some of Brundle's questions are contained in the comment on O $1s$ binding energy and surface heterogeneity in the remarks communicated by Tapping and myself.

It would certainly appear that the electronic environment of the oxygen atom changes very little, if at all, in the process of changing from an adlayer structure to nickel oxide, but until one can reliably separate the various contributions to chemical shifts or indeed put the reference level problem on a sounder basis, I feel that it is hazardous to draw definite conclusions as to the electronic state of the oxygen atoms from core level binding energies. LEED and low energy ion-scattering data locate the atoms above the surface plane but somewhat buried—certainly different from an NiO structure.

We agree with Brundle that our 300 K data show the presence of $\sim 10\%$ of the high binding energy O $1s$ line (531.4 eV). With regard to his comments that it may represent " true chemisorption ", the techniques mentioned above indicate that the surface is not fully reconstructed to the oxide, i.e., the oxygen atoms may still be regarded as chemisorbed, although the distinction, it seems to me, is rather vague. One can produce an oxygen species with this binding energy on Ni(111) by adsorption at 400 K, surely a temperature region that would not favour stabilization of " true chemisorption ".

In summary, I feel that it is premature to discuss the physical and electronic state of the adsorbed or incorporated oxygen atoms on the basis of data on polycrystalline surfaces.

Dr. S. Evans (*Aberystwyth*) (*communicated*): Brundle has frequently suggested that the electronic environment of the oxygen atom changes very little throughout the

[1] K. L. Kim and N. Winograd, *Surface Sci.*, 1974, **43**, 625.
[2] T. J. Chung, unpublished work.
[3] C. R. Brundle, *Faraday Disc. Chem. Soc.*, 1974, **58**.
[4] C. R. Brundle and R. I. Bickley, to be published.

chemisorption/oxidation process. We think this is unlikely: our previous criticisms [1] (which remain valid) are reinforced by the recently published [2-4] soft X-ray appearance potential spectra for $MO_{ads.}$ (M = Ti, Cr, Fe, Ni) and surface oxide. The 0 K spectrum, unlike the 0 1s XPS, changes radically between the two situations, showing only a single peak for $MO_{ads.}$ but a complex multipeaked structure in the later stages of the oxidation. This difference between the two techniques suggests that there may be substantial changes in the chemically important unoccupied sections of the valence bands even though the net potentials [1] as reflected in the 0 1s chemical shifts change very little. These results add significantly to the reservations outlined by Norton about drawing conclusions concerning the electronic environments of adatoms solely from XPS BE data.

Dr. C. R. Brundle (*IBM*) (*communicated*): I think Norton's caution is perhaps appropriate, but there are two points which ought to be borne in mind. First, the rapid decrease in sticking probability of O_2 on Ni to a minimum and its subsequent rise (300 K and above) is now well-established [5] as due to the onset of NiO nucleation and the subsequent island growth across the surface. These data were originally obtained for *films* [6] and have subsequently been observed for single crystal surfaces.[5, 7] At 80 K the minimum *does not occur*. If this can be considered as implying that NiO nucleation does not occur, it is significant that at 80 K the XPS 530 eV O(1s) peak is strongly inhibited and the 531.4 eV peak grows rapidly to saturation.

The second point concerns the LEED data. It is a misconception to think that the published LEED analyses of the Ni/O_2 system *prove* that only atoms chemisorbed above the Ni surface plane are present during the early stages of adsorption (less than half a monolayer). Under the adsorption and annealing conditions used to obtain the sharp LEED patterns it is not known how much of the total oxygen uptake is represented by the LEED patterns ($p(2 \times 2)$ and $c(2 \times 2)$) which are subsequently intensity analyzed. It is admitted that as much as $\frac{2}{3}$ of the uptake could be below the Ni surface and therefore only $\frac{1}{3}$ in the LEED intensity calculated position of 0.9 Å above the Ni surface.[8] If this is so, a *minor* component of the XPS O(1s) signal could easily correspond to the oxygen shown by LEED to be above the surface.

Dr. S. Evans (*Aberystwyth*) said: During the last year we have been examining the interaction of argon-ion bombarded polycrystalline, (100), (110), and (111) nickel surfaces with molecular oxygen using an XPS/UPS/$\Delta\phi$ combination.[9] Our experiments are generally in agreement with those reported by Norton and Tapping (NT), and our interpretation is broadly similar. However, we find difficulty in reconciling some of NT's coverage estimates with our own, which have been obtained by a rather different approach. We measured [10] the inelastic mean free path (λ) for electrons in nickel (14±3 Å at 1 185 eV, giving 8.4±2 Å at 400 eV, assuming $\lambda \propto E^{\frac{1}{2}}$) and the $Ni2p_{\frac{3}{2}}$ and O1s photoionisation cross-sections (3.2 and 0.55 respectively relative to

[1] S. Evans, M. J. Tricker and J. M. Thomas, *Faraday Disc. Chem. Soc.*, 1974, **58**, discussion remarks.
[2] S. Andersson, H. Hammarqvist and C. Nyberg, *Rev. Sci. Instr.*, 1974, **45**, 249.
[3] C. Nyberg, *Surface Sci.*, 1975, **52**, 1.
[4] S. Andersson and C. Nyberg, *Solid State Commun.*, 1974, **15**, 1145.
[5] P. H. Holloway and J. B. Hudson, *Surface Sci.*, 1974, **43**, 123.
[6] A. M. Horgan and D. A. King, *Surface Sci.*, 1973, **23**, 526.
[7] P. H. Holloway and J. B. Hudson, *Surface Sci.*, 1974, **43**, 141.
[8] S. Andersson, private communication.
[9] S. Evans, J. Pielaszek and J. M. Thomas, submitted for publication.
[10] S. Evans, R. G. Pritchard and J. M. Thomas, in preparation.

F1s = 1.00 for MgKα radiation), using a method described elsewhere.[1] These figures enabled us to estimate the O1s : clean Ni2$p_{\frac{3}{2}}$ peak intensity ratio to be expected following chemisorption of a monolayer of oxygen atoms (conveniently defined as one O atom for each Ni atom in the (100) plane, since we found similar uptake behaviour for the single-crystal and polycrystalline materials). Our estimate for the coverage at 5L exposure ($\theta = 0.5 \pm 0.1$) derived from these data is, gratifyingly, in excellent agreement with NT's estimate ($\theta = 0.43$). The final O1s (oxide) intensity shown in fig. 1 or NT's paper is, however, 2.5 to three times the monolayer intensity : our results, for (probably) somewhat more reactive Ar$^+$-bombarded surfaces, show ~ 3.7 times the monolayer intensity. If we take one layer of NiO to refer to a single (100) plane of the material, 3 O-monolayers as defined above are equivalent to over 4 NiO layers, making no allowance for inelastic losses : and the actual oxide layer thickness must therefore be greater than this. We have used two methods to estimate the true layer thickness. Observing that the total (parent and shake-up satellite) Ni2p intensity did not decrease on oxidation of the surface gives us (cf. Carlson and McGuire [2]) $\lambda_{NiO}(400 \text{ eV}) \sim 16$ Å and, again assuming $\lambda \propto E^{\frac{1}{2}}$, $\lambda_{NiO}(720 \text{ eV}) \sim 21$ Å. Using a simple exponential model, similar to that described previously,[3] these data, together with the initial clean Ni2$p_{\frac{3}{2}}$: final O1s intensity ratio, yield an estimate of 18.6 Å for the thickness of the nickel oxide layer. To confirm this result, we examined the decline in the Ni metal $2p_{\frac{3}{2}}$ signal on oxidation : an approximate deconvolution indicated that the residual metal signal (I_r) was ~ $\frac{1}{3}$ of the original signal, (I_c). (Deconvolution of the very similar spectra reported by NT gives a final metal signal ~ $\frac{1}{2}$ the original, for faster (AlKα) photoelectrons.) Using the relationship

$$I_r = I_c \exp [(-\text{oxide layer thickness})/\lambda_{NiO} \text{ for } 2p \text{ photoelectrons}]$$

we find the layer thickness to be 17.6 Å in our experiments and 14 Å in NT's. We conclude that the nickel oxide overlayer thickness averages 7-9 " layers " as defined above (the layer spacing is 2.09 Å), rather than 2 layers. In view of this, we feel it would be helpful if the authors would clarify their definition of " one layer " in respect of both the intial oxygen overlayer and the final nickel oxide layer, especially in view of the apparent internal inconsistency (outlined above) in their quoted coverage estimates.

Dr. P. Norton (*A.E.C.L.*) (*communicated*): We wish to made several points in reply to Evans. Firstly, our main intent in the paper was to establish self-consistently the point at which oxide nucleation occurs and to confirm the absence of effect on the *d*-electrons in the chemisorption regime, rather than to establish an absolute coverage calibration.

As stated in the paper, our coverage estimates are based upon O1s peak areas and thus fig. 1 cannot be used to reconstruct our coverage calibration as Evans has obviously tried to do. We do not try to estimate the absolute number of oxygen atoms in the final oxide layer but rather, *assume* that it contains two monolayer equivalents of oxygen (i.e., ~ 3×10^{15} O atoms cm^{-2}). We would agree that some confusion could arise from our terminology " 2 NiO layers ". The assumption of $\theta = 2$ is close to the experimental results quoted in ref. (1) and (2) in our paper and also those of Quinn and Roberts.[4] Recent results of Sewell and Mitchell [5] on single

[1] P. Cadman, S. Evans, J. D. Scott and J. M. Thomas, *J.C.S. Faraday II*, 1975, **71**, 1777.
[2] T. A. Carlson and G. E. McGuire, *J. Electr. Spectr.*, 1972, **1**, 161.
[3] S. Evans, *J.C.S. Faraday II*, 1975, **71**, 1044.
[4] C. M. Quinn and M. W. Roberts, *Trans. Faraday Soc.*, 1964, **60**, 899; 1964, **61**, 1775.
[5] P. B. Sewell and D. F. Mitchell, private communication.

crystal nickel surfaces indicate that, if our surfaces contain equal amounts of the (100), (110) and (111) planes, the actual number of oxygen atoms in the final oxide would correspond more nearly to $\theta \sim 2.3$ (where $\theta = 1$ is equivalent to $\sim 1.5 \times 10^{15}$ O atoms cm^{-2}). It is, however, pointless to argue whether $\theta = 2.3$ or 2.0 without an absolute calibration method or a knowledge of the crystallography of our surfaces. Having once made the assumption that $\theta = 2$ (or equivalently, that there are $\sim 3 \times 10^{15}$ O atoms cm^{-2} in the surface region), the use of the O1s peak *areas* together with an approximate estimate of the average escape depth enables us to correct our data for attenuation. The correction is small and depends on the various assumptions and for these thin films is within the uncertainty of the original assumption of $\theta = 2$. We would argue that our coverage calibration is entirely self-consistent and could be put on an absolute basis only if the structure of the surface layer were known.

With regard to the discrepancy in oxide film thickness, I can only refer Evans to the work contained in ref. (1) and (2) in our paper, to the results of Quinn and Roberts [1] and more recently to the work of Mitchell and Sewell [2] in whose paper (and references therein) the calibration procedure relies on gravimetric and/or manometric procedures. The latter authors also note the anisotropy in the oxidation of nickel with crystal orientation. I would therefore regard the final room temperature oxide film thickness on (100), (110) and (111) nickel as established as lying in the range 0.6 to 0.8 nm.

Some of the present difference could be attributed to the differences in surface morphology. In the comment of Tapping and myself on surface heterogeneity, we point out that considerable differences exist in the O1s line shapes on surfaces with different inherent heterogeneity, and certainly Evans' ion-sputtered and unannealed surfaces may be expected to behave differently from annealed polycrystalline of single crystal surfaces.

If, for the sake of comparison, we use only the O1s peak heights in fig. 1, we calculate that the ratio of O1s intensities at saturation and $\theta = 1$ is ~ 2.3 (not 2.5 to 3 quoted by Evans). Taking his ratio of ~ 3.7 without any correction for attenuation, we may calculate an effective coverage in monolayer equivalents of $(3.7/2.3) \times 2 = 3.2$ (in our units). The minimum thickness for this layer would thus lie in the range 1-1.3 nm. Upon correction for self-attenuation, the final thickness would lie in the range 1.5-2.0 nm, close to that determined by Evans.

We would therefore stress that it is not reliable to base even relative coverage estimates upon XPS peak heights.

Dr. S. Evans (*Aberystwyth*) (*communicated*): In general, we would echo Norton's strictures on the use of peak heights: we did, however, use peak *areas* to establish all our numerical estimates. Height data were only taken, *faute de mieux*, to confirm the rough comparability of his data and ours. Our estimate of his oxide thickness as ~ 14 Å (from the Ni 2p spectra) is independent of height measurements from fig. 1. We used the same mean free path data to obtain this estimate and that for 5 L exposure: and the inconsistency we remarked on thus remains unresolved, though the welcome clarification of NT's terminology reduces its magnitude considerably and its origin is now clear.

There is, as Norton indicates, much evidence that on many nickel surfaces the oxygen uptake at saturation approximates to 1.5-2 oxygen monolayers. However, θ estimates based on measured surface areas in excess of the geometric area will, in

[1] C. M. Quinn and M. W. Roberts, *Trans. Faraday Soc.*, 1964, **60**, 899; 1964, **61**, 1775.
[2] D. F. Mitchell and P. B. Sewell, *Thin Solid Films*, 1974, **23**, 109.

general, be smaller than our estimates, which must refer to the geometric area of the sample. Estimates of θ made by established techniques for two rather different surface types may thus agree where XPS data would differ. There are, moreover, substantial differences between XPS data reported by different groups, one of which [1] found *no* detectable Ni $2p$ shakeup structure following saturation exposure to oxygen. This result, unlike NT's and our own, *is* consistent with a very small average oxide thickness. We agree with Norton that the difference in uptake between his films and our etched samples is probably due to differences in the nature of the surface: might not some of the apparent differences between his data and the single-crystal results referred to also have a similar origin?

Dr. C. R. Brundle (*IBM*) said: Bickley and I [2] have also studied the oxidation of Zn films by XPS and UPS, both at 300 and 80 K, but at a fixed emission angle (45°). The 300 K results are in excellent agreement with those presented here by Briggs, including the fact that we observe the two O($1s$) peaks to co-exist even at the lowest exposures (cf. Ni). It is also gratifying to see that we are in agreement in assigning the 530 eV O($1s$) peak to oxide oxygen. There are by now so many examples of metal oxides (both grown from films and in bulk form) which have this O($1s$) B.E. that it is quite nonsensical for anyone to continue to argue that such a value is characteristic of a chemisorbed species only and not an oxide species.

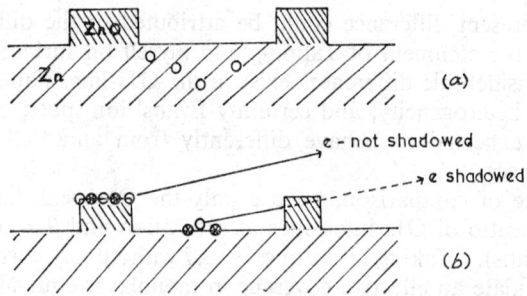

Fig. 1.—(*a*) Model of patched oxide surface proposed by Briggs. (*b*) Model of patched oxide surface with chemisorbed ○ in an overlayer position. ○ represents oxygen atom position. ⊗ represents zinc atom position.

The model developed by Briggs for the interpretation of the angle dependence of the two O($1s$) intensities may be compared to that proposed by Carley and myself to explain the presence of the two O($1s$) peaks in the Ni+O$_2$ systems.[3] In both cases the 530 eV O($1s$) peaks is considered to represent oxygen in oxide islands. In the Ni+O$_2$ system we have argued that the 532 eV peak may represent chemisorbed oxygen atoms remaining in an overlayer position, possibly involving d-orbital bonding. Briggs considers from his angular studies that the 532 eV peak for Zn+O$_2$ represents oxygen atoms embedded in patches of metal surface (fig. 1*a*), so that going to low angle emission will enhance the 532 eV/530 eV O($1s$) intensity ratio. LEED calculations on Ni [4] have shown that an overlayer oxygen atom can be as little as 0.9 Å above the plane of the metal surface. If one considered a model for Zn which had ZnO islands (which would be at least 3 Å thick, and could be much thicker if more than

[1] P. J. Page, D. L. Trimm and P. M. Williams, *J.C.S. Faraday I*, 1974, **70**, 1769.
[2] R. I. Bickley and C. R. Brundle, to be published.
[3] C. R. Brundle and A. F. Carley, *Chem. Phys. Letters*, 1975, **31**, 40.
[4] J. E. Demuth, P. W. Jepsen and P. M. Marcus, *Phys. Rev. Letters*, 1973, **31**, 540.

two ZnO layers were involved) and oxygen atoms in an overlayer separated by about 1 Å from the Zn metal substrate (fig. 1b), it would seem that the shadowing effect of the islands on going to low emission angles could also account for the enhanced 532/530 eV O(1s) ratio.

Our 80 K XPS results also show a marked similarity to those for Ni in that the 532 eV peak becomes relatively more important at this temperature (fig. 2), a fact which I would interpret as implying the stabilization of chemisorption with respect to oxidation (and island growth). One can also see from fig. 2 that a third oxygen state is observable at 80 K (cf. Ni) which is only stable at low temperature.

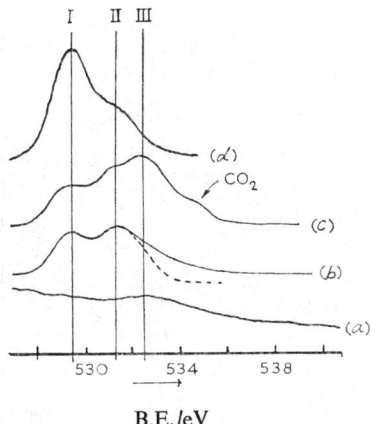

FIG. 2.—XPS O (1s) region for Zn film plus O_2. (a) Clean surface; (b) exposure at 10^{-6} Torr at 77 K. The dotted trace is obtained on warming to 300 K; (c) exposure at 5×10^{-4} Torr at 77 K; (d) exposure at 10^{-6} Torr at 300 K.

Dr. D. Briggs (*I.C.I. Runcorn*) said: The main reason for locating the 532 eV oxygen within the exposed Zn patches is the absolute variation in intensity of this peak with angle of emission which approximates to that expected from a homogeneous surface. It seems unlikely that a model of the type proposed involving islands of ZnO above the Zn surface could produce this invariance in intensity by means of shadowing, as Brundle suggests, since fortuitous compensation effects would have to operate over a large angular range. A more serious possibility is that island formation gives rise to a surface which is rougher than the evaporated film and this can lead to artefacts in the angular dependence curves as discussed by Fadley (ref. (4) of our paper). It may well be that, within the escape depths involved, the " embedded " oxygen species form a concentration gradient into the metal since our data show that the O(532) : Zn intensity ratio decreases at θ increases. On the other hand, the escape depth of the Zn *LMM* electron (~ 990 eV KE) is greater than that of the O1s electron (~ 720 eV) which could exaggerate this effect.

Brundle's observation of an O1s peak at ~ 533 eV at 77 K for Zn strengthens the arguments advanced in connection with his observation of a similar peak for the O_2/Ni system under similar conditions in the present paper. I expand on this theme in the reply to Roberts' comments (see below).

Dr. P. Pianetta (*Stanford*) said: The paper on oxidation of zinc as well as the other papers on chemisorption have concentrated on looking at changes in the valence band upon adsorption of gases. We have been studying oxidation by looking at chemical shifts in core levels, using photon energies between 25 and 325 eV from the Stanford Synchrotron Radiation Project.

The system we have chosen to study is GaAs (110). There has been quite a bit of interest on the oxidation of GaAs especially as to whether oxygen binds to the gallium or the arsenic atoms. This prompted us to investigate the possibility of chemical shifts in the Ga and As 3d levels. We also performed detailed studies of the valence band from 9 to 25 eV.

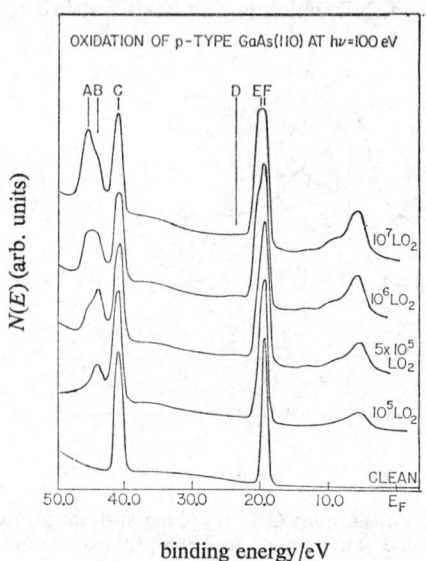

FIG. 1.—UPS spectra of clean and oxidized GaAs(110) for $h\nu = 100$ eV.

Fig. 1 shows a series of spectra taken after controlled oxidation of GaAs(110). The lowest curve, labelled " clean " was taken on GaAs(110) freshly cleaved *in situ* at a pressure less than 1×10^{-10} Torr. Oxygen was then admitted to give the exposures shown in fig. 1. The clean spectrum shows only two very sharp peaks: Ga 3d at 19 eV and the As 3d at 41 eV where the energy scale is taken relative to the Fermi level as determined from a gold sample. Upon exposure to 10^5 LO_2 (1 L = 10^{-6} Torr s), a second peak is clearly visible at about 44 eV. Since this is on the higher binding energy side of the As 3d, we interpret this as the chemically shifted As 3d due to charge transfer to the oxygen. Notice that the Ga 3d is as yet unchanged, indicating that there has been no effect on the Ga atom at this exposure. Upon further oxidation a second distinct peak at a binding energy of about 46 eV appears. This peak grows with oxidation until it dominates the first oxide peak, which seems to stop growing very soon in the oxidation process. This indicates that there are two binding states on the arsenic, one of which saturates at relatively low coverages between 1 and 5×10^5 L and another which continues to grow even at very high coverages. It is of interest to note that as soon as this second peak appears, the Ga 3d level starts to broaden until at 10^7 L, there appears to be a shift in the Ga 3d of about 1 V. The fact that the Ga 3d peak broadens only when the second oxide peak on the arsenic forms seems to show that the shift in the Ga 3d is a secondary effect, not due to the oxygen binding to the gallium but due to the fact that some of the gallium atoms are bonded to oxidized arsenic atoms which are now in different chemical environments.

In fig. 2 we show spectra of heavily oxidized GaAs(110) as a function of photon energy. By observing the ratio of the As 3d main peak and shifted peak, we see that the sensitivity for the oxide peak is largest at about 80 to 100 eV. This illustrates on important advantage of a tunable radiation source.

The photon energy can be used to change the kinetic energy of the outgoing electrons and thus change their escape depth. This allows studies of the electronic structure of surfaces as a function of depth into the material.

From our data we have been able to determine that oxygen binds to As in GaAs(110) in at least two different ways and that oxidation (as well as other gas adsorption) can be studied effectively by looking at core level electrons with small escape depths for a given photon energy.

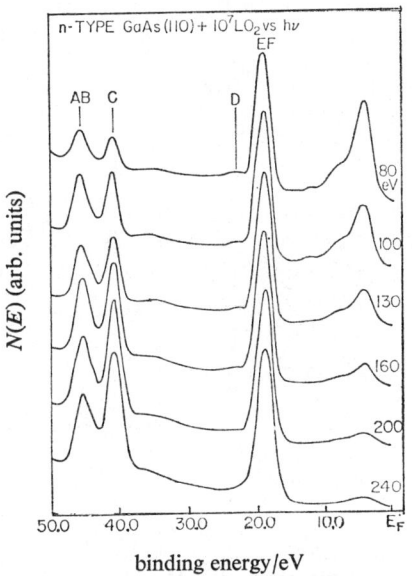

FIG. 2.—UPS spectra of GaAs(110) oxidized with 10^7 L O_2 for photon energies from 80 to 240 eV.

Prof. C. S. Fadley (*Hawaii*) said: Briggs has asked if there are any procedures by which effects on angular distributions due to patching of the surface layers could be distinguished from those due to surface roughness. Such a distinction could be very difficult, as both of these phenomena, as well as electron refraction in crossing the surface, tend to reduce the amount of surface enhancement relative to a simple uniform-surface-layer flat-surface model. Furthermore, the modelling of any of these effects in trying to make comparisons with experimental data will probably involve two or more adjustable parameters (as for example, surface coverage and effective layer thickness in the patched layer case), a number of degrees of freedom that may permit at least a reasonable description of the data by more than one effect. Thus it may not be possible easily to choose between effects. As a possible method for making such a distinction in a more convincing way, I would suggest performing experiments on Zn films evaporated on a unidirectionally-polished substrate so as to obtain angular distributions parallel to and perpendicular to the grooves (as we have done for oxidized aluminium [1]). Depositions could also be made on ultra-smooth glass substrates (for example, Corning 7059). Roughness effects should then be

[1] Paper by C. S. Fadley in this Discussion.

variable according to perpendicular > parallel > glass, and their maximum magnitudes might be estimable from such data.

We have also noticed absolute intensity curves with shapes similar to those in fig. 4(a) of Briggs' paper,[1] in particular as regards the fall-off of intensity for $\theta \gtrsim 60°$ (compare figs. 20 and 21 of ref. (1)). Our explanation of the effect was not off-axis rotation as Briggs has suggested, but rather that the broadly illuminated Henke-type anode in the AEI spectrometer begins to be partially shaded by the specimen itself as rotation proceeds to these higher angles.

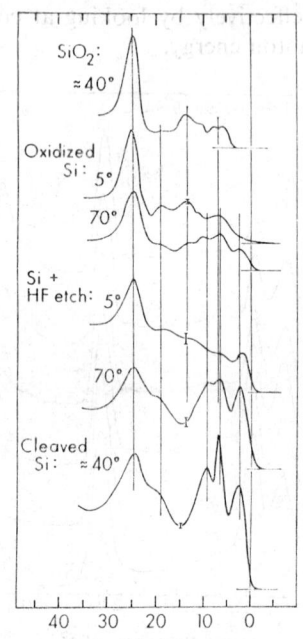

FIG. 1.—Valence-level XPS spectra for SiO_2 (ref. (4)); atmospherically oxidized Si at two angles of electron exit (this work); Si cleaned by means of an HF etch immediately before placement in the spectrometer, again at two angles of electron exit (this work); and a cleaved Si single crystal (ref. (4)). Prominent features in the SiO_2 and cleaved Si spectra are indicated by vertical lines.

This paper represents a very nice illustration of the kind of information that can be obtained from the analysis of core-level XPS angular distributions. I would like to point out similar possibilities for *valence-level* XPS studies based upon some recent work by Baird, Wagner and myself.[2] The figure shows the valence-level regions for: SiO_2, as obtained previously by DiStefano and Eastman [3]; a polished silicon specimen exposed to atmospheric oxidation; a similar specimen cleaned by means of an HF etch and immediately inserted into a vacuum of $\sim 5 \times 10^{-9}$ Torr for analysis; and a cleaved single crystal, as obtained previously by Ley, Kowalczyk and Shirley.[4] The oxidized specimen should possess a surface layer of oxide approximately 35 Å

[1] C. S. Fadley, R. J. Baird, W. Siekhaus, T. Novakov and S. Å. L. Bergström, *J. Electron Spectr.* 1974, **4**, 93.
[2] R. J. Baird, L. F. Wagner and C. S. Faldey, to be published.
[3] T. H. DiStefano and D. E. Eastman, *Phys. Rev. Letters*, 1971, **27**, 1560.
[4] L. Ley, S. Kowalczyk and D. A. Shirley, *Phys. Rev. Letters*, 1972, **29**, 1088.

thick.[1] The etched specimen had only a very thin oxide layer as judged by the Si2p (oxide)/Si2p (element) ratio of 0.02 at $\theta = 50°$; this ratio was thus lower than that found in the prior cleaved-crystal study.[2] The SiO_2 and cleaved Si valence-band profiles can be used as references for the two basic kinds of features that might be expected to occur in oxidized silicon, and certain peaks characteristic of both can be found in both the oxidized and etched specimens. (Note the accidental coincidence in position at 25 eV of the broad silicon plasma-loss peak and the narrower O2s-derived oxide peak.) The etched specimen, when viewed at high angle, looks very much like pure silicon, but at low angle, the overall spectrum changes considerably in such a way as to accent SiO_2-like features, especially the O2s peak at 25 eV. The more heavily oxidized specimen on the other hand only weakly retains certain silicon valence-band features such as the peak at 2.5 eV when observed at high angle; at low angle the 2.5 eV peak disappears completely and the overall structure is very similar to that of SiO_2. We thus conclude that selective enhancement of surface-associated valence-level features is also possible in variable-angle XPS measurements. Such enhancements could by very useful in distinguishing overlapping surface- and bulk-valence levels.

Dr. R. W. Joyner (*Bradford*) (*partly communicated*): It would be helpful if Briggs can clarify which model he proposes for that oxygen which has $B.E._{(F)} \simeq 532$ eV. He refers to this oxygen as being both " chemisorbed " and " incorporated " although these terms are mutually exclusive.[3] Also, can he tell whether the relative intensities of the two oxygen peaks (530.0 and 532.0 eV) are quantitatively compatible with his suggested model or with that introduced into the discussion by Brundle?

It is useful to recognize that the assignment of oxygen with $B.E._{(F)}$ 532 eV to oxygen chemisorbed on the oxide cannot be general. Thus, in the case of oxygen interaction[4] with a heavily contaminated lead surface, after prolonged oxygen exposure ($> 10^4$ Torr s) the peak at 532 eV is substantially bigger than that at 530 eV. Also, Briggs' angular variation data are the inverse of that predicted by assigning the 532 eV peak to chemisorbed oxygen. Norton's point in this discussion regarding the influence of film structure is likely to be important in this context. It must be remembered, however, that, as is the case on copper at 80 K,[5] weakly held oxygen can have B.E. in the range 532-534 eV. As happens for the ubiquitous peak at 530 eV[6] it seems likely that the simplicity of the XPS data conceals the complex nature of the oxygen interactions. It is therefore dangerous for Briggs to use results on zinc to adjudicate between the suggestions of Williams and Brundle on nickel—which is much more resistant to oxidation in any case.

Prof. M. W. Roberts (*Bradford*) said: Fadley raises the question as to whether Brigg's angular distribution data may reflect surface roughness rather than patches of oxide and zinc metal. Briggs gives no information on the extent of oxygen uptake and to fill this gap I would like to report some data obtained in Bradford by Metcalfe.[7,8] From physical adsorption studies of krypton, estimates of the surface

---

[1] F. Lukes, *Surface Sci.*, 1972, **30**, 91.
[2] L. Ley, S. Kowalczyk and D. A. Shirley, *Phys. Rev. Letters*, 1971, **29**, 1088.
[3] M. W. Roberts, *Quart. Rev.*, 1962, pp. 71.
[4] S. Evans and J. M. Thomas, *J.C.S. Faraday II*, 1975, **71**, 313.
[5] M. J. Braithwaite, R. W. Joyner and M. W. Roberts, this discussion.
[6] R. W. Joyner and M. W. Roberts, *Chem. Phys. Letters*, 1974, **29**, 447.
[7] L. Metcalfe, *Ph.D. Thesis*, (University of Bradford, 1970).
[8] R. I. Bickley, L. Metcalfe and M. W. Roberts, (unpublished).

area of evaporated zinc films were possible. At 77 K a 35 mg Zn film had an area which was five times its geometric area, at 290 K this surface had sintered to close to its geometric area. Assuming that each zinc surface atom is bonded to an oxygen " atom " than the number of oxygen molecules required to form a monolayer ($\theta = 1.0$) can be determined.

At 77 K the oxygen uptake at $\sim 10^{-6}$ Torr is equivalent to about 5.6θ, this oxygen is irreversibly adsorbed at a very fast rate. On increasing the pressure to $\sim 10^{-2}$ Torr a further, but reversible, uptake of oxygen (0.4θ) occurs. We are, therefore, concerned with a metal+oxygen system where oxidation is very extensive at low temperature. Furthermore thermal treatment *in vacuo* led to regeneration of the oxidised surfaces' ability to interact with oxygen. The regeneration was by oxide dissolution into the metal and, for example, was complete at 400 K. Any model of zinc oxidation must, therefore, take into account (*a*) the extensive oxidation that occurs at room temperature and low oxygen pressure; (*b*) the dissolution that is likely to occur of this oxide in the metal at room temperature, this is very obvious at somewhat higher temperature; (*c*) the average *oxide* thickness at 10^{-4} Torr is at least 12 Å, this is relevant to the " escape depth " with different photon sources (e.g., UV and X-ray); (*d*) the Zn+O_2 system will have distinctly different features from the Ni+O_2 system where θ is between 1 and 2 in the temperature range 77 to 290 K. Chemically the systems are, therefore, very different and conclusions based on a comparison of B.E. values will not be obvious. However the growth of " oxide pillars " on zinc is not an unrealistic model in view of the uptake data.[1, 2] Is this what Briggs is proposing?

Dr. D. Briggs (*I.C.I. Runcorn*) said: Roberts' additional uptake data are interesting, but his conclusions a little confusing at first sight. Thus, without resource to other experimental data except that adsorption at 77 K is equivalent to 5.6 monolayers at 10^{-6} Torr, he concludes the average oxide thickness at 10^{-4} Torr is $\geqslant 12$ Å (at an unspecified temperature) even though stressing that oxide dissolution is likely at room temperature. Our data (at room temperature) suggest oxide islands of at least double layer thickness (presumed necessary to produce well defined XPS and Auger spectra characteristic of ZnO) and oxygen incorporated into the exposed Zn patches. The two types of oxygen " atoms " are present in similar concentrations. Hence a total uptake of 6 monolayers is not in conflict with this model, but does not imply oxide islands 12 Å thick.

Even at 77 K Brundle's data show the presence of the two types of oxygen described in our paper. The type we ascribe to incorporate oxygen cannot be very different from that formed when ZnO dissovles in Zn at elevated temperatures. Thus the balance of oxygen in ZnO islands, or incorporated (dissolved) into Zn metal, can be looked upon as being thermodynamically controlled. The very close similarities noted in our paper and re-emphasised by Brundle in the behaviour of Zn and Ni suggests that similar models apply over the temperature range 77-300 K, although differences in the equilibria parameters would be expected for the two metals. On this basis I would not agree with Roberts' that a difference in uptake of $\times 3$ between Ni and Zn means that the systems are " chemically very different ".

Prof. M. W. Roberts (*Bradford*) (*communicated*): I confine my remarks to Briggs's last sentence and remind him that the chemical reactivity of nickel which has interacted with oxygen to form the equivalent of about one unit cell of " oxide " is very different from that which has formed many oxide layers.

[1] L. Metcalfe, *Ph.D. Thesis*, (University of Bradford, 1970).
[2] R. I. Bickley, L. Metcalfe and M. W. Roberts, (unpublished).

GENERAL DISCUSSION

Dr. D. Briggs (*I.C.I. Runcorn*) said: The paper seems to be quite clear in assigning the 532 eV oxygen type to an "atomic" species located within the metal which I described as "incorporated" and following the chemisorption process.

At this stage the analysis of the various dependences of the peaks with angle of emission in terms of a patched surface model is not complete, thus the second point awaits an answer. As noted above, and also by Fadley in this Discussion, the separation of effects due to patching and induced roughness is difficult and may turn out to be impossible without further experimental data.

Joyner's points in relation to the observation of O1s peaks with $BE_F = 532$ eV seem only to confirm our view that this cannot represent oxygen adsorbed on the oxide, although we do not postulate it to be weakly held in any form. We rely on the angular variation data to locate it beneath the Zn surface. The Zn/Ni comparison has been dealt with in reply to Roberts' comments.

Dr. C. R. Brundle (*IBM*) (*partly communicated*): XPS results [1] for the interaction of O_2 with Cu films at 300 K are in reasonable agreement with those for Cu (100) given by Braithwaite *et al.*, but are in complete disagreement with the results of Evans *et al.*[2] for polycrystalline Cu. The latter's exceptionally high reaction rate and extent of reaction, leading to the formation of bulk CuO at low O_2 pressures, therefore remains anomalous and cannot be considered generally characteristic of polycrystalline Cu.

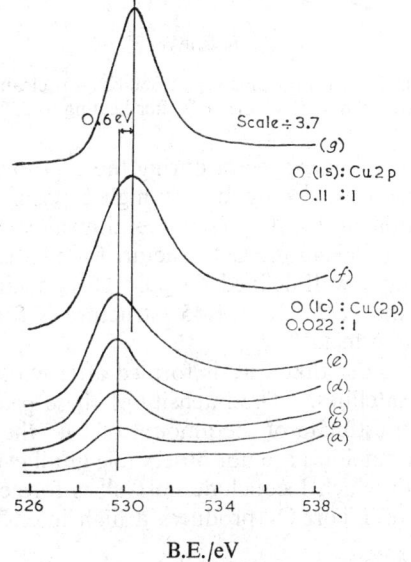

FIG. 1.—O(1s) XPS region for Cu film plus O_2 at 300 K. (*a*) clean; (*b*), (*c*), (*d*), and (*e*) after 3, 10, 100, at 2000 L exposure respectively; (*f*) after 10^5 L at 1×10^{-4} Torr exposure; (*g*) after exposure at 1 Torr for 5 min.

Fig. 1*a* shows an O(1s) uptake sequence for the Cu film which may be compared directly with fig. 3(*a*) of the present paper. There is a slight discrepancy in the B.E. scale which may be genuine, but is more likely a calibration difference. The Cu(2p)/O(1s) ratios at a given exposure are in reasonable agreement with those given by

[1] C. R. Brundle, to be published.
[2] S. Evans, E. L. Evans, D. E. Parry, M. J. Tricker, M. J. Walters, and J. M. Thomas, *Faraday Disc. Chem. Soc.*, 1974, **58**.

Braithwaite *et al.* The 0.6 eV O(1s) shift at high coverage is also observable for the film data (traces *f* and *g* of fig. 1*a*), but I would consider it to be possibly due to the growth of a second overlapping peak at higher B.E., rather than a simple shift to high B.E. with coverage.

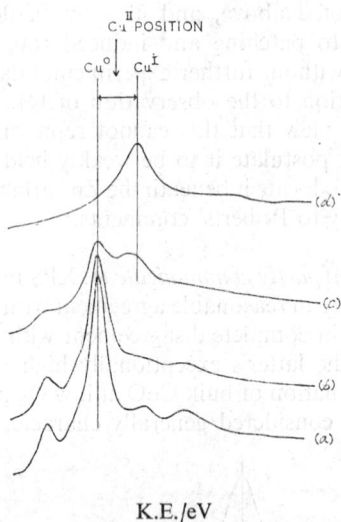

FIG. 2.—Cu Auger spectrum for Cu film plus O_2 at 300 K. (*a*) clean; (*b*) saturation coverage at 1×10^{-4} Torr; (*c*) after 5 min at 1 Torr; (*d*) after heating to 250°C in 1 Torr of O_2.

The polycrystalline Cu Auger spectra during the exposure sequence (fig. 2) show evidence for the presence of Cu^I only at very high exposure (5 min at 1 Torr) with very little after saturation at 1×10^{-4} Torr. Braithwaite *et al.* claim that for the (100) surface the $(\sqrt{2} \times 2\sqrt{2})$ R45° LEED structure found after exposures of 5×10^6 L at 1×10^{-4} Torr represents well-defined Cu_2O. Does their Auger data show the presence of Cu^I when the $(\sqrt{2} \times 2\sqrt{2})$ R45° structure is formed, or like mine still indicate predominantly Cu metal?

The authors also consider that Cu^{II} is formed after exposure to 5 Torr O_2 from the detection of Cu2*p* satellites. The intensity of these peaks is very low (fig. 4*a*). Do the authors have an estimate of the amount of Cu^{II} they consider to have been formed? For films the satellites are not observable at all even after exposure to one atmosphere of O_2, from which I conclude no Cu^{II} is formed. Heating the film to 250°C in the presence of 1 Torr O_2 produces a high intensity Cu^I Auger spectrum (fig. 2).

Finally, I would like clarification on how the authors reconciled some of the LEED sequences and XPS sequences. They consider that they have firm evidence that reconstruction of the surface only occurs when the $(\sqrt{2} \times 2\sqrt{2})$ R45° structure is formed (i.e., after high exposure only). The argument given for believing this is that the $(\sqrt{2} \times \sqrt{2})$ R45° structure is formed on exposure at 80 K and does not change on warming to 300 K, and that since one does not expect reconstruction at 80 K the $(\sqrt{2} \times \sqrt{2})$ R45° structure must, therefore, represent an un-reconstructed surface. However, their XPS O(1*s*) data at 80 K show 2 equal intensity peaks at saturation (fig. 6*a*). On warming to 300 K the higher B.E. component disappears (desorption) and the remaining peak moves from 532.0 eV to 530.8 eV (table 2)—a value characteristic of both chemisorbed O and Cu_2O according to the authors.

Thus, one has a situation where non-alteration in LEED pattern, which may represent only a fraction of the total oxygen present, is being taken to imply no change in the relative positions of the O and Cu atoms, whereas rather considerable changes in the O (1s) energies, representing the total oxygen content, are presumably being considered as of no significance.

Dr. S. Evans (*Aberystwyth*) (*communicated*): We agree that some of our early results [1] are somewhat anomalous as, in fact, we indicated in the Experimental section of our paper [2] for this Discussion. Nevertheless, our more recent experiments [2] are in agreement with the results contemporaneously reported by Braithwaite et al.[3] in that Cu^{II} satellites were again obtained at oxygen pressure well below an atmosphere; and He I spectra characteristic [1] of Cu^{II} have also been obtained under similar conditions by Richardson.[4] The use of photon-excited Auger spectra (previously suggested in the present context by Briggs [5]) suffers the drawback that the Auger structure is much more complex than is the 2p XPS. Moreover, with a rather high electron mean free path (~ 24 Å[6]) the contribution from a surface layer sufficient to give " oxide " UPS and LEED spectra is small, and consequently it is difficult to identify such small quantities of surface oxide.

Calculations of the extent of oxygen uptake using recently determined cross-section and mean free path data [6] suggest that the uptake in both our recent experiments [2] and in those of ref. (3) occurred at roughly comparable rates, the absolute coverages being about a factor of three greater than our earlier, more empirical, estimates.[1,2] The revised estimates are in reasonable agreement with the maximum coverage derived in ref. (3) from LEED data.

Prof. F. S. Stone (*University of Bath*) said: I would question whether Braithwaite, Joyner and Roberts have sufficient evidence to conclude that in the further stages of copper oxidation at 290 K and 5 Torr there is formation of both Cu_2O and CuO. Satellites in the Cu 2p XPS spectrum may certainly imply the presence of Cu(II), but that does not necessarily mean formation of CuO.

There is also some magnetic [7] and optical [8] evidence that Cu(II) is present in cuprous oxide formed by slow oxidation, but nucleation of CuO does not have to be invoked. The oxidation of copper has a unique coordination chemistry. The Cu(I) atoms in Cu_2O have a coordination number of two, and the initial stages of oxygen incorporation can be expected to reflect this, as is acknowledged in the paper. Further oxidation will involve the diffusion of copper atoms outwards into the array of adsorbed oxygen and this necessitates an increase in the coordination number of the metal atoms. The Cu(II) oxidation state is more suited to this situation than Cu(I), being characteristically stable in square planar, tetrahedral and Jahn-Teller distorted octahedral environments, so the occurrence of some Cu(II) in metastable oxide layers is to be expected. In the terminology of oxidation theory, one would

---

[1] S. Evans, E. L. Evans, D. E. Parry, M. J. Tricker, M. J. Walters and J. M. Thomas, *Faraday Disc. Chem. Soc.*, 1974, **58**, 97; S. Evans, *J.C.S. Faraday II*, 1975, **71**, 1044.
[2] S. Evans, D. E. Parry and J. M. Thomas, paper at this Discussion.
[3] M. J. Braithwaite, R. W. Joyner and M. W. Roberts, paper at this Discussion.
[4] N. V. Richardson, personal communication.
[5] D. Briggs, *Faraday Disc. Chem. Soc.*, 1974, **58**, discussion comments.
[6] S. Evans, R. G. Pritchard and J. M. Thomas, in preparation: see also Discussion comment on the paper by Clark *et al.*
[7] M. O'Keeffe and F. S. Stone, *Proc. Roy. Soc. A*, 1962, **267**, 501.
[8] H. Wieder and A. W. Czanderna, *J. Chem Phys.*, 1962, **36**, 816; A. W. Czanderna and H. Wieder, *J. Chem. Phys.*, 1963, **39**, 489.

say that the positive holes are trapped and that at the low temperature concerned the holes and the associated cation vacancies do not migrate sufficiently freely to the metal/oxide interface to be annihilated. An ordered superstructure of cation vacancies and Cu(II) in Cu_2O has in fact been proposed.[1] With due control of oxygen exposure, temperature and vacuum anneal variables, XPS should provide a good opportunity for understanding better the detail of this behaviour. More attention to the multilayer region of oxidation would be very valuable.

Dr. M. J. Tricker and **Prof. J. M. Thomas** (*Aberystwyth*) (*communicated*): Stone has commented that as well as the monolayer regime the later stages of the oxidation of metals are of interest and that the study of such systems should not be neglected. Recently at Aberystwyth we have been exploring the potential of ^{57}Fe conversion electron Mössbauer spectroscopy (CEMS) for studies of the progression of oxidation of iron-containing materials beyond the monolayer region. In essence the experiment involves the detection of the back-scattered iron conversion electrons (rather than the reduced flux of transmitted γ-radiation) following resonant absorption of Mössbauer γ-photons. Although the technique is necessarily restricted to iron and its compounds, it is now clear that the method is extremely useful in the study of overlayers [2, 3] in the thickness range of *ca.* 10 to 500 nm. A combination of photoelectron spectroscopy and CEMS therefore enables a more complete picture to be built up of the entire oxidation of the substrate.

Dr. M. J. Tricker (*Aberystwyth*) said: I would like to draw attention to an experiment performed by Porter and Freedman (*Phys. Rev. C*, 1971, **3**, 2285). These workers were able to measure the kinetic energies (*ca.* 7.3 keV) of the iron 1*s* conversion electrons resulting from the decay of the $I = \pm\frac{3}{2}$ excited spin state of ^{57}Fe populated by decay of ^{57}Co. A chemical shift was detected in the iron 1*s* binding energies between ^{57}Co implanted into and onto the surface of a graphite target. A potential advantage of the method is that there is no contribution to the spectral line width from a stimulating X-ray. It would be of interest to extend the method to other iron containing compounds where shake up and/or exchange effects might be present.

Dr. U. Landman (*Rochester*) said: I would like to comment on the structural assignments made in Roberts' paper. As is well known LEED spot patterns contain information only about the two-dimensional space-group in planes parallel to the surface plane. Determination of layer spacings and registries can be made via the analysis of the (intensity, voltage) data of the diffraction beams. Since intensity data were not employed by the authors in their analysis, their conclusions concerning overlayer structures and substrate reconstruction are rather intuitive and should be complemented by a proper detailed analysis of the intensities.

Dr. C. R. Brundle (*IBM*) (*communicated*): Roberts suggests that the " different molecular interactions " involved with the weakly bound oxygen adlayer present at 80 K, but desorbed at 290 K, are responsible for shifting the O(1*s*) B.E. of the strongly bound and non-desorbing oxygen from 532 eV at 80 K to 530.8 eV at 290 K. This

---

[1] H. Wieder and A. W. Czanderna, *J. Chem. Phys.*, 1962, **36,** 816; A. W. Czanderna and H. Wieder, *J. Chem. Phys.*, 1963, **39,** 489.
[2] M. J. Tricker, J. M. Thomas and A. P. Winterbottom, *Surface Sci.*, 1974, **45,** 601.
[3] J. M. Thomas, M. J. Tricker and A. P. Winterbottom, *J.C.S. Faraday II*, 1975, **71,** 1708.

O(1s) the authors associate with oxygen atoms which have invariant positions on warming from 80 K to 290 K (associated with the ($\sqrt{2} \times \sqrt{2}$) R45° LEED pattern).

The association of a chemical shift of ca. 1 eV with weak additional molecular interactions is hard to believe and in any case implies reversibility; i.e., re-cooling to 80 K and re-adsorbing oxygen should shift the O(1s) at 530.8 eV back to 532 eV. If it does not, or if re-adsorption of oxygen does not occur, then one is lead to the conclusion that the 530.8 → 532 eV O(1s) shift represents an irreversible change from an adsorption state of the strongly bound oxygen at 80 K to one of distinctly different electronic structure at 290 K.

Prof. M. W. Roberts (*Bradford*) said: I agree with Andersson's general comment that multi-techniques are an advantage in studying details of surface processes. In fact we chose the Cu(100)+O_2 system by LEED, AES and XPS since there was a strong possibility that we would be able to differentiate between Cu(metal), Cu_2O and CuO. This we have done, and it is the evidence from all three that adds weight to the interpretation. Furthermore at 77 K oxygen incorporation was not anticipated so that LEED data at this temperature would be easier to interpret. The background surface chemistry is therefore very important in determining the experimental strategy to be adopted. For example I would not chose the Fe+O_2 system mentioned by Andersson, as a " chemisorption system " since there is substantial evidence in the literature for oxygen incorporation even at low oxygen exposure at 77 K. Similarly I would not, in the first instance at least, choose a high index plane of Cu since " oxidation " is much more likely and we may miss entirely the chemisorption stage even at 77 K.

Stone raises the interesting point of the *mechanism* of the oxidation of Cu(100). Although our LEED and Electron Spectroscopic data indicate Cu_2O (probably the (110) plane) the manner in which the copper atoms move from their original positions in the Cu plane has not been considered. Clarke and Czanderna's [1] work and Stone's own studies with O'Keeffe [2] come to mind and have made us somewhat hesitant about proposing a detailed mechanism of oxidation at this stage. In particular Clark and Czanderna report that epitaxially grown Cu(100) oxidizes to give $CuO_{0.67}$, a defect structure of Cu_2O. How this affects a model for oxidation mechanism we have not considered.

One point that Stone's paper with O'Keeffe brings to mind, however, is the role that vacancies play in determining the nucleation of CuO. In particular the rate of production of cation vacancies as a function of oxygen pressure and their rate of diffusion will control whether CuO " nucleates ", and if it does, where it nucleates, i.e., at the surface itself or at the metal/oxide interface. Clearly the intensity of the satellites associated with CuO will be dependent on such factors and may account for the variable results reported. I am sure that the defective nature of oxides formed on bulk single crystals will be different from polycrystalline films, moreover different films will have different morphological characteristics.

In reply to Brundle's first point both we and the Aberystwyth group are very much aware of the real differences that exist between our two sets of data. Furthermore Evans *et al.* in a footnote to their paper draw attention to variation in their own experimental results over a period of a few months.

Evidence for Cu^{2+} in our work is based on data from a number of experimental parameters: (*a*) the presence of Cu(2p) satellites, albeit weak at 290 K, (*b*) LEED diffraction rings, (*c*) the Auger spectrum (fig. 4) and (*d*) the increase in width of the

[1] E. G. Clarke and A. W. Czanderna, *Thin Solid Films*, 1971, **12**, 443.
[2] M. O'Keeffe and F. S. Stone, *Proc. Roy. Soc. A,* 1962, **267**, 501.

Cu $2p_{\frac{3}{2}}$ level. The presence of Cu^+ is based on (a) the absence of $Cu(2p)$ satellites, (b) the presence of a ($\sqrt{2} \times 2\sqrt{2}$) R45° structure which is disordered at 290 K but well ordered when formed at high temperature and cooled *in vacuo* and (c) the predictions of thermodynamic calculations particularly with relation to the formation of CuO at high temperature in the presence of oxygen and its reduction to Cu_2O, e.g., on cooling *in vacuo* to 290 K. Brundle's data obtained by "heating the film to 250°C in the presence of 1 Torr" is intriguing in that he claims that only Cu_2O is formed (at 250°C). Presumably the spectra were taken at room temperature and a possible explanation is that if the oxidized film was cooled *in vacuo* rather than in oxygen, then the CuO formed initially at the higher temperature would be reduced to Cu_2O. The thermodynamic data reported in our paper would support this viewpoint.

With regard to the ($\sqrt{2} \times 2\sqrt{2}$) R45° structure observed at room temperature after an exposure of 10^6 L at 10^{-4} Torr it should be emphasised that it was not well ordered and, therefore, we would not claim it to be "well defined Cu_2O". It is, however, the same structure as that observed by "decomposing" CuO at high temperature *in vacuo* and known to give Cu_2O. In the latter case the O/Cu ratios indicate that oxygen is present to the extent of many monolayer equivalents.

Finally Brundle asks why we apparently attach less significance to the O(1s) data at 290 and 80 K than the LEED data at these temperatures. The important point to emphasise is that there is little to be gained in a comparison of an O(1s) value(s) at "saturation" at 80 K with one at 290 K. The "final O(1s) value" at 80 K which Brundle refers to will undoubtedly reflect very different molecular interactions from those occurring within the adlayer at 290 K. In other words oxygen present at 80 K, but desorbed on warming to 290 K, will influence the O(1s) value associated with the ($\sqrt{2} \times \sqrt{2}$) R45° structure at 80 K. With this in mind it should perhaps be pointed out that for comparable oxygen coverages as given by the $O(1s)/Cu(2p_{\frac{3}{2}})$ ratios, 0.011 at 80 K, cf. 0.017 at 290 K, the O(1s) peak positions are not too different (530.5±0.2 eV). The peak profiles are however different in that the FWHM value at 80 K is somewhat greater than at 290 K and this is entirely in keeping with some weakly held oxygen being also present at 80 K even at low θ but which at higher coverage is reflected in a "characteristic peak". Weakly adsorbed oxygen at 80 K will also be reflected in a disordered LEED pattern which we observe.

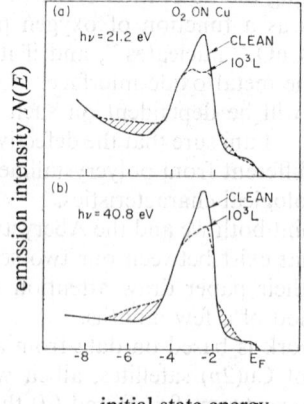

FIG 1.—He I and He II spectra showing the adsorption of oxygen on polycrystalline Cu at room temperature. The shaded regions are the observed resonance levels.

Mr. K. Y. Yu (*Stanford*) said: We have also looked at the adsorption of oxygen on Cu by photoemission (He I and He II). The data were taken by a double pass cylindrical mirror analyzer. The adsorption of oxygen on Cu gives rise to two resonant levels at -1.8 eV and -5.5 eV below E_F (fig. 1). On the clean Cu spectra, the *s-p* band intensity decreases, with respect to the *d*-band intensity, on going from $h\nu = 21.2$ eV to 40.8 eV. This is a well known effect; that the photoionization cross-section of *s-p* electrons decreases faster than *d* electrons with increasing photon energy, and it enables us to deduce the atomic origin of the -1.8 eV peak. We note that the magnitude of the -1.8 eV peak grows with respect to the *s-p* band as one goes from 21.2 eV to 40.8 eV. Therefore, this peak cannot be derived from the Cu *s-p* electrons, even though it is sitting on top of the *s-p* band. Tentatively we assign the level as arising from oxygen *p* electrons and Cu *d*-electrons, and, since the level is closer to the Cu *d*-band than the $O_2 p$ levels, it could be mainly *d*-like. Thus, this level at -1.8 eV can be viewed as a split off anti-bonding *d* state following oxygen adsorption, and the level at -5.5 eV is the usual bonding $O_2 p$ state.

Mr. N. V. Richardson (*Birmingham*) said: Some recent photo-emission work by Lloyd, Quinn and myself on the (001) face of a copper single crystal indicates that there are strong angular dependencies, suggesting, therefore, that one should be wary of making detailed assignments on fixed angle information. Fig. 1*a* shows the He I photoelectron spectra of copper for analyser angles between 0° (normal emission) and 80° in the arc projecting on to the [100] crystal axis (the ΓX direction of the Brillouin zone). Fig. 1*b* is the corresponding data for the arc projecting on to the [110] crystal axis (ΓK direction of the Brillouin zone).

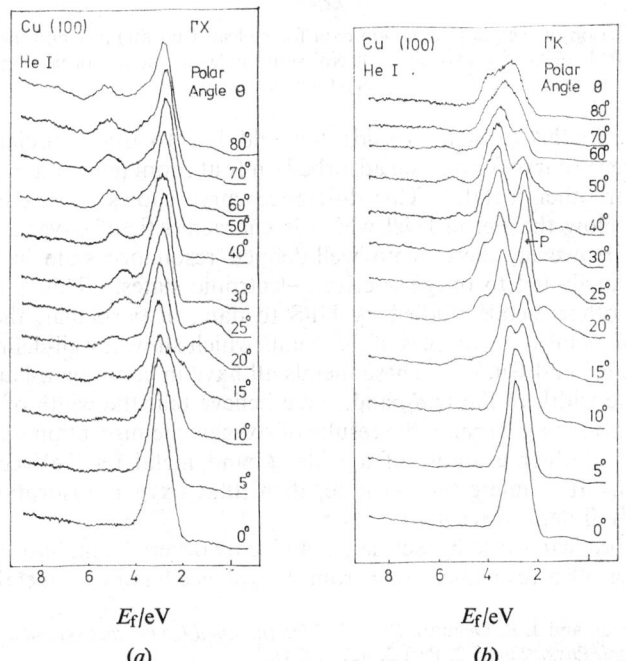

Fig. 1.—He I photo-emission spectra from the (001) face of a Cu single crystal. Emission direction varies from normal ($\theta = 0°$) to 80° off normal in two different arcs: (*a*) projecting on to the [100] crystal axis (*b*) projecting on to the [110] crystal axis.

The band labelled P in fig. 1b is found, under higher resolution, to have two components each with a half-width of *ca.* 150 meV.

Exposure of this crystal face to *ca.* 10^6 L of oxygen produced two significant effects which are exemplified in fig. 2.

(1) The marked suppression of band P.

(2) The appearance of the three bands A, B and C observed by Evans *et al.* Band B has two components whose relative intensity shows some angular variation. The deeped lying component is relatively more intense for emission close to normal.

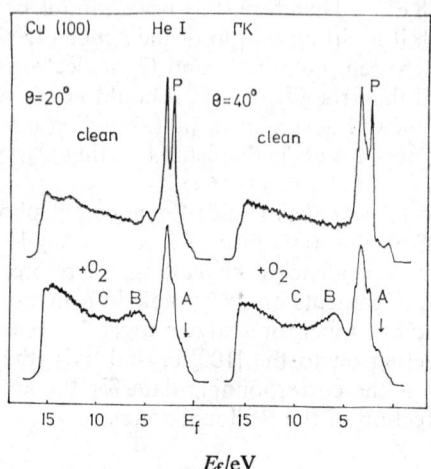

FIG. 2.—A comparison of the photo-emission data for a clean Cu (001) face and the same face after exposure to *ca.* 10^6 L of O_2 for two different emission angles in the arc projecting on to the [120] crystal axis.

Mr. D. M. Collins (*Stanford*) said: I would like to draw special attention to Brodén's difference curve for oxygen adsorbed on Ir at room temperature and compare it with data for other metals. This difference curve shows a preferential drop in emission just below the Fermi level which is characteristic of oxygen adsorption on most metals. However, there is no well-defined resonance state in the spectrum which can be attributed to oxygen-derived electronic states. This is in contrast to data for most other metals studied by UPS to date, in particular, the nearly free-electron and 3d metals. Examples of 3d metals which show a well-defined resonance state are Ni,[1] Fe,[2] and Cu.[2,3] These metals all have in common d-bands which are roughly half the width of the Ir d-band. We believe that the width of the d-band is an important factor in governing the results of oxygen chemisorption on these metals.

Platinum is another example of a wide d-band metal ($\sim$ 7 eV compared with $\sim$ 8 eV for Ir). It is interesting to note, then, that oxygen adsorption on Pt also fails to give a well-defined resonance state.[4]

Recent theoretical work by Schrieffer and co-workers[5] emphasizes the relative ease of splitting off a resonance state from the valence band of a metal when the so

---

[1] D. E. Eastman and J. E. Demuth, *Proc. 2nd International Conference on Solid Surfaces* (1974), *Japan J. Appl. Phys., Suppl.* 2, Part 2, 827 (1974).
[2] K. Y. Yu and W. E. Spicer, to be published.
[3] S. Evans, D. E. Parry and J. M. Thomas, paper at this Discussion.
[4] D. M. Collins, J. B. Lee and W. E. Spicer, to be published.
[5] J. R. Schrieffer and P. Soven, *Phys. Today*, 1975, **28**, 24 and references therein.

called " atomic " state of the adsorbed atom lies near the bottom of the valence band. This is a condition met by the $3d$ metals due to their relatively narrow valence bands but which is not met by Pt and Ir.

This supports the earlier conclusion that the widths of the metal d-bands have a strong influence on whether or not an adsorbed oxygen atom will result in a well-defined resonance state.

This rather unique behaviour of Pt and Ir has important bearing on their properties in catalyzing oxidation reactions. For example, Pt and Ir are two of a select few materials which catalyze the reaction of hydrogen and oxygen to form water at room temperature. There are no reports of this reaction occurring on the $3d$ metals.

Prof. M. W. Roberts (*Bradford*) said: Following Collins' remarks, I would like to mention a somewhat analogous situation which Kishi and I have observed,[1] namely the removal of an adsorbed layer of oxygen on lead by H_2S at 77 K. Just above this temperature the XPS data indicate the formation of H_2O (ads) and at somewhat higher temperature this is desorbed and replaced by a surface " sulphide ". At 290 K the adsorbed oxygen is completely removed after an exposure to H_2S of $\sim 10^{-4}$ Torr s. The surface reduction is, therefore, in one sense (i.e., it is complete) efficient but we should not lose sight of the fact that once the temperature is high enough for mobility to be induced in the adlayer there will be a large number of collisions between hydrogen adatoms and oxygen adatoms and this is what determines the *efficiency* of the process. This argument is also applicable to the $Pt+O_2$ system, the sticking probability of hydrogen may be low but the important parameter is the collisional frequency within the adlayer and this will be dependent on the number of surface sites visited before desorption.

Dr. U. Landman (*Rochester*) said: It is important to know the error bounds associated with the difference, and difference of differences, curves shown in the lower part of fig. 4.

Dr. G. Brodén (*Cornell*) said: It is true that the changes in the photo-emission curves due to oxygen adsorption on Ir(100) are small. However, in fig. 4 of our paper there is a feature labelled D which we ascribe to the $2p$-states of oxygen. In measuring our photoemission spectra we are using a pulse counting technique where the data are stored in a mini-computer. Consequently all difference curves are calculated by subtraction within the computer. We think this is essential in order to resolve weak structure such as peak D in fig. 4. We, therefore, conclude on the basis of our method that the observed fine structure in the Ir(100) spectra is significant.

Another question is why the oxygen peak on platinum group metals is so weak, in particular, in comparison to the results obtained for Ni and Fe. One reason for this could be related to oxygen coverage. On Pt(100), which is also a reconstructed surface, the saturation coverage of oxygen has recently been determined to be 0.25. On Ni(100),[3] on the other hand, the observed $c(2 \times 2)$ LEED pattern for oxygen is consistent with a coverage of 0.5. The oxygen coverage, therefore, seems to be smaller on Pt and Ir than on Ni and Fe. Another reason for the weakness of the oxygen features on platinum group metals might be related to the quantum yield. If the photoelectron emission from the valence band of Ir and Pt is larger than for Ni and Fe then the relative contribution to the spectra from a specific amount of

[1] K. Kishi and M. W. Roberts, *J.C.S. Faraday I*, 1975, **71**, 1721.
[2] G. Kneringer and F. P. Metzer, *Surface Sci.*, 1975, **49**, 125.
[3] J. E. Demuth, D. W. Jepsen and P. M. Marcus, *Phys. Rev. Letters*, 1973, **31**, 540.

adsorbed oxygen would be smaller for Ir and Pt than for Ni and Fe. If the spectra are not normalized to the quantum yield it would then *appear* as if the oxygen peak is smaller on Ir and Pt than on Ni and Fe.

Finally, as pointed out by Collins, the particular bonding mechanism for oxygen on different metals could affect the peak area and in this context the d-band width could be an important parameter.

Mr. D. M. Collins (*Stanford*) said: Mason has demonstrated the use of UPS to study the adsorption properties of certain organic molecules on Pt under various conditions.

We have extended the use of UPS in studying surface reactions by using it in conjunction with thermal desorption techniques to monitor the progression of the reaction of hydrogen with oxygen adsorbed on Pt to form water at room temperature.

Thermal desorption was used to determine the coverage of oxygen on our predominantly (111)-oriented Pt ribbon as a function of oxygen exposure. Through the use of thermal desorption techniques [1] we determined that the highest energy binding state of oxygen on Pt, designated the β state by Alnot and co-workers,[2] saturates at an exposure of 5 L (1 L = 10^{-6} Torr s). Alnot and co-workers [2] determined that the saturation coverage of the β state was $\sim 2 \times 10^{14}$ atoms cm^{-2} for a sample similar to ours.

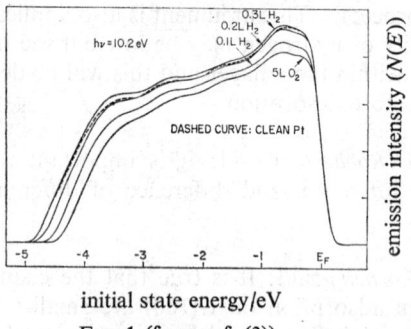

FIG. 1 (from ref. (3)).

At this coverage the changes in the Pt UPS—Energy Distribution Curve (EDC) had saturated. In the EDC's shown in fig. 1 we see that the 5 L oxygen exposure, corresponding to the filling of the β state, resulted in a relatively uniform decrease in the electron emission throughout the entire spectrum as well as an increase in the work function of ~ 0.45 eV. At this stage the sample was exposed to 0.1 L of hydrogen, resulting in the changes shown in fig. 1. Separate control experiments identified the reaction product as H_2O. After an additional 0.1 L of hydrogen exposure, the EDC had reverted to nearly that of the clean Pt surface (fig. 1). With another 0.1 L hydrogen exposure the emission increased still further. The increase beyond the emission for the clean surface was attributed to hydrogen chemisorption. The sticking probability for hydrogen chemisorption was measured to be $\sim 10^{-3}$ in separate experiments.

At this stage, thermal desorption was performed with the result that only hydrogen

[1] D. M. Collins, J. B. Lee and W. E. Spicer, to be published.
[2] M. Alnot, A. Cassuto, J. Fusy and A. Pentenero, *Proc. 2nd International Conference on Solid Surfaces* (1974), *Japan J. of Appl. Phys.*, Suppl. 2, Part 2, 79 (1974).
[3] D. M. Collins, J. B. Lee and W. E. Spicer, *Phys. Rev. Letters*, 1975, **35**, 592.

and no oxygen or water desorbed. This indicates that all the oxygen had reacted to form water which subsequently desorbed.

An exposure of 0.2 L of hydrogen corresponds to 3×10^{14} hydrogen molecules cm^{-2} striking the Pt surface. The removal of 2×10^{14} oxygen atoms, corresponding to saturation coverage of the β state, by 3×10^{14} hydrogen molecules leads to a strikingly high reaction probability of nearly one oxygen atom removed per incident hydrogen molecule. A reaction probability approaching unity has strong implications as to the mechanism of the reaction. Since the sticking probability for hydrogen chemisorption is small, it is clear that the reaction must occur between a physically adsorbed hydrogen atom which has high enough mobility and residence time on the surface to encounter an adsorbed oxygen atom before desorbing. The reaction mechanism is thus analogous to the fast non-activated chemisorptions with the chemisorption step being replaced by the formation of water and the site being on adsorbed oxygen atom.

Dr. R. W. Joyner (*Bradford*) said: At 295 K the initial adsorption of hydrogen chloride on evaporated nickel films is dissociative. At higher coverages molecular adsorption commences and predominates close to the monolayer point.[1] This result parallels that of Clarke *et al.* for vinyl fluoride and vinyl chloride on Pt(100). Both results are susceptible to explanation by assuming that the collective electronic properties (and thus the ability to dissociate an incoming molecule) of the metal+ adsorbate system change as the adsorbed layer is built up. The ability of CO pre-adsorption to prevent dissociation of vinyl halides supports this explanation, which may also be appropriate to the case of CO chemisorption on iron, where preadsorption of H_2S prevents carbon monoxide dissociation.[2] It is, therefore, not clear why Clarke *et al.* suggest that " only certain sites " on the homogeneous single crystal surface are active in dissociation.

Prof. R. Mason (*University of Sussex*) said: Joyner may well be right in suggesting that there is a close analogy between the associative-dissociative behaviour of hydrogen chloride and the vinyl halides. So far as his main point is concerned, we chose carbon monoxide for its minimal electronic perturbation of surface states, its relatively small molecular volume and our (limited) understanding of the metal-ligand bond as a function of surface coverage. The data we present in our paper cannot rule out, however, the possibility that electronic changes at the surface, following the preadsorption of carbon monoxide, may be still sufficient to affect the rate of alkene dissociation. But the results of further experiments add weight to the sense of original argument.

Preadsorption of, say, a quarter-monolayer of carbon monoxide on the Pt(111) surface again provides for associative sorption of the vinyl halides up to a total coverage of three quarters or so of a monolayer; at that point *dissociative* sorption of the alkene begins. Secondly, and much more illuminating, we have separately adsorbed up to 25 % monolayer-trifluorochloroethylene and -*cis* 1,2 difluoroethylene (both of which are associatively bound) and subsequently ethylene up to a total coverage of one monolayer. The sorption of ethylene produces dissociation of the haloalkenes, a feature which would not be explained by Joyner's suggestion since the haloalkenes provide a considerable perturbation of the surface states. Rather, the results establish conclusively the importance of carbon-hydrogen bond fission and

[1] R. W. Joyner, *Ph.D. Thesis*, (University of Bradford, 1969). R. W. Joyner and M. W. Roberts (unpublished work).
[2] K. Kishi and M. W. Roberts, *J.C.S. Faraday I*, 1975, **71**, 1715.

the subsequent intermolecular dehydrohalogenation reaction. They do not, of course, add to our suggestion that activation occurs only at bridging sites.

Prof. R. Mason (*University of Sussex*) said: In reply to Thomas, our assignment of core binding energies has followed from a systematic survey of a large number of substituted alkenes chemisorbed on Pt(100) and Pt(111). The important points to mention are that the C(1s) binding energy of the (non co-ordinated) trifluoromethyl groups in adsorbed trifluoropropene and trifluoropropyne are ca. 6 eV different from those of the olefinic and acetylenic carbons—a value similar to those observed in the gas phase spectra of these molecules; the C(1s) binding energies in 1,1 difluoroethene and 1,1 dichloroethene differ, however, by only 2.0 eV or so; with our instrumental resolution (~ 1.2 eV) and with their dissociative sorption, we cannot distinguish separate binding energies for the carbon atoms of chemisorbed vinyl fluoride and vinyl chloride which our data on the dihaloalkenes would suggest would be separated only by a little more than 1 eV if they were associatively bound at the surface.

Dr. P. Pianetta (*Stanford*) said: In addition to the interesting results that have just been presented by Linnett and Perry, there has been a considerable amount of work done on tungsten in the past few years. As part of a larger program on the study of tungsten carbide, we have also studied the carburization of clean W(100) surfaces. Part of this work was done using He I resonance radiation and the rest using synchrotron radiation at the Stanford Synchrotron Radiation Project.

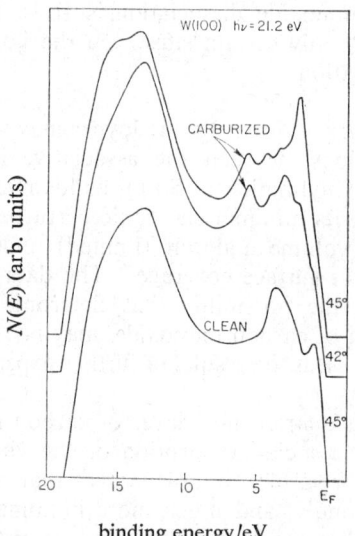

FIG. 1.—UPS Spectra of clean and carburized W(100) as a function of angle.

After cleaning the sample by repeated heating in oxygen, we adsorbed C_2H_2 on the surface of the W(100) sample and heated it to 1000 K. This temperature is high enough to dissociate the C_2H_2 and drive off the hydrogen, leaving a carbon layer on the surface of the sample. Fig. 1 shows the results of this carburization. The most interesting feature of these spectra is the very angular dependent peak about 1 eV below the Fermi level. The angles in the figure refer to the angle between the analyzer axis and the sample normal; a change of 3° is sufficient to make the leading peak disappear. No such angular dependence was observed on the clean sample nor on the adsorbate covered sample. It should be noticed that our spectrum for carburized W(100) is almost identical to Linnett's spectrum for H_2 on W(100). Furthermore, this is also different from the spectra for H_2 on W(100) reported by Feuerbacher and Adriaens [1] in which the hydrogen peak is 2 eV rather than 1 eV below the Fermi level.

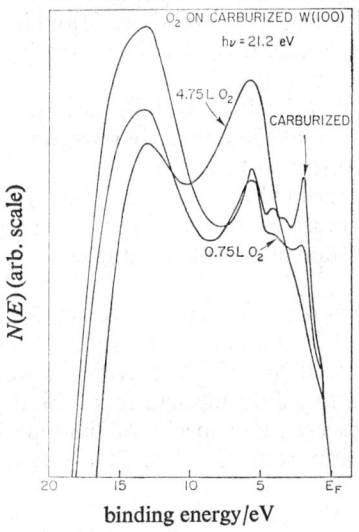

FIG. 2.—UPS spectra of oxygen adsorbed onto the carburized W(100) surface.

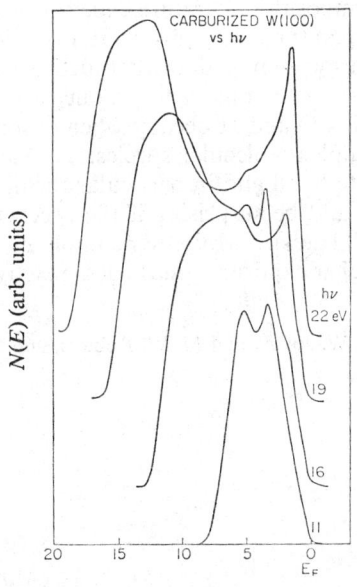

FIG. 3.—UPS spectra of carburized W(100) as a function of incident photon energy.

[1] B. Feuerbacher and M. R. Adriaens, *Surface Sci.*, 1974, **45**, 553.

In fig. 2 we show the oxidation of the carburized W(100). The most striking feature of this spectrum is again associated with the angularly sensitive peak. The peak almost completely disappears at an exposure of as little as 0.75 L O_2 (1 L = 10^{-6} Torr s); if the oxidized surface is heated to 1000 K, the angularly sensitive peak reappears, but is less intense than for the original carburized surface. In fact, this reduction in peak height is proportional to the amount of oxygen adsorbed on the carburized surface. Heating after repeated oxygen exposure brings back the clean W(100) surface. The angular sensitivity of this peak and its behaviour during oxygen adsorption suggest that some sort of ordered carbon layer, possibly graphitic, is formed during the C_2H_2 decomposition.

Fig. 3 shows spectra for carburized W(100) as a function of incident photon energy. It should be noticed that the angularly sensitive peak has a strong photon energy dependence and disappears below 16 eV.

The data that we have presented here are preliminary in nature and were shown mainly to stress that caution must be exercised in interpreting data taken with electron energy analyzers that have different acceptance angles.

Dr. R. W. Joyner (*Bradford*) said: At 295 K we have noted [1] that methanol decomposes rapidly at a nickel film surface to yield apparently, only chemisorbed carbon monoxide, as monitored by UPS and XPS. I wonder if Egelhoff, Linnett and Perry can give any indication of the mechanism of this decomposition on tungsten, and indicate if they have observed any species of the type $H_NC=O$. In particular, would they expect the spectrum from adsorbed CO to be very different to that from, say, adsorbed HCO?

Dr. D. L. Perry (*Cambridge*) said: At room temperature the decomposition of methanol on a clean tungsten surface is very fast so that we are unable to speculate about the mechanism of decomposition. Future experiments with a tungsten crystal cooled to liquid nitrogen temperatures may give us information on this aspect of the reaction. The UPS, flash desorption and work function experiments all indicate that, at low exposures, methanol decomposes to produce a mixed adlayer of CO and H, in the binding states which CO and H occupy when adsorbed separately. There is no evidence for a more complex molecular species. In view of the different bond order for the surface-adsorbate bond and in particular the dissociative nature of CO adsorption on tungsten, it would be surprising if the spectrum of the species HCO (ads) was identical to that of CO (ads). Molecular complexes, derived from methanol, are, of course, observed at higher exposures on the less reactive CO covered tungsten surfaces.

[1] M. J. Braithwaite, R. W. Joyner and M. W. Roberts, unpublished results.

Electronic Structure of Some Periodic Polymers Studied by ESCA and Band Structure Calculation

By Melvyn H. Wood,† Richard Lavery, Ian H. Hillier*
and Michael Barber‡

Chemistry Department, University of Manchester, Manchester M13 9PL

(Received 24th February, 1975)

The valence region of the ESCA spectra of polyethylene, poly(tetrafluoroethylene), poly(ethylene oxide) and polyglycine are reported, and discussed with the aid of band structure calculations using the NDO approximations. When allowance is made for the variation of the ionization cross-section of the crystal orbitals with their atomic orbital character, theoretical spectra, in quite good agreement with experiment can be constructed.

High energy photoelectron spectroscopy (ESCA) which allows ionization potentials to be obtained in the solid state may be used to study experimentally the band structure of solids. Although the resolution obtained using this technique is usually inferior to that achieved using low-energy photon sources (e.g., He(I) radiation), the ESCA spectra invariably show sufficient structure to allow this technique to yield useful results in such band structure studies. From a theoretical viewpoint a one-dimensional periodic solid presents a more tractable computational problem than a two- or three-dimensional one. For this reason we have investigated experimentally and theoretically a number of one-dimensional polymers, with the aim of interpreting the observed ESCA spectra as fully as possible using the methods of semi-empirical molecular orbital theory.

THEORY AND METHODOLOGY

Several semi-empirical methods have been used to calculate band structures of solids. We here use the tight-binding crystal orbital method with a basis of valence atomic orbitals, the computations being simplified by means of the complete neglect of differential overlap (CNDO) and intermediate neglect of differential overlap (INDO) approximations which are well known from studies of molecular electronic structure.[1] As such a scheme has been used by a number of authors[2] for band structure calculations we present only an outline of the method here.

In the tight binding, LCAO approximation the one-electron crystal orbitals (CO)(ϕ_{ks}) for a helical polymer chain may be written

$$\phi_{ks} = \left(\frac{1}{N}\right)^{\frac{1}{2}} \sum_{j=0}^{N-1} \sum_{t=1}^{n} C_{ks,jt} \chi_t(T^{-j}\mathbf{r} - j\mathbf{a}). \tag{1}$$

Here k and s denote the wavenumber and energy band respectively. $\chi_t(T^{-j}\mathbf{r}-j\mathbf{a})$ is the tth atomic orbital in the jth unit cell and $C_{ks,jt}$ the corresponding expansion coefficient, the summations being over all unit cells (N) and valence atomic orbitals

---

† present address: Theoretical Physics Division, A.E.R.E., Harwell, Oxfordshire OX11 0RA.
‡ present address: Chemistry Department, University of Manchester Institute of Science and Technology, Manchester M60 1QD.

per unit cell (n). The unit axial translational vector is given by **a**, and the rotational operator around the helical axis by T. Thus, the application of j rotations and j translations maps the a.o.s. of the first unit cell into the equivalent orbitals of the jth unit cell. Use of the Born–Von Karman periodic boundary condition,

$$\phi_{ks}(T^N\mathbf{r}+N\mathbf{a}) = \phi_{ks}(\mathbf{r}), \quad (2)$$

leads to the Bloch condition for the linear expansion coefficients,

$$C_{ks,jt} = \exp(2\pi i p j/N)C_{ks,t} \quad p = 0, 1, 2\ldots, (N-1). \quad (3)$$

For large N, we may effect the replacement of p by the wavenumber, then

$$k = 2\pi p/N,$$

giving

$$\phi_{ks} = \left(\frac{1}{N}\right)^{\frac{1}{2}} \sum_{j=0}^{N-1} \sum_{t=1}^{n} C_{ks,t} \exp(ikj)\chi_t(T^{-j}\mathbf{r}-j\mathbf{a}). \quad (4)$$

Incorporation of this expansion into the Roothaan equations in the usual manner leads to the set of N, coupled, matrix eigenvalue equations,

$$F(k)\,C(k) = C(k)\,\varepsilon(k). \quad (5)$$

Here $\varepsilon(k)$ is a diagonal matrix of orbital energies, $C(k)$ is the matrix of expansion coefficients ($C_{ks,t}$), and $F(k)$ the Fock matrix elements which are evaluated using the CNDO or INDO approximations. These coupled equations are then solved iteratively for a limited number of values of k distributed uniformly between 0 and π to give a calculated band structure which is invariant to an increase in this number. Stable results were achieved with approximately 15 such k values and with approximately 8 unit cells in eqn (4). Further details of the computational method are contained in ref. (3). The calculations were performed using standard geometric parameters, and the CNDO parameters given by Pople,[1] and by Sichel and Whitehead (denoted R2 in ref. (4)).

Our calculations yield a number of solutions of eqn (5) at each point in k-space, but, as McCubbin has pointed out,[5] the determination of the band structure from these eigenvalues may not be straightforward due to the problem of band crossings. In this work we have calculated the point differentials [6] of the band energy at the various points of solution of eqn (5) and used these to predict the correct band structure by means of linear extrapolation. To interpret the intensities of the observed ESCA peaks it is necessary to obtain density of states curves from the calculated valence band structures and to take into account the dependence of the photoionization cross-section on the atomic orbital character of the CO from which ionization occurs. To obtain the density of states $D(E)$ at energy E, from the calculated band structure, we use the formula [7]

$$D(E) = \sum_{s=1}^{\text{occ.}} \sum_{j=0}^{\pi} \delta(j,s). \quad (6)$$

Here

$$\delta(j,s) = 1, \text{ if } (E-0.125)\,\text{eV} \leqslant \varepsilon(j,s) \leqslant (E+0.125)\,\text{eV}$$

otherwise

$$\delta(j,s) = 0,$$

where $\varepsilon(j,s)$ is the energy of the sth band at point j in k-space, and the summations in eqn (6) are over all occupied bands and over $\sim$ 1000 points in k-space for each band. Such a procedure resulted in a smooth density of states curve, the values of $\varepsilon(j,s)$

which did not coincide with k-values for which eqn (5) had been solved being obtained by interpolation.

In order to predict the ESCA spectra from such density of states curves it is necessary to take account of the variation of ionization cross-section with the atomic orbital character of the band from which ionization occurs. We allow for such effects in an empirical manner, following Gelius and Siegbahn,[8] by evaluating the cross-section ($\sigma^{CO}(j, s)$) of a crystal orbital of the sth band at point j in k-space as

$$\sigma^{CO}(j, s) = \sum_A \sum_t P_{At}(j, s)\sigma_{At}. \qquad (7)$$

Here the summations are over all atoms (A) in the unit cell and all orbitals (t) on each atom. $P_{At}(j, s)$ is the gross population of orbital (A, t) in band s at point j in k-space, and σ_{At} are the atomic orbital cross-sections obtained from ref. (8). The ionization intensity at energy E ($I(E)$) is obtained by weighting eqn (6) with $\sigma^{CO}(j, s)$, thus

$$I(E) = \sum_{s=1}^{occ.} \sum_{j=0}^{\pi} \sigma^{CO}(j, s)\delta(j, s). \qquad (8)$$

EXPERIMENTAL

ESCA spectra were obtained on an A.E.I. ES200 spectrometer, of $C_{36}H_{74}$, poly(tetrafluoroethylene)(PTFE), poly(ethylene oxide) (PEO) and polyglycine (PG). The data for $C_{36}H_{74}$ was taken to be the same as for polyethylene (PE) and have been previously reported by us.[9] Commercial high molecular weight samples of the remaining polymers were used.

THEORETICAL BAND STRUCTURE AND INTERPRETATION OF EXPERIMENTAL SPECTRA

The results of the INDO calculations did not differ significantly from those using the CNDO approximations. For this reason only, the latter are described here.

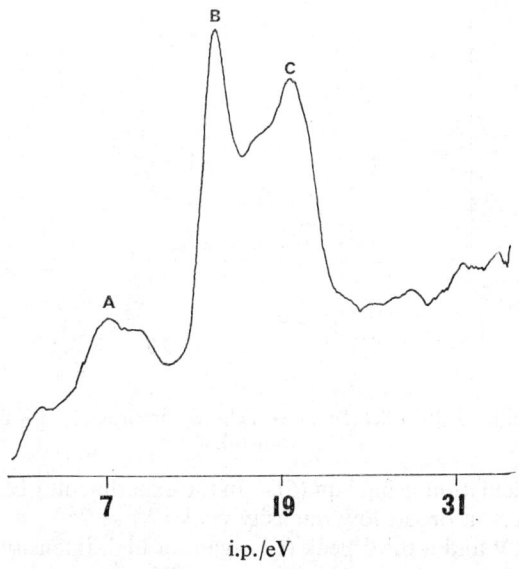

FIG. 1.—Experimental ESCA spectrum of $C_{36}H_{74}$.

POLYETHYLENE

We first discuss the interpretation of the experimental spectrum of PE, (fig. 1), examining the effect of the CNDO parameterization schemes of Sichel and Whitehead and of Pople, and the effect of including the atomic orbital cross-section in the calculation of the ESCA spectrum.

In fig. 2a we show the band structure calculated using the parameters of Sichel and Whitehead and in fig. 2b, the ESCA spectra predicted simply on the basis of a

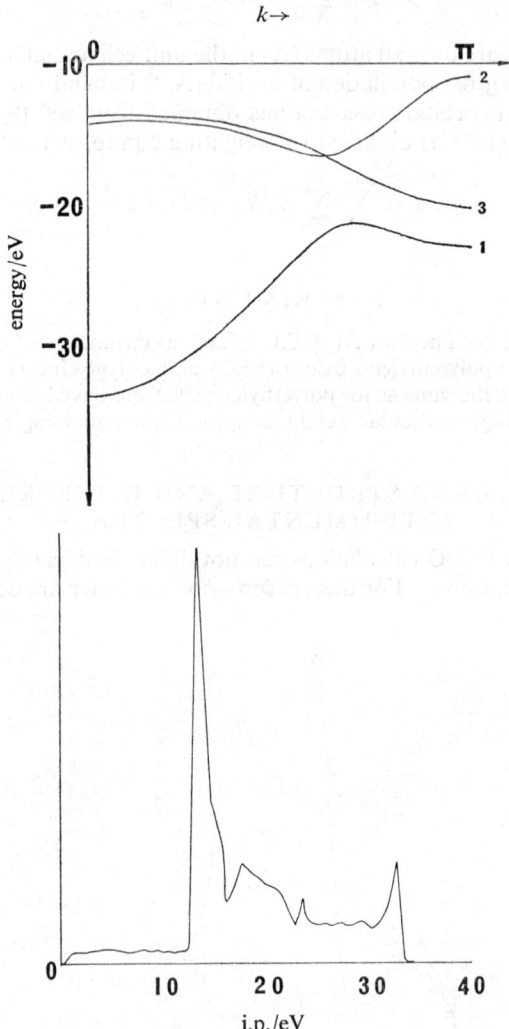

FIG. 2(a).—Band structure of PE (Sichel and Whitehead parameters). (b) Density of states curve from (a).

density of states calculation using eqn (6). In the experimental ESCA spectrum there are three main peaks, a broad low intensity peak (A) at 7 eV, a sharp peak of high intensity (B) at 14 eV and a third peak (C), again of high intensity at 20 eV. Fig. 2b can be roughly correlated with such a spectrum. The intense peak at ∼ 13 eV arises from the lower k-value regions of bands 2 and 3 (fig. 2a) and may be correlated with peak

A in the spectrum (fig. 1). The broad band near 20 eV arises from band 3 (k-values near π) and may be correlated with peak B in the ESCA spectrum. The final peak calculated to be near 30 eV which arises from band 1 (k-values near 0) may be correlated with peak C (fig. 1). Thus, although three peaks are indeed predicted in line with the ESCA spectrum, the energy separation and relative intensities of these peaks are poorly predicted using only a density of states curve. In fig. 3a we show the

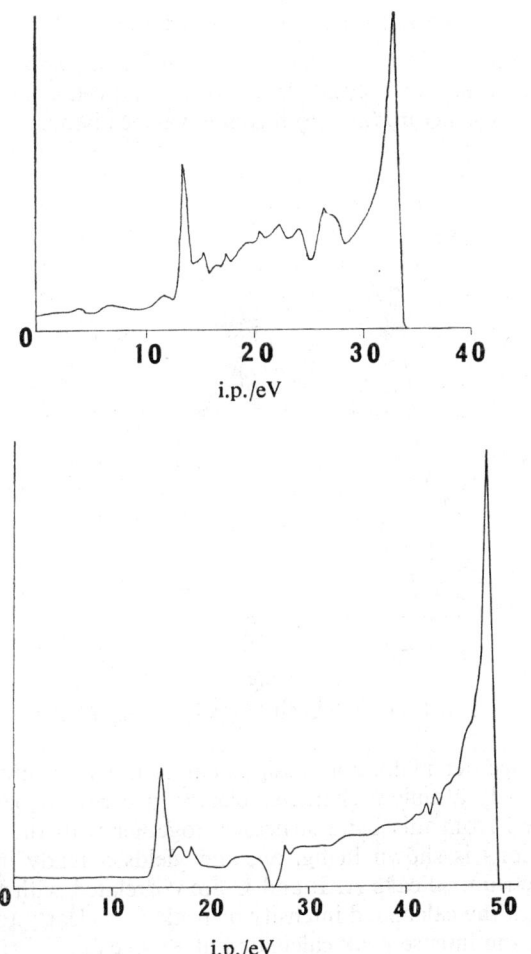

FIG. 3(a).—Predicted ESCA spectrum of $C_{36}H_{74}$, from 2(b). (b) as (a) using parameters of Pople.

effect of including CO cross-sections in our calculations (eqn (8)). As band 1 (fig. 2a) has much more C 2s character than the top two bands, and as the cross-section for s electron ionization is greater than for p ionization, the higher energy region of the predicted spectrum increases in intensity relative to that of the low energy region (fig. 3a). This predicted spectrum now bears a much closer resemblance to the experimental one than does fig. 2b. There is a distinct low-energy peak at ~ 14 eV, and a much more intense peak at ~ 34 eV. Between these peaks are a number of rather ill-defined features, but a distinct peak at ~ 28 eV is evident. This peak probably corresponds to peak B in the spectrum although it is still much too weak.

In fig. 3b we show a similar calculation to that of fig. 3a, but using the CNDO parameters of Pople.[1] Sichel and Whitehead[4] noted that for free molecules, molecular properties are generally reproduced better using their parameters. We find that the energy spread of the peaks is much larger than the experimental values using the parameters of Pople so that for calculations on the remaining polymers we show only the results for the parameterization of Sichel and Whitehead.

POLY(TETRAFLUOROETHYLENE)

The ESCA spectrum of PTFE (fig. 4) shows four main peaks, three of intermediate intensity centred at approximately 20, 25 and 31 eV and one of high intensity at 41 eV. The calculated band structure (fig. 5a) has nine valence bands, the upper seven being

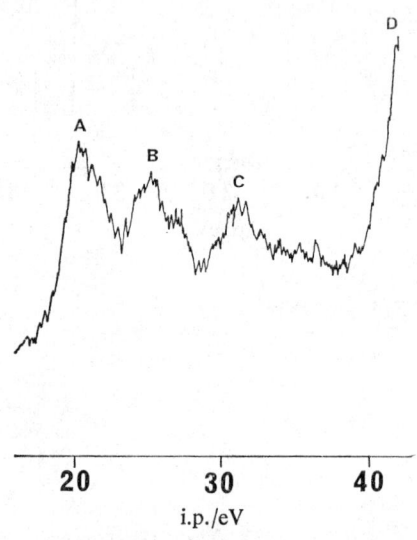

FIG. 4.—Experimental ESCA spectrum of PTFE.

closely grouped together making an assignment of the spectrum difficult. Bands 1 and 2, having mostly atomic s character clearly give rise to peak D (fig. 4). The spectrum predicted from this band structure, together with the inclusion of atomic orbital cross-sections is shown in fig. 5b, and yields a ready interpretation of the experimental spectrum. Peaks A, B and C are correlated with the first three peaks of fig. 5b, although the calculated intensity of peak A is clearly too low, and peak D is correlated with the intense peak calculated at ~ 45 eV.

POLY(ETHYLENE OXIDE)

The experimental ESCA spectrum of PEO (fig. 6) shows a broad peak envelope (A) centred at ~ 16 eV and three further peaks centred at 23, 26 and 35 eV. The calculated band structure (fig. 7a) shows nine bands whose energies are not strongly k dependent. Although the calculated spectrum (fig. 7b) shows a large number of peaks, it correlates quite well with the experimental data. The intense high energy peak calculated to be near 40 eV is correlated with experimental peak D. This peak arises mainly from band 1 (fig. 7a) which has mostly s character. The next two peaks found to lower energy (B,C) may be correlated with the next two calculated peaks at ~ 28 and ~ 22 eV and the broad peak (A) with the remaining calculated peaks.

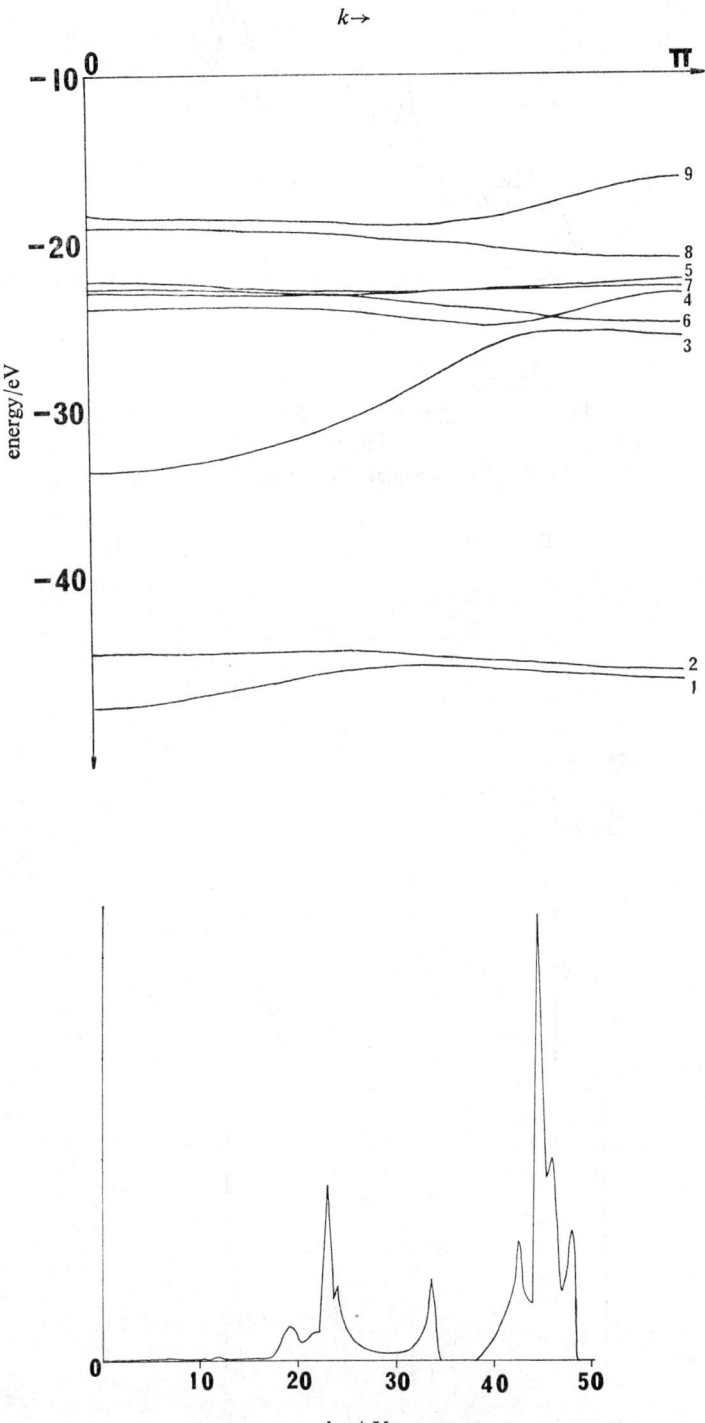

FIG. 5(a).—Band structure of PTFE. (b) Predicted ESCA spectrum of PTFE, from 5(a).

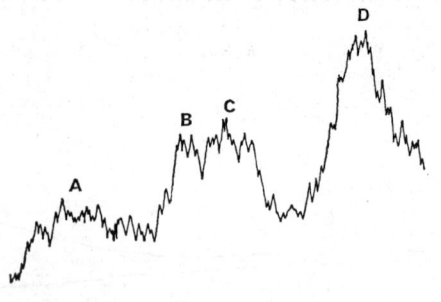

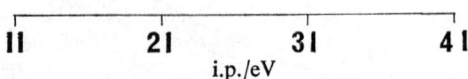

FIG. 6.—Experimental ESCA spectrum of PEO.

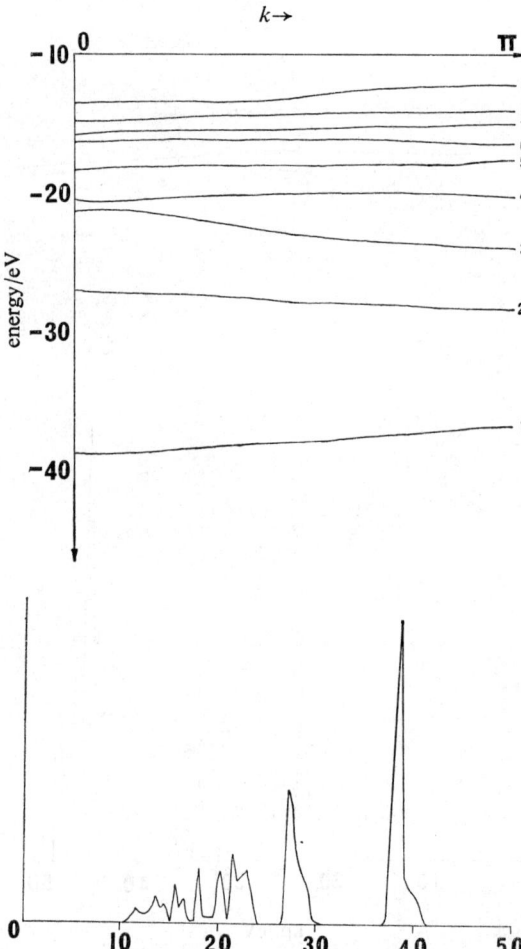

FIG. 7(a).—Calculated band structure of PEO. (b) Predicted ESCA spectrum of PEO from 7(a).

POLYGLYCINE

In view of the relative complexity of this polymer, it is expected that its ESCA spectrum is rather featureless. The spectrum shown in fig. 8 confirms this, there being a broad high energy peak (E) and a number of low energy peaks (A-D) which are just resolved. The use of the calculated band structure (fig. 9a) to interpret this spectrum is complicated by the large number of bands (11) although the high energy peak (E) is clearly associated with bands 1-3 of predominantly s character. The

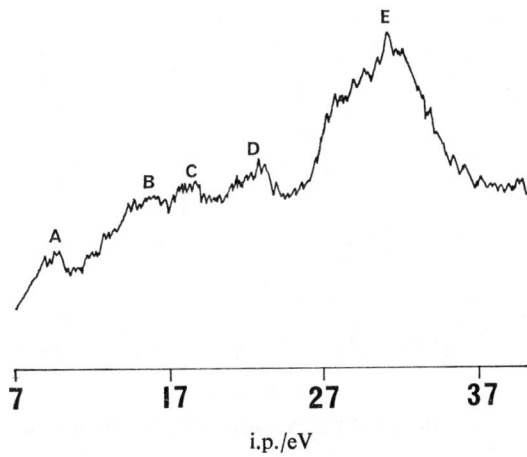

FIG. 8.—Experimental ESCA spectrum of PG.

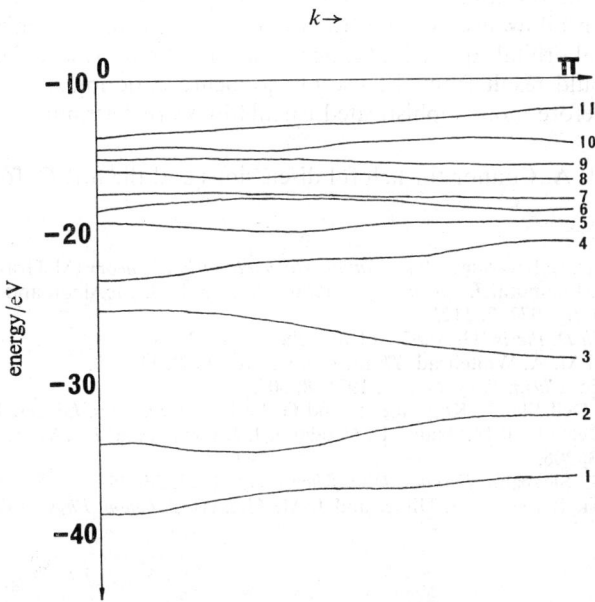

FIG. 9(a).—Band structure of PG.

calculated spectrum (fig. 9b) shows a number of closely-spaced high intensity peaks which give rise to the experimental peak E, however, between 10 and 30 eV an envelope of peaks are predicted which can only be assigned together as the experimental peaks A-D.

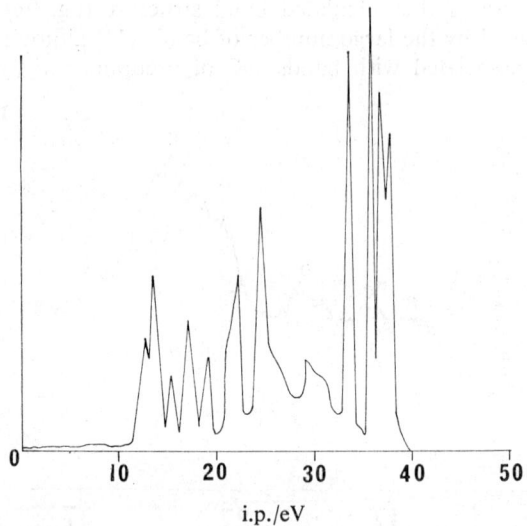

FIG. 9(b).—Predicted ESCA spectrum of PG, from 9(a).

CONCLUSIONS

The results presented here show that calculations at a semi-empirical level are quite successful in interpreting the ESCA spectra of periodic solids. This success is only achieved when allowance is made for the dependence of the ionization cross-section of the crystal orbitals on their atomic character. Better resolved experimental spectra, which should result from the use of monochromatic ionizing radiation are probably needed before more sophisticated calculations are warranted.

We thank Dr. J. A. Connor for helpful discussions and the S.R.C. for supporting this research.

[1] J. A. Pople and D. L. Beveridge, *Approximate Molecular Orbital Theory* (McGraw-Hill, 1970).
[2] H. Fujita and I. Imamura, *J. Chem. Phys.*, 1970, **53**, 4555. B. J. McAloon and P. G. Perkins, *J. C. S. Faraday II*, 1972, **7**, 1121.
[3] M. H. Wood, Ph.D. thesis (University of Manchester, 1972).
[4] J. M. Sichel and M. A. Whitehead, *Theor. Chim. Acta.*, 1968, **11**, 220.
[5] W. L. McCubbin, *Chem. Phys. Letters*, 1971, **8**, 507.
[6] J. M. Andre, J. Delhalle, G. Kapsomenos and G. Leroy, *Chem. Phys. Letters*, 1972, **14**, 485.
[7] J. Delhalle, S. Delhalle, J. M. Andre, R. Caudano, J. J. Pireaux and J. J. Verbist, *Chem. Phys. Letters*, 1973, **23**, 206.
[8] U. Gelius and K. Siegbahn, *Faraday Disc. Chem. Soc.*, 1972, **54**, 257.
[9] M. H. Wood, M. Barber, I. H. Hillier and J. M. Thomas, *J. Chem. Phys.*, 1972, **56**, 1788.

Application of ESCA to Studies of Structure and Bonding in Polymers

By D. T. Clark*, A. Dilks, J. Peeling and H. R. Thomas

Department of Chemistry, University of Durham, South Road, Durham City

Received 17th March, 1975

Shake-up phenomena in a series of *para*-substituted polystyrenes have been studied by ESCA. Comparison with other spectroscopic data and with theoretical calculations within the sudden approximation, equivalent cores model, and CNDO SCF MO formalism identifies the shake-up structures arising from $\pi^* \leftarrow \pi$ transitions involving the highest occupied and lowest unoccupied orbitals of the pendant phenyl groups. The relative intensities of different core levels and associated shake-up satellites are used to investigate the structure and surface morphology of some alkane styrene copolymers and polystyrene-polydimethylsiloxane AB block copolymers.

INTRODUCTION

In a series of papers [1] over the past few years, we have systematically investigated the application of ESCA to studies of structure and bonding in the polymer field. These studies have revealed the great potential of ESCA in many important fields of polymer chemistry and physics. In the study of surface phenomena, for example, it is apparent that ESCA has a unique role to play whilst in other areas the technique nicely complements the more established spectroscopic tools.[2] In general, however, ESCA provides data at a somewhat coarser level than most other spectroscopic tools such that short-range phenomena may be studied directly but longer-range phenomena may only be inferred indirectly. It is also apparent that in many areas of application ESCA does not compare particularly favourably in terms of resolution, sensitivity etc. with more established spectroscopic tools. The fact remains, however, that this is more than compensated by the great range of information available from a single ESCA experiment such that in the future one can envisage that ESCA will be the technique of choice for any initial investigation of a polymer sample. The ability to provide information straightforwardly on uncharacterized samples is unique to ESCA and gives the technique great potential (already exploited in some areas) for tackling not only academic problems but those of an applied " trouble shooting " nature.

The hierarchy of information levels available from ESCA experiments is detailed in table 1. The primary sources of data which have been predominantly utilized to date in the polymer field are absolute and relative binding energies and relative intensities for the core levels. The escape depth dependence on kinetic energy gives a ready means of analytical depth profiling and the advantages of studying the valence bands of polymers by ESCA have also been detailed.[1b] Much less emphasis has thus far been placed on angular and photon energy dependent studies or on multiplet effect and shake-up and shake-off phenomena. It is becoming increasingly apparent, however, that data derived from shake-up studies in particular can add a new dimension to the application of ESCA to studies of structure and bonding, particularly from the standpoint of removing ambiguities of interpretation arising from the primary sources of ESCA data.[1a] In this paper we outline some of the systematic studies of shake-up phenomena in polymers which we have recently undertaken which effectively illustrate the great potential for such investigations. These studies range from fundamental

TABLE 1.—INFORMATION FROM ESCA

CORE ELECTRONS

(1) Binding energy characteristic of a given level of a given element therefore useful for analysis. Different KE dependencies for different core levels provides means for analytical depth profiles

(2) Absolute BE may be characteristic of particular structural features (e.g., CF_3, C—O, —NH_2 etc.). " Shifts " can be related to electron distribution.

VALENCE ELECTRONS

(1) Can study valence energy levels of insulators. Densities of states for conduction bands of metals (of interest in study of metallised polymers).

(2) Studies of differential changes in cross section with photon energy provides information on symmetries of orbitals (σ, π etc.). Angular dependence studies.

(3) MULTIPLET SPLITTINGS
For paramagnetic species observation of multiplet splittings provides information on spin states of atoms or ions and distribution of unpaired electrons.

(4) Shake-up and shake-off satellites. Information on excited states of hole states.

examinations of simple substituted systems where assignments of transitions have been made both by comparison with other available spectroscopic data and by complementary theoretical studies, to the application of shake-up studies to the determination of copolymer compositions and to the investigation of the surface morphology of block copolymers.

2. SHAKE UP STUDIES IN *para*-SUBSTITUTED POLYSTYRENES

(*a*) INTRODUCTION

An area where the additional data derived from the observation of shake-up satellites can resolve ambiguities is for systems for which shifts in core binding energies are insufficient to obtain the required information. A typical example would be for hydrocarbon-based polymer systems since it is known that the absolute binding energies for carbon $1s$ levels corresponding to —$\overline{CH}$—, H—$\overline{C}$= and H—$\overline{C}$≡ type environments are closely similar.[3] Fig. 1 shows the $\overline{C}$ $1s$ spectra for typical saturated polymers, polyethylene and polydimethylsiloxane and polystyrene and polydiphenylsiloxane which represent prototype systems with saturated backbones and unsaturated pendant groups. For the latter, well developed shake-up structures are apparent which clearly distinguish them from the saturated systems although the differences in lineshape and linewidths for the main photo-ionization peaks are closely similar.

(For the siloxanes of course, the relative intensities of the C 1s with respect to the O 1s and Si 2p levels may be used to effect a ready distinction between the two siloxanes).

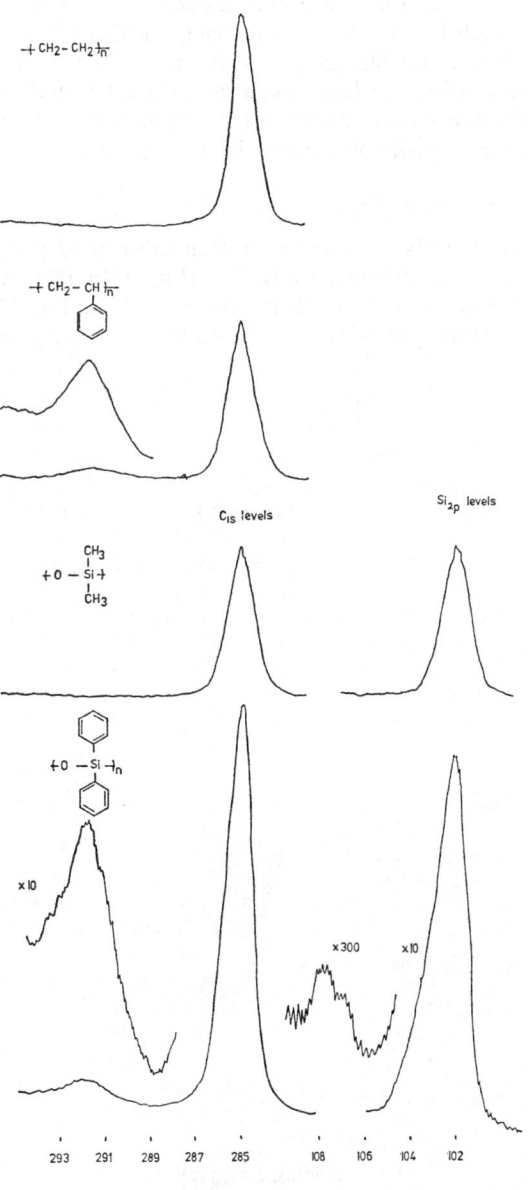

Fig. 1.—ESCA spectra of polyethylene, polystyrene, polydimethylsiloxane and polydiphenylsiloxane showing shake-up structure.

As a point of practical detail it is worthwhile emphasising that in studying solids by ESCA with an unmonochromatized X-ray source, low intensity transitions arising from shake-up or shake-off process of energy > 10 eV are often largely masked by the tail arising from inelastic scattering events. Low energy shake-up transitions arising from, e.g., $\pi^* \leftarrow \pi$ transitions (energies typically < 10 eV) are therefore particularly favourable for study. It will become apparent that the transitions giving rise to the satellite structure in PS and PDPS are indeed due to $\pi^* \leftarrow \pi$ transitions and it may also be shown that such transitions are effectively localized in a given phenyl group. To elucidate the nature of the shake-up transitions, as a preliminary to utilizing such data for structural studies, we have made a systematic study of para-substituted polystyrenes and the experimental studies have been complemented by a theoretical investigation of shake-up probabilities within the sudden approximation.

(b) RESULTS AND DISCUSSION

The measured core levels for a series of para-substituted polystyrenes are shown in fig. 2 and 3 where the substituents are H, Me, tBu, OMe, NH_2, Cl and Br. Several features are evident from these data. First, the overall band profiles for the shake-up satellites for the C 1s spectra are similar. Secondly, the separation between the shake

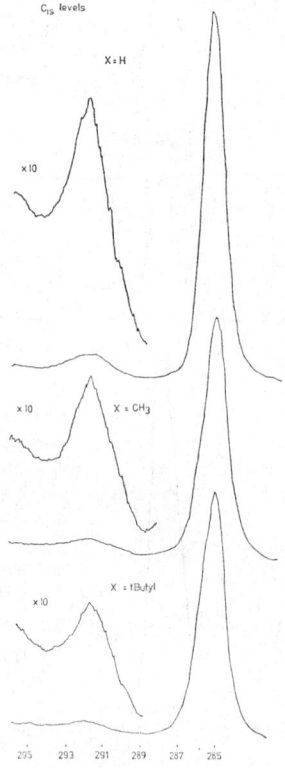

FIG. 2.—ESCA spectra for *para*-substituted polystyrenes (X = H, CH_3, t-butyl) showing shake-up structure.

up satellites and the main photoionization peaks are within a very narrow range (6.6±0.2 eV) for all core levels. Finally, the most striking feature is the very strong dependence of the shake-up intensities on the substituent. Table 2 shows the relevant data and it is clear that the shake-up intensity for the C 1s levels decrease as a function of the increasing π electron donating power of the para-substituent. The trend for the shake-up satellites associated with core levels of the substituent on the other hand exhibit a clear trend in the opposite sense. This becomes more evident on consideration of the ratio of shake-up intensities for the different core levels as indicated in

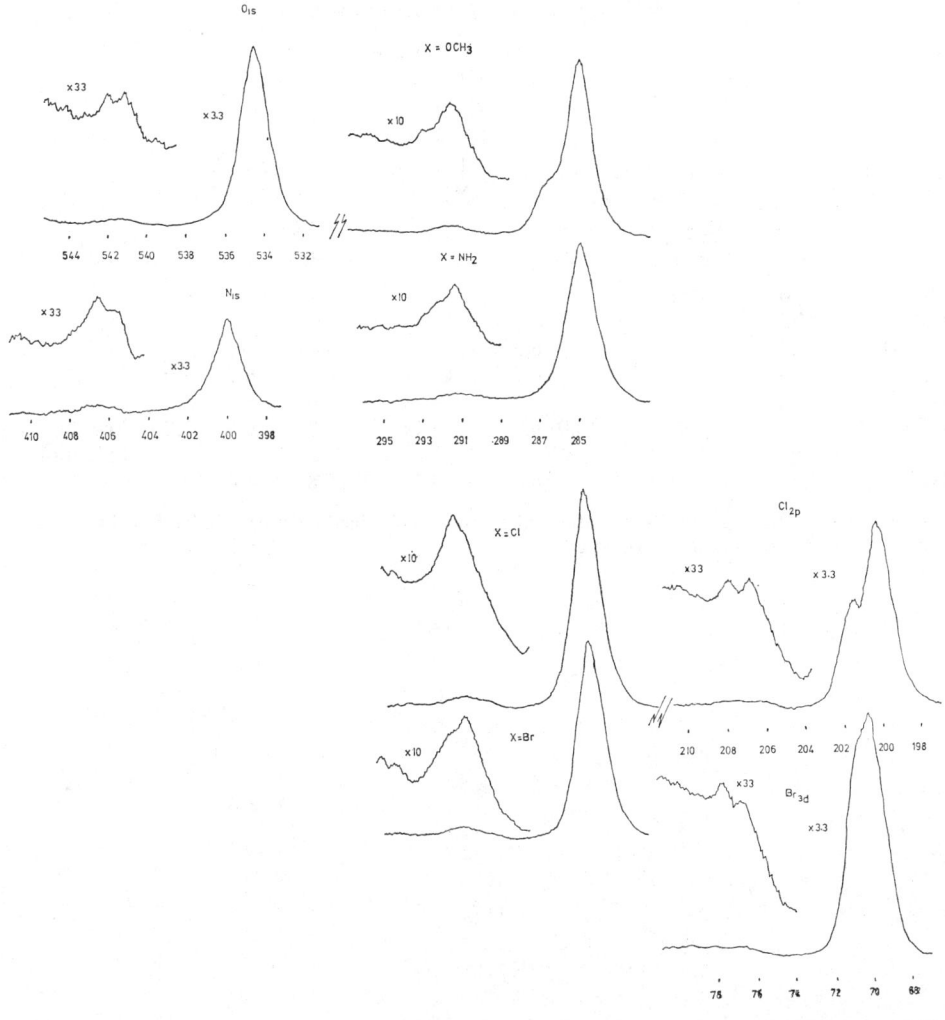

FIG. 3.—ESCA spectra for *para*-substituted polystyrenes (X = OCH$_3$, NH$_2$, Cl, Br) showing shake-up structure.

the final column of table 2. The intensity ratio for the methoxy compound is almost certainly a lower limit since there is some evidence from the O 1s spectra that there is a small amount of surface oxidation.

The most obvious interpretation of these data is that the shake up satellites arise from $\pi^* \leftarrow \pi$ transitions accompanying core ionization. This is strongly reinforced when direct comparison is made of data pertaining to the corresponding transitions for the neutral molecule (which follow dipole as opposed to monopole selection rules).

Fig. 4 shows such a comparison with Platts' spectroscopic moment and with the coulomb integral of the substituent (appropriate to a localized orbital model for discussing the electronic spectra of substituted benzenes).[4] The correlations leave little room for doubt that the satellites are in fact due to $\pi^* \leftarrow \pi$ transitions.

TABLE 2.—SHAKE-UP PHENOMENA IN POLYMERS

$-(-CH_2-CH-)_n$
|
(ring)
|
X

X	area ratios	C 1s levels shake up % intensity [a]	ΔE [b]		X core levels shake up intensity	ΔE [b]	relative shake up intensities (%X/%C) [c]
H	13.0	8.1	6.6	—			
CH_3	14.9	7.0	6.5	—			
t-butyl	21.7	6.5	6.5	—			
Cl	17.0	6.3	6.7	Cl 2p	2.7	6.7	3.0
Br	18.2	5.9	6.6	Br 3d	1.9	6.8	2.3
OCH_3	33.3	3.7	(8.0) 6.7	O 1s	2.0	6.6	3.8 ← low due to oxidation
NH_2	33.3	3.3	6.6	N 1s	5.8	6.4	12.3

a with respect to ring and directly attached carbons, *b* obtained from centroids of peaks (in eV), *c* standardized to equal numbers of core levels.

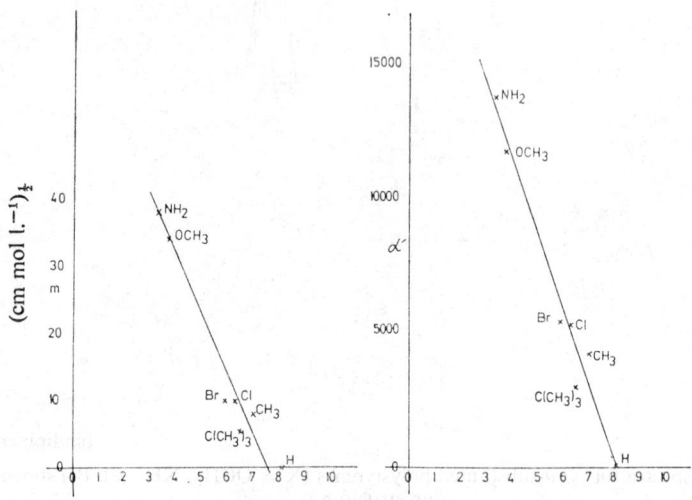

FIG. 4.—Plot of Platts' spectroscopic moment and Coulomb integral of substituent versus % shake-up intensity of C 1s levels of *para*-substituted polystyrenes.

The theoretical interpretation of the data pertaining to the intensity of the shake-up transitions has been within the sudden approximation and CNDO SCF MO formalism. Within this approximation the probability of observing a transition from occupied orbital ψ_i corresponding to an unrelaxed valence electron in the hole state to a virtual orbital ψ_f for the hole state is proportional to $|\langle \psi_i | \psi_f \rangle |^2$. The virtual orbitals for the hole states were simulated by taking the appropriate equivalent cores species.

It is a comparatively straightforward matter to establish that on the ESCA depth scale the surface morphology of the polymers studied are such that the repeat units are statistically sampled and also that the transitions are effectively localized within a given pendant group. This being the case the model systems for the theoretical computations were taken to be the para-substituted toluenes.

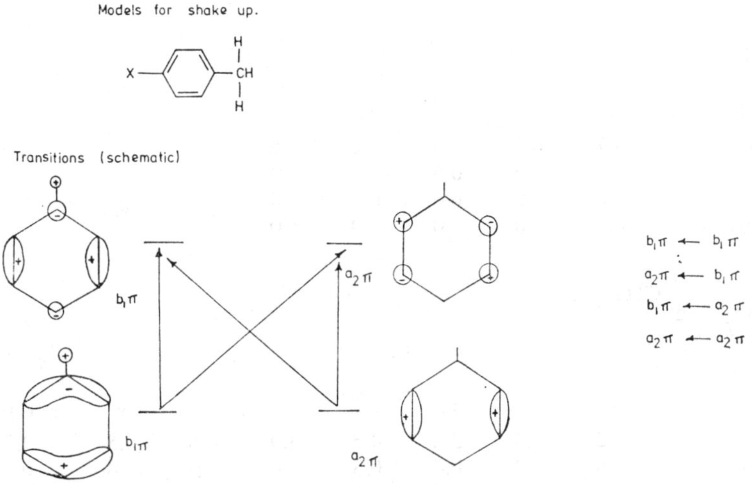

FIG. 5.—Schematic of shake-up transitions involving highest occupied and lowest unoccupied orbitals for model systems.

Fig. 5 shows the four one-electron transitions involving the two highest occupied and lowest unoccupied orbitals (symmetries designated with respect to approximate local C_{2v} symmetry of the pi electron system). Of the four transitions those arising from the $b_{1\pi} \leftarrow a_{2\pi}$ and $a_{2\pi} \leftarrow b_{1\pi}$ excitations are formally monopole forbidden. However, the strong perturbation consequent upon removal of a core electron effectively removes the symmetry restriction except for hole states corresponding to photo-ionization from core levels associated with atoms located on C_2 axes (e.g., C1, C2, C5 and X). The calculated shake-up probabilities for the two extremes in the series (X = H, OMe) are shown in table 3. Two features are evident. First, the intensity for the low energy shake-up satellites to the C 1s spectra largely derives from two transitions ($b_{1\pi} \leftarrow b_{1\pi}$ and $b_{1\pi} \leftarrow a_{2\pi}$) and indeed the distinct asymmetry of the satellites would tend to confirm that more than one transition is involved. Secondly, when hydrogen is replaced by a pi electron donating substituent, whilst the intensity of the $b_{1\pi} \leftarrow b_{1\pi}$ transition is predicted to remain essentially the same that for the $b_{1\pi} \leftarrow a_{2\pi}$ transition is calculated to decrease. The computed net overall decrease in shake-up intensity predicted by the model is therefore qualitatively in agreement with experiment. On an absolute scale the calculated transition probabilities are somewhat smaller than experimentally determined. This is, to some extent at least, a function of the parameterization particularly for the equivalent cores species and agreement could no doubt be improved by judicious optimization of parameters. Since our main concern has been to understand qualitatively the data no attempt has been made improve matters in this direction.

STUDIES OF STRUCTURE AND BONDING IN POLYMER

The calculated data for the other model systems is illustrated schematically in fig. 6. With the single exception of NH$_2$ as substituent the correlations involving $+M$, $+I_\pi$ substituents suggest that the $b_{1\pi} \leftarrow b_{1\pi}$ transition probability remains essentially constant in the series X = H, Me, tBu, OMe, Cl, Br whilst that associated with the $b_{1\pi} \leftarrow a_{2\pi}$ transitions decreases. As a matter of some interest calculations have also

TABLE 3.—CALCULATED SHAKE-UP PROBABILITIES FOR MODEL COMPOUNDS

X=H

transition	1	2	% shake-up 3,7	4,6	5	total C 1s
$b_{1u} \leftarrow b_{1\pi}$	3.7	6.2	2.0	2.8	3.6	2.6
$a_{2\pi} \leftarrow b_{1\pi}$	0	0	0.6	0	0	0.1
$b_{1\pi} \leftarrow a_{2\pi}$	0	0	6.0	6.6	0	1.8
$a_{2\pi} \leftarrow a_{2\pi}$	0.1	0	0	0	0	0

X = OCH$_3$

transition	1	2	% shake-up 3,7	4,6	5	8	9	total C 1s	O 1s
$b_{1\pi} \leftarrow b_{1\pi}$	3.2	7.3	2.2	4.2	3.9	0.6	0.5	2.7	0.6
$a_{2\pi} \leftarrow b_{1\pi}$	0	0	0.8	0.3	0	0	0	0.1	0
$b_{1\pi} \leftarrow a_{2\pi}$	0	0	5.2	4.4	0	0	0.1	1.2	0
$a_{2\pi} \leftarrow a_{2\pi}$	0.1	0	0	0	0	0	0	0	0

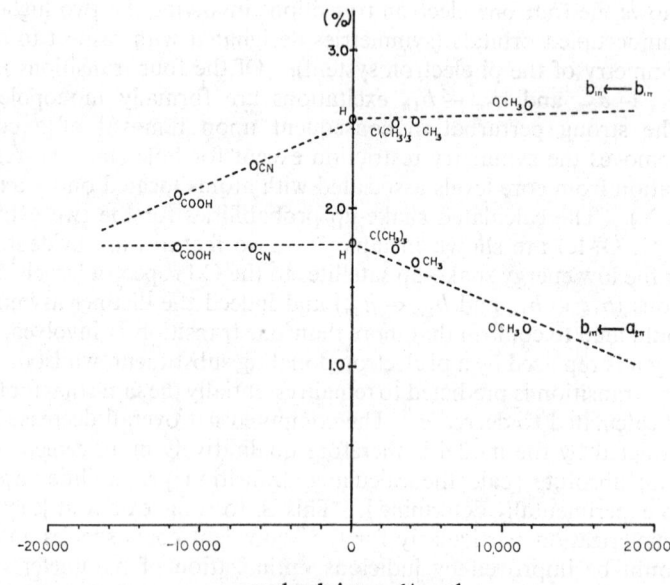

FIG. 6.—Calculated shake-up probabilities for model systems as a function of substituent.

been made on some typically electron accepting substituents (e.g., COOH) and the situation is now seen to be the reverse of that previously described for electron donating substituents. Thus the $b_{1\pi} \leftarrow b_{1\pi}$ transition decreases in intensity as the substituent increases in electron accepting power whereas the $b_{1\pi} \leftarrow a_{2\pi}$ transition is calculated to have roughly the same intensity as for the parent system. Although the corresponding polymers have not yet been studied, investigations of the relevant para-substituted toluenes confirms that the shake-up intensity for the C 1s levels decreases for substituents which are either good π electron donors or acceptors in complete agreement with the theoretical models.

The experimental observations and the correlation with other spectroscopic parameters taken together with the theoretical calculations are internally self-consistent in attributing the observed low energy shake-up satellites in the para-substituted polystyrenes to transitions involving the highest occupied and lowest unoccupied pi orbitals of the pendant substituted phenyl rings. This provides a sound basis for using such phenomena as an aid to structural studies in the polymer field of which two examples are described in the succeeding sections.

3. SHAKE UP PHENOMENA IN ALKANE-STYRENE COPOLYMERS; AN INVESTIGATION OF COPOLYMER COMPOSITIONS AND SURFACE MORPHOLOGIES

(a) INTRODUCTION

In this section we outline an investigation of some alkane-styrene copolymers of general formula,[5]

$$[CH-CH_2CH_2-CH-(CH_2)_n]_m$$
         | |
         Ph   Ph

$n = 0, 1, 3, 5, 6, 10$

which illustrate the utility of ESCA for studying copolymer compositions in systems for which the primary sources of information are of themselves insufficient and for which the extra dimension is provided by observation of shake-up satellites.

(b) RESULTS AND DISCUSSIONS

It should be evident from the previous discussion that a characteristic low energy shake up structure accompanying direct photoionization of the C 1s levels should be specifically associated with the styrene component. Comparison of the data for toluene (studied as a thin film) with that for polystyrene readily confirms that the shake-up transitions are effectively localised within a given pendant phenyl group and on this basis therefore the shake-up phenomena might be expected to be additive in the absence of any specific chain orientation effects.*

Fig. 6 shows the measured C 1s levels and shake-up satellites for the alkane-styrene copolymers and it is evident by visual inspection that the relative intensities of the shake-up satellites with respect to the main photo-ionization peaks decrease

---

* Our extensive studies of surface fluorination of polyethylene suggests [2] a typical electron mean free path of $\sim$10 Å for electrons of kinetic energy of $\sim$968 eV. This being the case $\sim$50% of the intensity of the elastic peak derives from the topmost 7 Å whilst $\sim$90% of the signal areas from the outermost $\sim$27 Å.

with increasing chain length of the alkane component. The measured intensities and energy separations are given in table 4.

A trend exists between shake-up intensity and the chain length of the alkane component and the structure of the shake-up satellites and the energy separations remain essentially constant. A least squares plot of the intensity ratio of the direct photo-ionization peak to shake-up satellites versus n (the chain length of the alkane component) gives a correlation coefficient of 0.997 the slope being 1.91 and intercept 12.91. The latter may be compared with the measured value of 13.4 for the parent system (i.e., polystyrene). Taking the measured shake-up intensities for polystyrene and the repeat unit for a given copolymer, the calculated slope assuming an additive model is 0.90. The fact that an additive model applies to the experimental data but with a much larger dependence of intensity on n than predicted theoretically would strongly suggest that there are specific orientation effects of the polymer chains in the topmost ~ 50 Å probed by ESCA studies of the C $1s$ levels. Two models can be envisaged which form a basis for rationalizing these data and are discussed in detail elsewhere.[6]

TABLE 4.—ESCA DATA FOR THE ALKANE-STYRENE COPOLYMERS

	binding energies/eV*			
n	C $1s$	C_{1s}^S	Δ	area ratios (C $1s/C_{1s}^S$)†
0	285.0	291.6	6.6	13.4
1	285.0	291.6	6.6	14.5
3	285.0	291.6	6.6	18.5
5	285.0	291.6	6.6	22.7
6	285.0	291.6	6.6	23.8
10	285.0	291.6	6.6	32.2

* relative to C $1s$ at 285.0 eV, Δ given with respect to centroid of asymmetric shake-up peak.
† ratio of main photo-ionization peak to shake-up peak.

4. SURFACE STRUCTURES OF AB BLOCK COPOLYMERS OF POLYDIMETHYLSILOXANE AND POLYSTYRENE

(a) INTRODUCTION

We have previously demonstrated that the surface sensitivity of ESCA can be turned to good advantage in delineating surface from sub-surface and bulk structures in inhomogeneous films.[2] An obvious extension of this work would be in the investigation of the surface structure of block copolymers in which the possibility of different morphologies for the bulk material exists. In the following section we show how a consideration of the relative intensities of the elastic peaks for core ionizations and of accompanying shake-up satellites for block copolymers of PS and PDMS may be utilized to provide detailed information on their surface structures.

AB block copolymers frequently exhibit phase separation which typically gives rise to a dispersed phase consisting of one block type in a continuous matrix of the second block type. The detailed morphology of the domain structure of block copolymers depends upon such factors as the relative proportions of the two components, the molecular weight, the thermal and physical history of the polymer and for solvent-cast films, upon the solvent and temperature. A variety of experimental techniques have been employed in investigations of the domain structures of block copolymers with the predominant emphasis being on the bulk morphology. By contrast, the detailed structure of the surface of such polymer systems has been subject to much less detailed consideration. Since many properties of a polymer, however, depend on the detailed structure of the surface and since the latter may be considerably different from the bulk, any technique which can in principle differentiate surface from bulk effects is likely to be of some considerable importance in this area.

Solvent-cast films of PDMS-PS AB block copolymers have been extensively studied by electron microscopy in recent years.[7, 8] The extreme incompatibility of the two molecular segments has been shown to result in the formation of discrete and continuous regions and a variety of domain structures have been observed for the

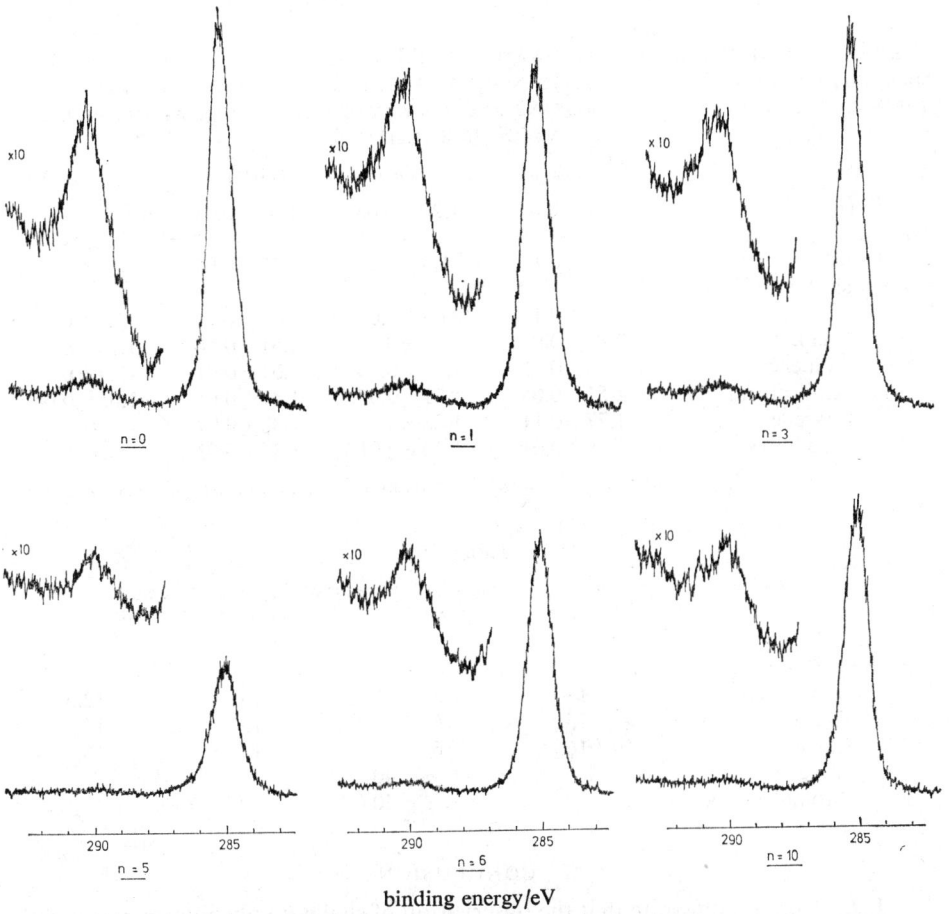

FIG. 7.—C $1s$ levels for some alkane styrene copolymers.

polymer bulk. The large differences in surface free energy of the two components makes this copolymer system of particular interest with respect to the surface structure.

Two block copolymers have been studied. Copolymer 1 contained 23 weight % PS and copolymer 2,59 weight % PS, with corresponding number average molecular weights of 121,000 and 124,000 respectively. Films were cast from a variety of solvents and the relevant data are given in table 5. It is clear both from the relative intensities of the direct photoionization peaks and the shake-up satellites (specific to the styrene component) that in all cases the immediate surfaces of the films independent of the bulk morphologies are essentially pure PDMS. With a knowledge of the escape depths appropriate to photo-ionization from the O $1s$, C $1s$ and Si $2p$ core levels [2] and data pertaining to the parent homopolymers, it is possible to establish the thickness of the surface PDMS layer for the PS-PDMS block copolymers. The data are shown in table 6. The results are internally self-consistent and show that the thickness of the polydimethylsiloxane outer layer varies between ~ 13 to > 40 Å depending on the method of preparation of the films of copolymer 2. The exact correspondence of the data for copolymer 1 and that for the PDMS homopolymer shows that the surface to $> \sim 40$ Å is essentially pure PDMS.

TABLE 5.—EXPERIMENTAL INTENSITY IN RATIOS THE ESCA SPECTRA OF FILMS OF POLYDIMETHYLSILOXANE, POLYSTYRENE AND POLYSTYRENE-POLYDIMETHYLSILOXANE AB BLOCK COPOLYMERS. THE VALUES FOR POLYDIMETHYLSILOXANE AND POLYSTYRENE ARE TAKEN AS THE REFERENCE VALUES (SEE TEXT)

polymer film	$I_{C\,1s}/I_{Si\,2p}$	$I_{C\,1s}/I_{O\,1s}$	$I_{O\,1s}/I_{Si\,2p}$	$(I_{shake\,up}/I_{C\,1s}) \times 100$
PDMS (I_∞ values)	1.66±0.05*	1.20±0.05	1.38±0.05	—
PS	—	—	—	6.94
copolymer 1	1.7±0.1	1.3±0.1	1.35±0.05	—
copolymer 2 cast from				
cyclohexane	3.1±0.1	2.05±0.05	1.5±0.1	1.8±0.1
benzene	2.83±0.03	2.87±0.10	1.51±0.10	1.75±0.10
toluene	2.63±0.05	1.73±0.04	1.51±0.01	1.4±0.1
chloroform	2.51±0.05	1.71±0.05	1.47±0.05	1.05±0.05
styrene	1.77±0.11	1.28±0.08	1.38±0.01	—
bromobenzene	1.68±0.01	1.24±0.01	1.36±0.03	—

* Errors are standard deviations in measurements on a number of spectra.

TABLE 6

polymer film	\multicolumn{4}{c}{thickness (in Å) of PDMS surface calculated from}			
	$I_{C\,1s}/I_{Si\,2p}$	$I_{C\,1s}/I_{O\,1s}$	$I_{O\,1s}/I_{Si\,2p}$	$I_{shake\,up}/I_{C\,1s}$
copolymer 1		≳ 40		
copolymer 2 cast from				
cyclohexane	13	13	13	12.5
benzene	15	15	13	14
toluene	16	16	13	16
styrene		≳ 40		
bromobenzene		≳ 40		

5. CONCLUSIONS

These examples illustrate that the observation of shake-up phenomena can greatly extend the scope of ESCA in studying structure and bonding in polymer systems, particularly those in which the primary sources of ESCA data are of themselves insufficient to provide the required information.

Thanks are due to S.R.C. for provision of equipment and for a research studentship to one of us (A. D.). Thanks are also due to the National Research Council of Canada for the award of a research fellowship (J. P.) and to Xerox Corporation, Webster, N.Y. for a maintenance grant (H. R. T.).

[1] (a) R. D. Chambers, D. T. Clark, D. Kilcast and S. Partington, *J. Polymer Sci.*, (*Polymer Chem. Edn.*), 1974, **12**, 1647 and references therein; (b) D. T. Clark, *The Application of ESCA to Studies of Structure and Bonding in Polymers* in *Structural Studies of Macromolecules by Spectroscopic Methods*, ed. K. J. Ivin, (J. Wiley and Sons Ltd.), in press.

[2] cf. ref. 1(b) and D. T. Clark, W. J. Feast, I. Ritchie and W. K. R. Musgrave, *J. Polymer Sci., Polymer Chem. Edn.* 1975, **13**, 857.
[3] D. T. Clark, I. W. Scanlan and J. Muller, *Theor. Chim. Acta*, 1974, **35**, 341.
[4] (a) J. R. Platt, *J. Chem. Phys.*, 1951, **19**, 263, 1418; (b) J. N. Murrell, *The Theory of the Electronic Spectra of Organic Molecules*, (Methuen and Co. Ltd., London, 1963); (c) H. H. Jaffe and M. Orchin, *Theory and Applications of Ultraviolet Spectroscopy*, (John Wiley and Sons Inc., N.Y., 1962).
[5] D. H. Richards, N. F. Scilly and F. J. Williams, *Polymer*, 1969, **10**, 603. (We are indebted to Dr. D. H. Richards, E.R.D.E., Waltham Abbey for providing these samples).
[6] D. T. Clark and A. Dilks, to be submitted for publication.
[7] J. C. Saam, D. J. Gordon and S. Lindsey, *Macromolecules*, 1970, **3**, 1.
[8] J. C. Saam and F. W. G. Featon, *Ind. Eng. Chem. Prod. Res. Dev.*, 1971, **10**, 10.

GENERAL DISCUSSION

Dr. T. A. Carlson (*Oak Ridge*) said: First, I should like to applaud the work of Clark and his co-workers for developing another practical use of ESCA, which is to say the evaluation of double bonding in polymers through the study of satellite structure due to electron shake up. As a supplement to their work on solids I should like to mention studies of satellite structure of a series of gaseous alkenes, which Dress, Haggerty and I have recently completed. Preliminary analysis of the data shows that the intensity of the low-energy satellite peak due to π orbital excitation is qualitatively explained by the number of carbon-carbon double bonds relative to the total number of carbon atoms. In the case of butadiene, two relatively sharp satellites of nearly equal intensity are found.

I should, however, like to give a warning when interpreting the satellite peaks in the photoelectron spectra of solids. In gases it is found that sharp peaks similar in energy and appearance are found to arise from characteristic energy losses. For gases, corrections arising from characteristic energy losses can be made in a straightforward manner; in solids, it is not so simple. Although I believe that the bulk of the data taken by Clark *et al.* can only be interpreted in terms of electron shake up, analysis of many of the lower intensity peaks may be observed by characteristic energy losses, and experiments ought to be designed for measuring electron loss spectra for the appropriate solids.

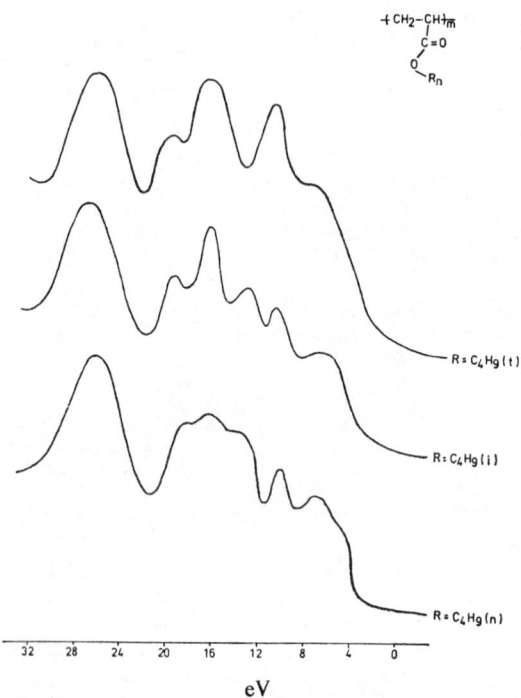

FIG. 1.—Valence levels for isomeric polybutyl acrylates measured with $MgK\alpha_{1,2}$ photon source. (Binding energies in eV).

GENERAL DISCUSSION

Dr. D. T. Clark (*University of Durham*) said: I would like to point out that for many polymer systems the interpretation and assignment of the valence bands may straightforwardly be accomplished by direct comparison with appropriate model systems incorporating the relevant repeat units.[1,2] For such model system UPS studies using He(I), He(II) are often available and the differential changes in cross section with photon energy may greatly aid in the assignment. For sufficiently simple model systems it is also possible to complete the assignment by reference to non-empirical calculations within the Hartree Fock formalism on the neutral molecule and relevant hole states.

It should also be emphasized that the valence energy levels are extremely useful in distinguishing structurally isomeric polymer systems for which the core level spectra in appropriate cases can appear identical. As a simple illustration of this, fig. 1 shows the valence energy levels of a series of polybutyl acrylates. Comparison with model systems allows a ready identification to be made of the structural isomers.

Dr. I. H. Hillier (*Manchester*) said: Care should be exercised in the use of CNDO and other semi-empirical molecular orbital methods to interpret shake-up states. Such calculations do not take explicit account of the unpaired core electron, and hence do not satisfactorily describe the *two* linearly independent doublets arising from each valence orbital transition which normally yield two peaks in the shake-up spectrum.[3]

Dr. S. Evans (*Aberystwyth*) said: We are concerned about the assumption in the paper by Clark *et al.* that a suitable value for the electron inelastic mean free path (λ) in polymers is ~ 10 Å at 968 eV. We are currently engaged [4] in measuring a number

TABLE 1.—INELASTIC MEAN FREE PATHS FOR 1 000 eV ELECTRONS [4]

material	λ/Å, relative to $\lambda_{Au} = 20.2$ Å [5]
stearic acid	103
graphite fluoride	80
polyethylene	100
2-bromostearic acid	94
polytetrafluoroethylene	92
silicon	~ 60
graphite	44
diamond	24
copper	24
silver	21
iron	17
nickel	14

[1] D. T. Clark and D. Kilcast, *Nature Phys. Sci.*, 1971, **233**, 77.
[2] D. T. Clark *Structure and Bonding in Polymers as Revealed by ESCA*, in *Electronic Structures of Molecular Solids and Crystals*, ed. J. M. Andre and J. Ladik, Proceedings of NATO Advanced Study Institute, Namur, September 1974 (D. Reidel, Dordrecht, Holland), in press.
[3] I. H. Hillier and J. Kendrick, *J. Electron Spectr.*, 1975, **6**, 325.
[4] S. Evans, R. G. Pritchard and J. M. Thomas, in preparation.
    P. Cadman, J. D. Scott and J. M. Thomas, *Chem. Comm.*, 1975, 654.
    P. Cadman, J. D. Scott and J. M. Thomas, in preparation.
[5] (*a*) M. Klasson, J. Hedman, A. Berndtsson, R. Nilsson, C. Nordling and P. Melnik, *Phys. Scripta*, 1972, **5**, 93;
    (*b*) Y. Baer, P. F. Heden, J. Hedman, M. Klasson and C. Nordling, *Solid State Comm.*, 1970, **8**, 1479;
    (*c*) B. L. Henke, *Phys. Rev. A.*, 1972, **6**, 94.

of mean free paths both to aid the quantitative interpretation of XPS adsorption experiments and, we hope, to gain a more fundamental insight into the various factors governing their magnitude. Our method is essentially similar to that described by Carlson and McGuire [1] and is fully described, with reference to the determination of λ in graphite and silver, elsewhere.[2] Other data obtained more recently are included in the table: it can be seen that for electrons of similar energy there is a variation in λ of almost an order of magnitude, hydrocarbon polymers having $\lambda \sim 100$ Å, rather than 10 Å. The effect of this on the data reported by Clark et al. (table 6) would be to increase the depths suggested approximately ten-fold. Although all our values are relative to the value for gold, any overall scaling-down would result in unacceptably short mean paths for the transition elements. Moreover, the data agree quite well with other determinations for the stearic acids,* silicon,[4] and (after extrapolation) silver,[5] although there is a discrepancy (discussed elsewhere [2]) for graphite. It is also clearly unwise to assume, as is often the case,† that λ is independent of the substrate, although there are classes of material, such as polymers, where variations may be rather small.

Dr. D. T. Clark (*Durham*) said: In general I would agree with Carlson that one must be aware of the likely contribution to satellite structure arising from characteristic energy losses when studying solids. In our particular work, however, it is clear that the dominant contribution to the low energy satellite structure arises from shake-up processes since direct comparison can be made with model systems studied in the gas phase.

Whilst we agree with Hillier that a model of shake-up intensities based on the sudden approximation, equivalent cores concept and employing the CNDO/2 SCF MO formalism cannot give a quantitative description of such phenomena, we have only sought to interpret gross trends and differences. The qualitative features of substituent effects are clearly encompassed by such a model and provide a basis for rationalizing the data. With the computational facilities currently available a series of higher level calculations encompassing multiplet effects would not be feasible. Our approach therefore follows the maxim of Occam's razor.

Arising out of comments made by Evans, I would point out that in ref. (2) of our paper we have set out in detail the logic of our analysis which leads us to the conclusion that in a typical polymer the mean free path for 968 eV electrons is ~ 10 Å. The analysis, of course, only provides ratios of mean free paths and our criteria for assigning this particular value are clearly set out in our previous papers. The data produced by Evans et al. do little to alter our current view of the situation and we believe that mean free paths in organic materials are close to our initial estimate, and it is interesting to speculate on the reason for this. Before considering this aspect we may note that mean free paths of ~ 100 Å would be difficult to reconcile with other observations. For example, mean free paths of this order imply that $\sim 90\%$ of the signal intensity in a polymeric film derives from the outermost ~ 300 Å. In a polymer sample 10 μm thick a " surface " coating of this order of magnitude would

[1] T. A. Carlson and G. McGuire, *J. Electr. Spectr.*, 1972, **1**, 161.
[2] P. Cadman, S. Evans, J. D. Scott and J. M. Thomas, *J.C.S. Faraday II*, 1975, **71**, 1777.
[3] K. Siegbahn et al., *ESCA*, (Almqvist and Wiksell, Uppsala, 1967).
[4] M. Klasson, A. Berndtsson, J. Hedman, R. Nilsson, R. Nyholm and C. Nordling, *J. Electr. Spectr.*, 1974, **3**, 427.
[5] C. J. Powell, *Surface Sci.*, 1974, **44**, 29.

* Analysis of the data in ref. (4) gives values of 40-130 Å for stearic and iodostearic acids.
† See, for example, Somorjai's paper at this Discussion.

be detectable by double beam infra-red spectroscopy and also by multiple attenuated total internal reflectance spectroscopy using single crystal germanium. As our investigation of the surface fluorination of polyethylene has shown, such is not the case. A similar problem would arise in attempting to interpret data on the CASING of polymers.[1] In our extensive studies [2] of fluorographite the detection of the tertiary CF sites and CF_2 sites at the prismatic edges compared with the tertiary CF sites arising from fluorination in the basal plane is readily understandable in terms of the distribution of crystallite sizes for the shorter mean free paths. ESCA data relating to oriented surfactant systems at surfaces would also be difficult to understand in terms of mean free paths some order of magnitude larger than our own work would tend to indicate.[3] In this connection the data in table 1 of Evans' comments are largely meaningless, notwithstanding the method by which they were established, solely on the basis that the samples are not characterized. For example do the data for polyethylene refer to high or low density material? Was any effect observed depending on the porosity of the PTFE samples employed? What were the compositions of the graphite fluorides used? The results in table 1 would imply a drastically different overall signal level for graphite with respect to graphite fluorides which is not observed in our own work.[2]

It seems clear, therefore, that this is an area of some controversy and we have now completed the construction of the requisite instrumentation to measure escape depths in polymers directly. From a study of the method described by Evans et al. the most obvious source of error would seem to be from an over-simplification of the basic intensity model and from an incomplete knowledge of the effect of surface topography and angular dependence.* It has been established that with optically flat (viz. topographically uniform) films the basic intensity model follows a simple relationship of the form

$$I = I_\alpha e^{-d/\Omega}$$

with the usual notation. The method of choice for direct measurement of mean free paths would, therefore, seem to be the production of polymer films of known thickness on optically flat surfaces and the observation of relative signal intensities. The apparatus we have constructed consists of a reaction chamber in which paralene films of known thickness are generated by reaction of the appropriate paraxylylenes produced in an attached pyrolysis apparatus. Continuous films are laid down on an appropriate substrate on a sample probe maintained at the requisite temperature and this is held in a symmetrically related position (with respect to the inlet jet from the pyrolysis apparatus) with a quartz crystal of a deposition monitor, the surface of the crystal having an evaporated film identical to that of the sample probe. At this time we can report that all of the essential parts of the apparatus work and all of the preliminary work has been completed on thick polymer films. We are currently engaged, therefore, in elaborating the variables necessary to ensure the production and accurate measurement of films in the range < 100 Å and we will report on these measurements in due course.

[1] cf. D. T. Clark, *The Application of ESCA to Studies of Structure and Bonding in Polymers*, 5A, 241, plenary lecture ACS International Symposium on Advances in Polymer Friction and Wear, Los Angeles, March 1974, ed. L. H. Lee, (Plenum Press, New York, 1974).
[2] D. T. Clark and J. Peeling, *J. Polymer Sci.*, 1976.
[3] D. T. Clark et al., unpublished data.

* Subsequent to the conference Dr. Stephen Evans has kindly supplied a preprint of his ref. (4). Our interpretation of the data differs somewhat from his and we obtain a mean free path of ~17Å for graphite from his data.

Dr. S. Evans (*Aberystwyth*) (*communicated*): We regard the characterisation of our samples as adequate. The polythene (sheet) had a density of 0.9 g cm^{-3}; we used two PTFE samples, (one sheet (density 2.17 g cm^{-3}) and one pressed pellet (density 2.08 g cm^{-3})), which gave very similar results. The graphite fluoride, of composition $CF_{1.13}$, gave a pellet of density 1.44 g cm^{-3}. Stearic acid was also examined as a pellet, of density very close to the literature value. It is nonetheless extremely doubtful whether samples of these materials, however prepared, would show variations in λ approaching an order of magnitude. Moreover, as Clark admits, his own previous work establishes only ratios of mean free paths: and a majority of the few previous absolute determinations of mean free paths in carbonaceous materials in the required energy range of which we are aware are in general agreement with our values.

In the absence of any explanation of its origin, we remain sceptical of his analysis of our data for graphite in terms of $\lambda \sim 17$ Å. Even this (in our view, excessively low) value is, however, about double the value he uses for non-conducting polymers of about half the density: whereas other things being equal, one would expect on theoretical grounds an approximately *inverse* relationship between the density and the mean free path. Our results for the carbonaceous materials do, however, follow the expected trend. The mean free path should also depend [1] on the valence electronic structure of the material, transition metals being expected in general to exhibit rather shorter mean free paths than insulating materials, and this expectation also is satisfied in our data.

As Clark's extensive studies (his ref. (2) and (3) above) have not been made available to us and will apparently not be published until at least 1976, we are unable to comment on them: all of our not insubstantial body of data on the surface chemistry of carbon and of several metals can be rationalised using values similar to those in our table. Further experimental work should ultimately resolve the problem and we await the results of Clark's proposed study with interest.

Dr. I. H. Hillier (*Manchester*) said: Lavery and I have performed *ab initio* and CNDO calculations on oxygen chemisorbed on lithium clusters. For adsorption onto the 4-coordinate site of a square-pyramical arrangement of lithium atoms (Li_5O), the CNDO calculations gave an optimum O–Li_4 surface distance of 3.87 a.u., the minimum being very shallow. The CNDO calculations for a series of clusters up to Li_{13} resulted in a small net positive charge (~ 0.1) on the oxygen atom, in contrast to the suggestion of Quinn and Richardson. Our conclusion is substantiated by a minimal basis *ab initio* calculation on Li_5O.

We further find that the higher filled valence orbitals progressively decrease in energy with increasing cluster size, and the cluster stability appears to be approaching convergence for the larger clusters.

[1] C. J. Powell, *Surface Sci.*, 1974, **44**, 29.

Roles of Lateral Interactions between Adatoms and Local Surface Geometry in Determining the Binding Energies of Core Electrons in Adsorbed Species

BY C. M. QUINN AND N. V. RICHARDSON

University of Birmingham, Birmingham, England

Received 12th March 1975

The MSXα cluster technique is used to calculate the energies of core and valence levels in models for oxygen chemisorbed on lithium. It is shown that it is possible to maintain traditional concepts of the chemisorbed state if the clusters are considered to be open electron systems taking due account of lateral interactions in the surface and bulk regions.

Recent advances in the techniques of surface chemistry, and especially the advent of electron spectroscopy in the study of surface species,[1,2] have emphasised the need for a satisfactory theory of adsorption based on quantum rather than classical principles. To this end, in the past few years, several groups have attempted calculations of adsorbate–adsorbent interactions using a variety of approximate procedures. Estimates of the bonding in adsorbed layer models have been made using extended Hückel theory,[3-6] the CNDO method,[7,8] model Hamiltonians,[9] and perturbation techniques.[10]

Such calculations are to be contrasted with model system approximations in one and two dimensions.[11-13] Their extension to complex adsorption systems and to detailed examinations of binding energies and bonding schemes on the exact environment of adsorbate fragments is limited, however, in two ways. First, the demands these procedures make on computer time is very high, unless the basis sets used are kept to a minimum. Secondly, the reality of the models used for adsorbate–adsorbent structure must be questioned if no special features are introduced into the models, which distinguish the fact that the adsorbed species contacts an effectively infinite structure, the bulk adsorbent. For metals, this predominant feature must be the ability to supply charge to, or accept charge from, an adatom without significantly altering the local electron density around any metal atom of the adsorbent.

In this paper, we report our progress to date in introducing this feature into calculations of the core and valence level binding energies for oxygen adsorbed on lithium metal using the multiple scattering X-alpha (MSXα) cluster technique of Johnson and Slater.[14-16] This technique has several advantages which seem to make it suited to surface applications. It is not limited by the gross approximations of the extended Hückel and CNDO methods and provides within the framework of the statistical approximation a fast calculation route to self-consistent results since the problem of evaluating large numbers of integrals is avoided. For our present purposes, its major advantage in surface applications is the ease with which the electron content can be varied in model clusters chosen to represent the local surface geometry and environment of adsorbed species.

Our model clusters for chemisorbed oxygen on lithium metal are shown in fig. 1. Lithium was chosen as the adsorbent in these initial calculations in order to simplify them by reducing the total numbers of electrons that had to be accounted for. It is true that oxygen–lithium interaction is unlikely to be restricted to a single layer and that the major effect even at low temperatures is likely to involve bulk reaction at all coverages. Our models should be viewed as stages in the initiation of this process.

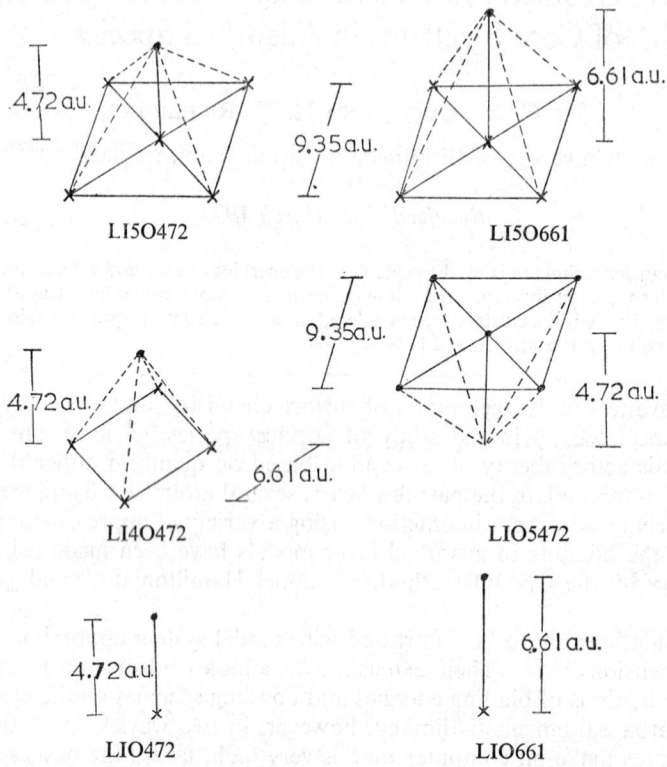

FIG. 1.—Clusters used in the multiple scattering x-alpha calculations as models for chemisorbed oxygen on lithium. Distances are given in atomic units.

In metallic lithium the lithium–lithium distance along the b.c.c. unit-cell edge is 6.61 a.u. For a (100) surface and the clusters LI5O472 and LI5O661 centred around any one lithium atom in the surface, this is also the distance to neighbours in the surface plane. The distances of chemisorbed oxygen from surfaces is unknown. We have used two bond lengths in our calculations, 4.72 and 6.61 a.u. The distance 4.72 a.u. is about the lithium distance from the centre of the surface plane of the alternative cluster LI4O472 for a different siting of an oxygen species on the (100) plane. The other clusters of fig. 1 correspond to extreme situations that we have used for comparison purposes to determine the influence of lateral interactions in the surface layers. LIO5472 is intended to reveal the maximum possible degree of interaction between neighbouring oxygen species on the surface. LIO472 and LIO661 are the minimum models for an adsorbate–adsorbent interaction in which no lateral interactions can occur and in which the major influence on chemisorbed species is the length of the chemisorption bond. These last two models are limiting ones in the use of pseudomolecular clusters as models for the adsorbed state.

For the MSXα calculations, the α-exchange parameters chosen for the "muffin-tin" spheres around each atom of the clusters were the virial theorem values as calculated by Schwarz [17]: for the intersphere and outer sphere regions of the clusters the α-value for oxygen was used. All electron calculations were performed for each model cluster and the iteration procedure was continued until the self-consistency in the orbital emergies was better than 0.1 eV.

The ground-state SCF orbital energies for the clusters of fig. 1 are tabulated in table 1. The levels are labelled as the irreducible representations of the appropriate symmetry group and are filled as required by the number of electrons in each cluster. The spherical harmonic components contributing to the wave functions for each energy are also given in the table.

TABLE 1

MSXα-SCF eigenvalues for the clusters used as models for oxygen chemisorbed on lithium. The clusters are identified as in fig. 1. The symmetries of the orbitals, together with the spherical harmonics contributing to the wave functions in each sphere, are given in parentheses. Li_I denotes a central lithium sphere, Li_{II} a peripheral lithium sphere. A similar notation is used for the oxygen spheres in LIO5472. All energies are in rydbergs (1 rydberg = 13.6 eV).

LI5O472	LI5O661	LI4O472
-0.16 $(e, Li_{II}(s)-O(p_x))$	-0.13 $(e, Li_{II}(s)-O(p_x))$	-0.22 $(e, Li_{II}(s)-O(p_x))$
-0.17 $(a_1, O(p_z)+Li_I(s)-Li_{II}(s))$	-0.13 $(a_1, O(p_z)+Li_I(s)-Li_{II}(s))$	-0.23 $(a_1, O(p_z)-Li_{II}(s))$
-0.19 $(e, O(p_x)+Li_{II}(s))$	-0.14 $(e, O(p_x)+Li_{II}(s))$	-0.24 $(e, O(p_x)+Li_{II}(s))$
-0.24 $(a_1, O(p_z)+Li_I(s)+Li_{II}(s))$	-0.20 $(a_1, O(p_z)+Li_I(s)+Li_{II}(s))$	-0.31 $(a_1, Li_{II}(s)+O(p_z))$
-1.20 $(a_1, O(s))$	-1.16 $(a_1, O(s))$	-1.28 $(a_1, O(s))$
-3.85 $(a_1, Li_I(s))$	-3.79 $(a_1, Li_I(s))$	-4.05 $(a_1, Li_{II}(s))$
-3.96 $(a_1, b_1, e, Li_{II}(s))$	-3.87 $(a_1, b_1, e, Li_{II}(s))$	
-37.27 $(a_1, O(s))$	-37.15 $(a_1, O(s))$	-37.29 $(a_1, O(s))$

LIO5472	LIO472	LIO661
-0.79 $(b_1, O_{II}(p_x))$	-0.35 $(\pi, O(p_x))$	-0.88 $(\pi, O(p_x))$
-0.79 $(e, O_{II}(p_y))$	-0.36 $(\sigma^+, O(p_z))$	-0.89 $(\sigma^+, O(p_z))$
-0.79 $(e, O_{II}(p_z))$		
-0.79 $(a_2, O_{II}(p_y))$		
-0.79 $(a_1, O_{II}(p_x))$		
-0.80 $(b_2, O_{II}(p_y))$		
-0.84 $(a_1, O_I(p_z)+O_{II}(p_z))$		
-0.86 $(e, O_I(p_x)+O_{II}(p_x))$		
-1.86 $(a_1, O_I(s)-O_{II}(s))$	-1.39 $(\sigma^+, O(s))$	-1.97 $(\sigma^+, O(s))$
-1.87 $(b_1, O_{II}(s))$		
-1.87 $(e, O_{II}(s))$		
-1.93 $(a_1, O_{II}s)+(O_I(s))$		
-4.63 $(a_1, Li(s))$	-4.06 $(\sigma^+, Li(s))$	-4.19 $(\sigma^+, Li(s))$
-37.90 $(a_1, O_I(s))$	-37.40 $(\sigma^+, O(s))$	-38.13 $(\sigma^+, O(s))$
-37.93 $(a_1, b_1, e, O_{(II}s))$		

Photoelectron spectroscopy data [18-21] for oxygen in oxides or after metal-gas reaction has occurred, whether this has led to chemisorption or incorporation, suggest a general insensitivity in the energy of the O 1s level with the exact local environment of the oxygen species. This stability, it should be emphasised, is relative to the equilibrium Fermi-level position in the measuring equipment. Only for zero kinetic energy emission is the photo-electric measurement work-function sensitive. The experimental results, therefore, suggest that both the superposed screened nuclear potential due to neighbours around oxygen in oxides, or in the chemisorbed or incorporated states, and the charge on oxygen in these different situations, are such that the local electronic environment for O 1s is the same in all these cases. In molecular systems the two effects can be expected to complement each other. A transfer of charge from neighbours to a particular atom in a cluster will raise the levels on that atom. To some extent this decrease in stability will be compensated for by an increase in the nuclear component of the potentials due to surrounding neighbours.

For many gas–solid reactions the influence of screened potentials, due to neighbouring nuclei, on adsorbate energy levels is likely to be small because of the large mesh size in crystals both in the bulk and on their surfaces. In our preliminary non-SCF calculations this was indeed found to be the case, the maximum variation in the O 1s level in the model clusters being -514.82 ± 0.5 eV. At the interatomic distances set in the models, little superposition of potential occurs around any site, and the non-SCF values are close to those in isolated oxygen or lithium atoms.

In the SCF calculations, however, where the charge density in each cluster is allowed to become self-consistent, there is a relatively large variation in the energy levels. In particular, the large variation of the Li 1s energies is unsatisfactory since the variation reflects varying positive charges on the metal atoms. The differences in the valence region charge density distribution, which are reflected in the core level energies, we believe to be due to the deficiencies implicit in the use of pseudo-molecular clusters as models for the adsorbed or incorporated state. For an adsorbent we must expect that the environment around individual metal atoms can be little changed by chemisorption or even more substantial gas–solid interaction. The bulk Bloch states are relatively insensitive to the mixing-in of a small number of different orbitals due to adsorbed species and the mixing process in the valence region leads at most to virtual levels at the surface, filled or partially filled, dependent on their position with respect to the system Fermi level.

These effects can be illustrated in simple examples. Consider the model problem of a single adsorbate atom at the end of a chain of atoms of a different type, a model metal adsorbent. The general molecular orbitals are of the form,

$$\psi_k(x) = \sum_{n=0} C_{nk}\phi_n(x-na), \quad (1)$$

where a is the " lattice " spacing, k the wave vector identifying the energy of the wave, and $\phi_0(x)$ is an orbital of the adsorbate atom, the extra atom in the chain. The " coefficient guessing " method [12, 13] is to assume the C_{nk} equal to $\exp(ikna)$ as in the ideal crystal under periodic boundary conditions, so that

$$\psi_k(x) = \sum_{n=0} \exp(ikna)\phi_n(x-na). \quad (2)$$

Using Hückel-type approximations the energy-level spectrum can be obtained from the difference equations

$$\begin{pmatrix} \gamma-E & \beta & 0 & 0 & \ldots \\ \beta & \alpha-E & \beta & 0 & \\ 0 & \beta & \alpha-E & \beta & \ldots \\ \cdot & \cdot & \cdot & \cdot & \cdot \\ \cdot & \cdot & \cdot & \cdot & \cdot \\ 0 & \ldots & \beta & \alpha-E & \beta \end{pmatrix} \begin{pmatrix} 1 \\ \exp(ika) \\ \cdot \\ \cdot \\ \cdot \\ \exp(inka) \end{pmatrix} = \quad (3)$$

with

$$\langle \phi_i | \hat{H} | \phi_j \rangle = \alpha \delta_{ij} = \beta \delta_{i, i \pm 1} \tag{4}$$

and

$$\langle \phi_i | \hat{H} | \phi_0 \rangle = \gamma \delta_{i,0} = \beta \delta_{i,1}. \tag{5}$$

The bulk and surface conditions then are

$$(\alpha - E) + 2\beta \cos ka = 0, \tag{6}$$

and

$$(\gamma - E/\beta) + \exp(ika) = 0. \tag{7}$$

The bulk band is defined by $-2 \leqslant \cos ka \leqslant +2$ which yields real k, real E values and so Bloch functions approximated by transcendental modulation of the atomic orbitals. The surface condition which isolates the adsorbate orbitals and mixes in the chain orbitals under an exponentially decaying envelope to a degree dependent on the initial separation of the level in energy, occurs for complex k. This can be examined by setting

$$p = \frac{\gamma - E}{\beta} - \frac{\alpha - E}{\beta} = \frac{\gamma - \alpha}{\beta} = \exp(-ika). \tag{8}$$

For real k, $-1 < p < 1$. For $p > 1$, i.e., $(\gamma - \alpha)/\beta > 1$ then the wave vector is complex and an isolated surface state appears below the adsorbent valence band. For $p < -1$, $(\gamma - \alpha)/\beta < 1$ and the surface state is empty above the band. For p within these limits the Coulomb and resonance integrals in the surface are little different to those for the atoms of the bulk, and no isolated state leaves the band.

This last possibility is the one most likely to be the normal situation in chemisorption. Even, in this simple model, if we change the localization condition, by altering the resonance integrals $\langle \phi_1 | H | \phi_0 \rangle$ and $\langle \phi_2 | H | \phi_1 \rangle$, the generation of localized orbitals characteristic of isolated surface species depends on the energies of such orbitals lying outside the band edges of the bulk material. But the transfer of electrons from the bulk to surface levels will give rise to Coulomb repulsions in these levels and most probably raise them into the band structure. As Grimley [9] has pointed out, the positions of surface levels and their occupancy must be determined together in a self-consistent way, the surface regions being viewed as an open system in which the number of electrons is not fixed on a local basis but rather is determined by the Fermi level of the whole. This does not mean that such delocalized states cannot be distinguished in photoelectron spectra. Modern photo-electric techniques specifically sample only the surface regions of emitters. Delocalized wave functions extending through the chemisorbed layer, due to strong mixing of adsorbate orbitals with similar energy Bloch functions, can be expected to contribute more to densities of states determined photo-electrically than those bulk wave functions terminating in the surface of the adsorbent.

Yet a localization condition is unavoidable in the cluster technique, when this is not modified to take account of the underlying bulk adsorbent, in surface applications. Lateral interactions in the bulk are so extensive that we cannot expect to reproduce the bulk situation with even relatively large clusters. Thus the lithium atoms in our clusters are generally positively charged and the adsorbate oxygen atoms are negatively charged, as judged by the core level energies at the end of the SCF calculations yielding the results in table 1, to a degree dependent on the numbers of each kind of atom present in the clusters and on the cluster structure. Substantial

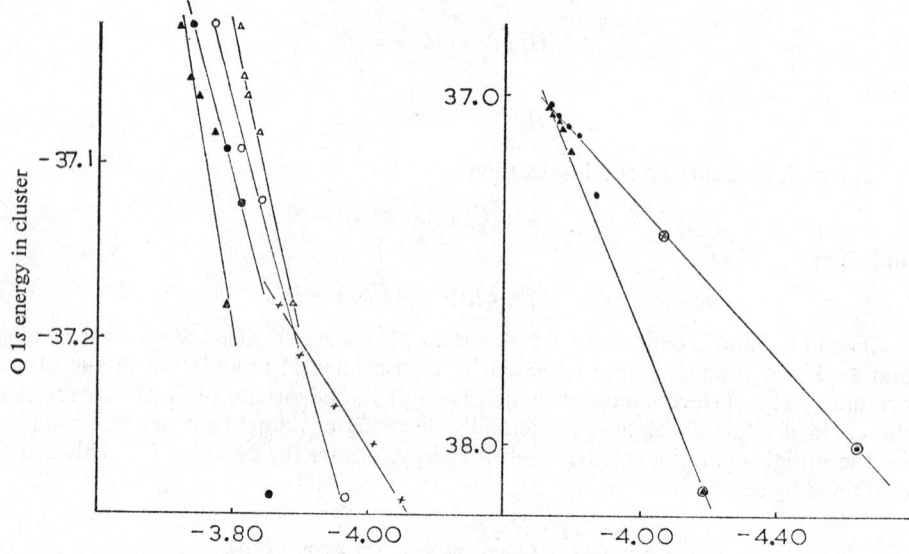

Fig. 2.—The variation in O 1s energies in the clusters with added charge reflected by the Li 1s values:
(a) ▲, △ O 1s in LI5O661 referred to centre and peripheral Li 1s values: ●, ○ O 1s in Li5O472 referred to centre and peripheral Li 1s values: × O 1s in LI4O472 referred to Li 1s value in cluster.
(b) extrapolations to include O 1s values in LIO661 ⊕, LIO472 ⊗ and LIO5472 ⊙ using Li 1s values observed in these clusters.

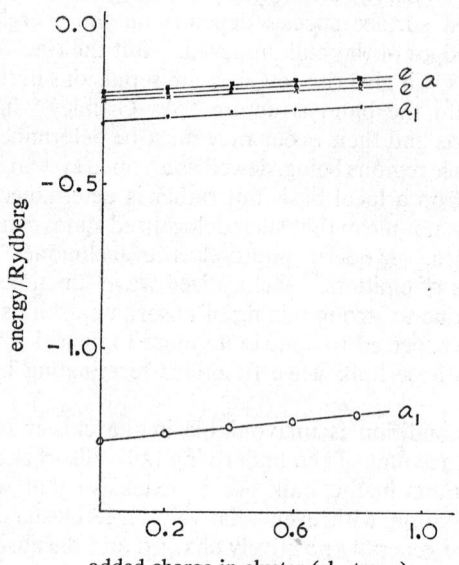

Fig. 3.—Variations in valence levels in LI4O472 with added charge. The symmetries are as marked in table 1.

local charge density differences around the atoms in the various clusters occur compared with the expected situation in the real adsorbate–adsorbent system. These differences we attribute to the closed nature of the clusters, the total number of electrons in a cluster being determined by the number of atoms included.

Accordingly we have attempted to make the model clusters more representative of the true physical situation in surface layers by adding electrons to the clusters and repeating the SCF calculations. We find a generally linear dependence in the energy levels of the clusters on this added charge. The results are shown in fig. 2 to 5. It can now be seen that the apparently anomalous results of the primitive clusters LIO472, LIO661 and the extreme model LIO5472 lie on the appropriate extrapolations of the central lithium atom data for the LI5O type clusters. These deficiencies in the models do appear to be overcome when the electron content of the clusters are suitably varied.

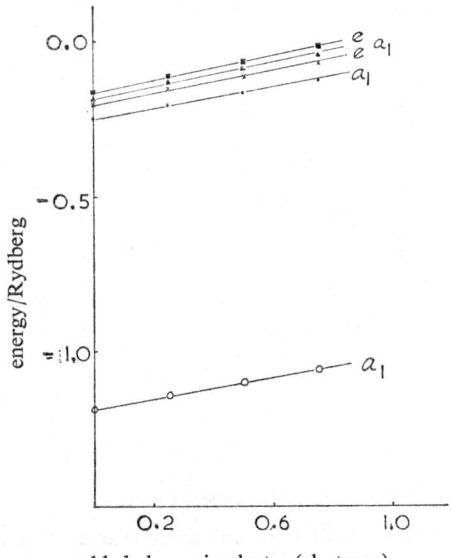

Fig. 4.—Variations in valence levels in LI5O472 with added charge. The symmetries are as marked in table 1.

This suggests that good representations of the chemisorbed state can be obtained with small model clusters if the matching to experimental data is made on the basis of an electron content in any cluster which takes account of any lateral interactions between the adsorbed species in the real situation and the attachment of the adsorbed species to a bulk adsorbent. Our suggestion conflicts with recent conclusions by Johnson and Messmer,[16, 22] based on MSXα cluster calculations in the nickel–oxygen system. These authors report that oxygen chemisorption data cannot be accounted for using clusters based on traditional chemisorption models involving linear, bridged or pyramidal structures and that a cluster NiO_6^{10-}, more representative of bulk oxide, must be used. In the case of nickel, it is difficult to accept these highly oxygenated clusters as good models for the chemisorbed or incorporated states. Oxygen interaction is limited for this metal at room temperature to at most two layers even at 10^3 L,[23, 24] yet the validity of the NiO_6^{10-} type clusters depend on the formation of a distinct surface film of nickel oxide even at much lower exposures.

In order to make a direct comparison of our data with experimental data it is necessary to know the electron content appropriate to each cluster type. An acceptable way of determining these quantities may be the requirement that the metal core levels be restored to the same energies as in the bulk metal or in isolated atoms. In the present calculations we have made the matching to the atomic Li $1s$ value (3.887 Ry) as the criterion. Currently [25] we are determining bulk core energy levels using the KKR technique [13, 26] and we intend in the future to use these as our standards. For exact comparisons with experimental photo-emission data two further factors need to be considered. It is necessary to add transition state relaxation energies [27] to the SCF-orbital energies and also to take account of the work-function which has to be added to the photo-electron spectroscopy data.

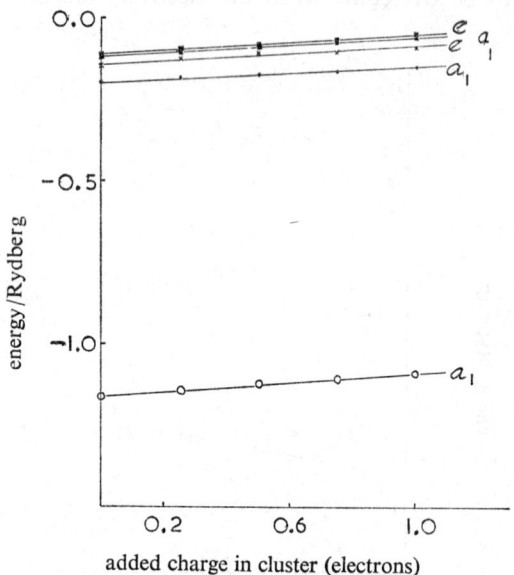

FIG. 5.—Variations in valence levels in Ll5O661 with added charge. The symmetries are as marked in table 1.

Average transition state energies in the region of the O $1s$ core levels in our calculations are about 28 eV. For the valence regions the transition state relaxation term is about 3 eV. This latter value is also likely to be close to the work-function of a hypothetical lithium surface covered with a chemisorbed layer of oxygen. The maximum variation in the O $1s$ levels in our model clusters is -506 ± 0.45 eV when these are adjusted to the electron content required to maintain the metal core levels at the atomic values. The O $1s$ value therefore is not sensitive to a specific location on the lithium and does not vary greatly with the length of the chemisorption bond.

In the valence regions, three principal structural features of photoemission data for a wide variety of surfaces, both oxide and chemisorbed layers, are observed.[18-21] In the region of 21 eV below the system Fermi level O $2s$ photoemission dominates the density of states. Some 16 eV higher in the spectrum a peak can be detected which is attributable to emission from O $2p$ levels, and this peak is within the valence band of the bulk emitter. At low oxygen coverage the underlying metal states can be clearly discerned little perturbed by the oxygen adsorption.

Our results agree with these general observations. For what is probably the most realistic model at low coverage, LI4O472, we find O $2p$ levels in the Li $2s$ band at -3 eV and O $2s$ levels at -17.6 eV. Even for the case LIO5472 a virtually identical energy gap is found between these levels, though, in this case, if the electron content of the cluster is not adjusted, these levels are considerably more stable and deeper lying because of the substantial electron deficiency of this extreme cluster.

Two features of our results require more explanation and will be considered elsewhere.[25] First, the individual results obtained for the central lithium atoms of the clusters LI5O472 and LI5O661 are slightly different from those for the peripheral atoms on which we base the matching to experiment. A similar difficulty in treating the central atoms in clusters has been reported by Gyorffy et al.[28, 29] in density of states calculations. Secondly, the simple linear relationship between the energies and the charge added for the clusters presumably occurs because the charge is being added to only one level with both lithium and oxygen character. In principle, situations may arise in which excess charge has to be added first to one level and then another of different symmetry in order that the core levels of cluster metal atoms attain the values in the bulk metal. We expect that this type of situation would lead at least to a change of slope in curves of energy against added charge for a cluster.

[1] M. W. Roberts and J. M. Thomas, *Surface and Defect Properties of Solids*, vol. 1 and 2 (The Chemical Society, London).
[2] Photo-effects in adsorbed species, *Disc. Faraday Soc.*, 1974, **58**.
[3] A. B. Anderson and R. Hoffmann, *J. Chem. Physics*, 1974, **61**, 4545.
[4] R. C. Baetzold, *J. Catalysis*, 1973, **29**, 129.
[5] J. C. Robertson and C. W. Wilmsen, *J. Vac. Sci. Techn.*, 1972, **9**, 901.
[6] A. J. Bennett, B. McCarroll and R. P. Messmer, *Surface Sci.* 1971, **24**, 191.
[7] G. Blyholder, *J. C. S. Chem. Comm.*, 1973, 625.
[8] A. J. Bennett, B. McCarroll and R. P. Messmer, *Phys. Rev. B*, 1971, **3**, 1397.
[9] T. B. Grimley and M. Torrini, *J. Phys. Solid State Phys. C*, 1973, **6**, 868.
[10] T. B. Grimley, in *Molecular Processes on Solid Surfaces*, ed. E. Drauglis, R. D. Gretz and R. J. Jaffee (McGraw Hill, New York, 1969).
[11] S. G. Davidson and J. D. Levine, *Solid State Phys.*, 1970, **25**, 1.
[12] J. D. Levine and P. Mark, *Phys. Rev.*, 1969, **182**, 926.
[13] C. M. Quinn, *Introduction to the Quantum Chemistry of Solids* (Clarendon Press, Oxford, 1972).
[14] K. H. Johnson and J. C. Slater, *Phys. Rev. B*, 1972, **5**, 844.
[15] K. H. Johnson and F. C. Smith, Jr., *Phys. Rev. B*, 1972, **5**, 831.
[16] R. P. Messmer, C. W. Tucker, Jr. and K. H. Johnson, *Surface Sci.*, 1974, **42**, 341.
[17] K. Schwarz, *Phys. Rev. B*, 1975, **5**, 2466.
[18] S. Hufner and G. Wertheim, *Phys. Rev. B*, 1973, **8**, 4857.
[19] D. E. Eastmann and J. K. Cashion, *Phys. Rev. Letters*, 1971, **27**, 1520.
[20] R. W. Joyner and M. W. Roberts, *Chem. Phys. Letters*, 1974, **28**, 246.
[21] C. R. Brundle, *Disc. Faraday Soc.*, 1974, **58**.
[22] R. P. Messmer and K. H. Johnson, *J. Vac. Sci. Techn.*, 1974, **11**, 236.
[23] C. M. Quinn and M. W. Roberts, *Trans. Faraday Soc.*, 1964, **60**, 899.
[24] C. M. Quinn and M. W. Roberts, *Trans. Faraday Soc.*, 1965, **61**, 1775.
[25] C. M. Quinn and N. V. Richardson, to be published.
[26] F. S. Ham and B. Segall, *Phys. Rev.*, 1961, **124**, 1786.
[27] J. C. Slater, *Adv. Quantum Chem.*, 1972, 6, 1.
[28] B. L. Gyorffy, D. House and P. V. Smith, *J. Phys., Metal Phys. F*, 1973, **3**, 745.
[29] B. L. Gyorffy, D. House and G. M. Stocks, *J. Physique*, 1974, **35**, C4.

The Structure of Clean Crystalline Surfaces and Chemisorbed Overlayers

By F. Jona

Department of Materials Science, State University of New York, Stony Brook, New York, 11794, U.S.A.

Received 17th March, 1975

Low-energy electron diffraction (LEED) is being successfully applied to the determination of atomic arrangements in the surface region of crystalline solids. Some recent work in this field is presented. Partial results are discussed for the structure analyses of: (1) the clean Mo{001} surface, which is found to be contracted by approximately 11 % with respect to the bulk; (2) the clean Ti(0001) surface, which is only very slightly contracted (by about 2 %) with respect to the bulk; (3) the 1 × 1 structure of an overlayer of silicon on Mo{001}; and (4) the $c(2 \times 2)$ structure of an overlayer of nitrogen on Mo{001}. The confidence level for the latter is considerably lower than that for the former structures. Possible reasons for situations of this kind are discussed, and an assessment of the *status quo* in surface crystallography is given.

The past four or five years have witnessed considerable advances in the development of theoretical and experimental methods for the determination of atomic arrangements within solid crystalline surfaces. The experimental probe is a beam of low-energy electrons that is allowed to diffract from the surface under study, while the theoretical tool is the dynamical theory of diffraction that describes the interactions of the probing electrons with the atoms located in the top few layers of the surface. Thus, low-energy electron diffraction (LEED) is supposed to do for crystal surfaces what X-ray diffraction has done for bulk solids.[1] So far, the method has been mainly applied to two types of surface structures, viz., the clean low-index surfaces of pure metals and the ordered arrangements of up to one monolayer of foreign atoms adsorbed on such clean metallic surfaces—commonly called overlayer structures. The study of clean surfaces has in fact generally a dual purpose: to determine atomic positions within the surface (in particular, the eventual expansion or contraction of the top atomic layer with respect to the bulk) and to prepare for the study of overlayer structures. In the group of clean surfaces, work has been concentrated predominantly on surfaces of face-centered cubic metals (Al, Ni, Ag and Cu being most prominent). In the overlayer group, attention has been devoted almost exclusively to superstructures with periodicities twice as large as that of the substrate, mostly with a centred unit mesh—the $c(2 \times 2)$ structure.[1]

The present paper reviews some recent work done at Stony Brook in both groups of structures. In the clean-surface group, the results of structure analyses of the {001} surface of molybdenum and the (0001) surface of titanium are presented. The former is the first analysis of the surface structure of a body-centred cubic metal and the latter is the first rigorous study of the atomic arrangement on the basal plane of an hexagonal close-packed material.[2] The results indicate that Mo{001} is contracted by about 11 %, and Ti(0001) by approximately 2 % with respect to the bulk. For f.c.c.{001} surfaces either no distortion or only a slight expansion of the bulk interplanar spacing were reported.[1] In the group of overlayer structures, the present paper reviews current work on overlayers on Mo{001}, namely, the (1 × 1) structure of a monolayer of silicon on Mo{001} and the $c(2 \times 2)$ structure of an overlayer of nitrogen on Mo{001}. The former has been convincingly determined, and the model is depicted in fig. 1(a), while the latter is doubtful, the most probable model being the one, shown in fig. 1(b), which has been proven correct for all $c(2 \times 2)$ structures of overlayers on cubic {001} surfaces investigated so far.[1]

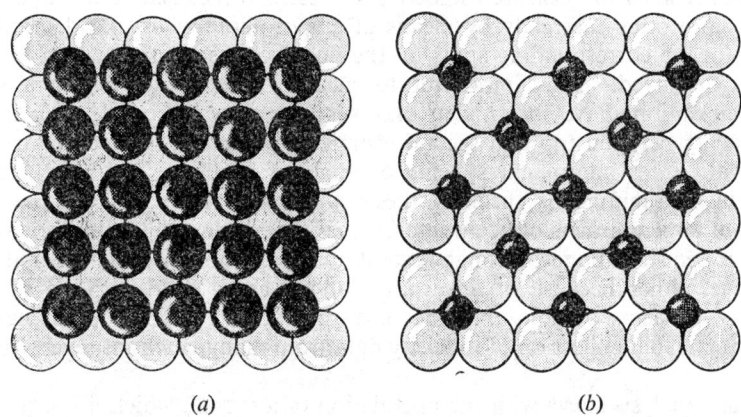

(a) (b)

FIG. 1.—Models of overlayer structures on cubic {001} surfaces (large circles represent substrate atoms, small circles represent adsorbed overlayer atoms): (a) The 1 × 1 structure of silicon on Mo{001}; (b) The $c(2 \times 2)$ structure of nitrogen on Mo{001}.

In this discussion, the author acts as the spokesman for a number of people who have collaborated on the projects described herein: A. Ignatiev, D. W. Jepsen, K. O. Legg, P. M. Marcus and H. D. Shih. Section 1 briefly reviews the experimental and calculational procedures followed at the present time for surface structure analysis. Section 2 presents the results and discusses the conclusions. Section 3 attempts an assessment of the situation in surface crystallography at the present time.

1. PROCEDURES FOR SURFACE-STRUCTURE ANALYSIS

The surface to be studied is polished parallel to the chosen crystallographic plane ({001} for the Mo sample, (0001) for the Ti sample) within approximately 0.5°, and is then cleaned in the ultra-high vacuum system by means of ion bombardments followed by anneals and occasionally by means of heat treatments in suitable reactive atmospheres (e.g., oxygen to burn out carbon or sulphur impurities). The chemical composition of the surface is monitored with Auger-electron spectroscopy (AES) and the crystallinity with LEED. For the study of clean surfaces, purity and good crystallinity are the two most important requirements. Purity of the surface is established by the absence of impurity lines in the AES spectra at maximum sensitivity —with retarding-field detection schemes the impurity concentration on the surface is then at most 0.1 % of a monolayer. Good crystallinity of the surface is established by the observation of a low-background, high-contrast LEED pattern with sharp diffraction spots. For the study of overlayer structures, preparation of the overlayer is at least as critical as that of a clean surface but tests of chemical composition (in particular, stoichiometry) and crystallinity are less quantitative and less reliable. Both AES and LEED spectra are monitored to establish *that* range of exposure of the surface to the overlayer atoms, at which changes are either zero or minimal as functions of exposure. Under these conditions, the overlayer structure attained is more stable and thus presumably better defined than otherwise. Heating of the sample, usually an efficient way to improve order on the surface, should be used only very cautiously, as it often causes diffusion of the adsorbed atoms into the substrate and thereby complicates the structure analysis.

For structure analysis, the intensities of several diffracted beams are needed as functions of energy for two or three angles of incidence of the primary beam. The diffracted intensities are measured sometimes directly by means of a movable Faraday cage and more often indirectly by means of a spot-photometer that determines the brightness of the corresponding spots on the fluorescent screen on which the LEED pattern is displayed. It is important to measure several non-degenerate beams because calculations have shown that only very few vary noticeably with the interplanar spacing between top and second atomic layer—most, in fact, change imperceptibly with either interplanar spacing or angle of incidence and are therefore of little use in determining or checking these parameters. The price to pay for the collection of large numbers of beam intensities is that during collection (often extending, with present techniques, through several hours) the surface structure under study may change, either because of adsorption of unwanted or desorption of wanted impurities. It seems obvious that the next generation of LEED equipment must allow for very rapid collection of intensity data from many beams at several angles of incidence.

Structure analysis starts with the postulation of a model, which, in turn, requires specification of a number of parameters. We call *structural* those parameters that have to do with atomic positions, interatomic or interplanar distances, and *nonstructural* those parameters that relate to the potential chosen to describe the scattering surface, to the adsorption and refraction of electrons upon entering the sample and to the isotropic or anisotropic vibrations of surface and bulk atoms. After postulation of a model, calculations are performed of the diffracted intensities that would be expected from such a model, and the results are compared with experiment. The calculations are done with the layer KKR-procedure (KKR stands for Korringa–Kohn–Rostocker) which has been described elsewhere [3] and applied successfully to clean surfaces of f.c.c. metals [4] and overlayers thereon.[5] The calculations make use of 8 phase shifts over the energy range 0-150 eV and most often 29 (or 58) beams in the representation of the wave function.

2. RESULTS AND DISCUSSION

A. CLEAN SURFACES

Partial results of the study of clean Mo{001} and of that of clean Ti(0001) are presented in fig. 2 in the form of experimental and theoretical LEED spectra. Complete and more detailed accounts of both studies than can be presented here will be published elsewhere.[6, 7]

For Mo{001}, we show in fig. 2(a) the 00 (or "specular beam"), 10 and $\bar{1}0$ spectra for an angle of incidence $\theta = 8°$ and azimuth $\phi = 0°$ (incident beam in the {010} plane). Location of these beams in the corresponding LEED pattern can be identified, e.g., in fig. 3(a) to be discussed below. We observe in fig. 2(a) that the correspondence between observed and calculated spectra is satisfactory—we refer, in particular, to peak positions, peak width, peak shapes and relative intensities. There is no agreement between absolute intensities: for the 00 and the 10 beams the calculated intensities are, on the average, 1.5 times larger than the observed, for the $\bar{1}0$ beam the same ratio is about 5.5; for other beams (not shown here) it varies between 0.8 and 15. It is

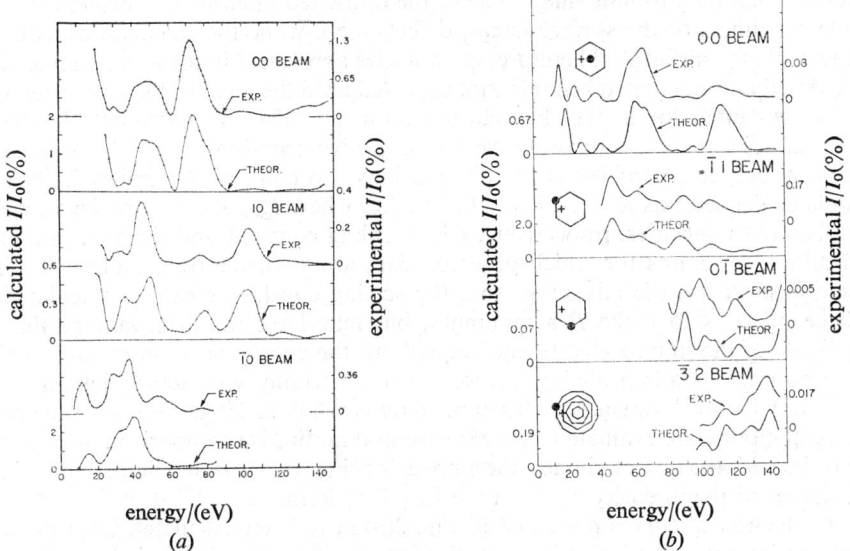

FIG. 2.—Observed and calculated LEED spectra for: (a) clean Mo{001} at angle of incidence of 8° and {010} azimuth; (b) clean Ti(0001) at angle of incidence of 20° and (01̄10) azimuth. The dot in each of the inscribed hexagons indicates the position of the beam considered in the LEED pattern; the plus sign shows the approximate position of the electron gun.

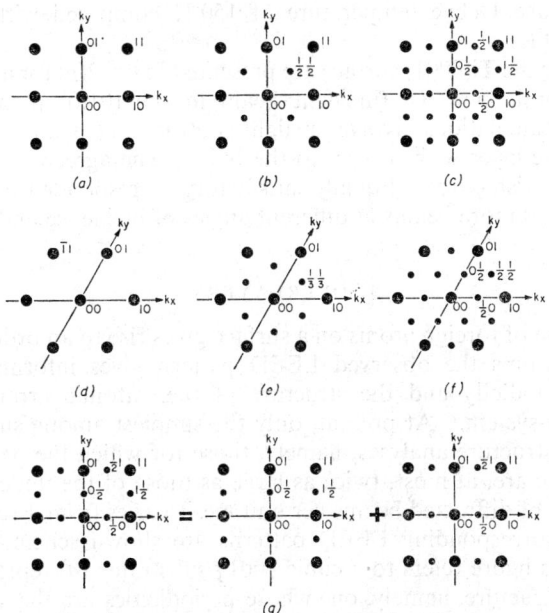

FIG. 3.—Schematic LEED patterns for simple overlayer structures: large circles represent integral-order, small circles, fractional-order beams; some representative indexing is given for both: (a) 1×1 structure on cubic {001}; (b) $c(2 \times 2)$ structure on cubic {001}; (c) $p(2 \times 2)$ structure on cubic {001}; (d) 1×1 structure on close-packed surface; (e) $\sqrt{3} \times \sqrt{3} - 30°$ structure on close-packed surface; (f) $p(2 \times 2)$ structure on close-packed surface; (g) pattern resulting from superposition of two types of domains of a (2×1) structure on a cubic {001} surface.

believed [8] that the absolute magnitude of the diffracted intensities is strongly affected by the roughness of the surface (steps, defects etc.), while the calculations assume a perfectly planar surface. Therefore, quantitative agreement in the absolute intensities of observed and calculated beams is not expected until the effects of surface roughness can be accounted for in the theoretical treatment. This is unfortunate because it is very difficult to find a quantitative and objective criterion by which to judge the agreement between calculations and observations. A convenient " reliability factor " R could be defined [9] as $R = \Sigma(cI_{calc} - I_{obs})^2/\Sigma I_{obs}^2$, where I_{calc} and I_{obs} are the calculated and observed intensities, respectively, c is a scaling constant and the sums are taken over all energy points for which observed data are available (say, every eV). This definition of R has the advantage that the scaling constant c can be calculated for each beam so as to make R a minimum, but may have the disadvantage that the magnitude of the intensity is weighed equally to the energy position of each peak in the spectrum, which is probably unwise. Thus, a wholly satisfactory definition of a " reliability factor " for surface-structure analysis is still lacking. For the time being, structural models are evaluated by making semi-quantitative comparison of calculated intensities with the observed ones, the emphasis being on positions, shapes and relative magnitudes of the intensity peaks in the LEED spectra.

The theoretical curves presented in Fig. 2(a) have been calculated for a model in which the interplanar spacing between the first and the second atomic layers is about 11 % smaller than that between any two successive {001} planes in the bulk—this considerably improves the agreement between theory and experiment over the model of a surface obtained from simple termination of the bulk structure.[6] The atoms in top atomic layer are found to vibrate more vigorously than those in the bulk, corresponding to a surface Debye temperature of 150 K compared with a bulk Debye temperature of 360 K.

Some results for the Ti(0001) surface are presented in fig. 2(b) for angle of incidence $\theta = 20°$ and azimuth $\phi = 30°$ (incident beam in the (01.0) plane). The model postulated for the calculations involves a slight contraction (by about 2 %) of the top close-packed atomic layer with respect to the bulk. The agreement between theory and experiment is satisfactory. Equally satisfactory correspondence with experiment is found for about 20 more beams at different angles of incidence and azimuths.[7]

B. OVERLAYERS

If the adsorption of foreign atoms on a surface gives rise to an ordered distribution of the adsorbed atoms the observed LEED pattern gives information about the symmetry, the periodicity and the structure of (i.e., atomic arrangement in) the overlayer-substrate-system. At present, only the simplest among such systems have been subjected to structure analyses, namely, those for which the periodicities in the plane of the surface are, at most, twice as large as those of the underlying substrate, so that the number of diffracted beams per unit mesh varies from 1 to a maximum of 4. Some of the corresponding LEED patterns are shown schematically in fig. 3. The top row in this figure refers to a cubic {001} substrate: (a) represents the LEED pattern of a 1 × 1 structure, namely, one whose periodicities are the same as those of the underlying substrate (although the intensities are, in general, different); (b) is the pattern of a $c(2 \times 2)$ structure, in which the periodicities are twice as large as that of the substrate and the unit mesh is centred; (c) represents a $p(2 \times 2)$ structure, in which the periodicities are again twice as large as that of the substrate but the unit mesh is primitive. We discuss below an example of (a) and one of (b). The second row in fig. 3 depicts patterns of equivalent structures on close-packed surfaces, (d) representing a 1×1; (f) a $p(2 \times 2)$ structure, while (e) represents a $\sqrt{3} \times \sqrt{3} - 30°$ structure, fairly common on f.c.c.{111} or h.c.p.(0001) surfaces. In the third row, the pattern sketched on the left is assumed to result from the contributions of two domains of a (2×1) structure, the two domains being rotated 90° with respect to each other as indicated by the two schematic LEED patterns drawn on the right in (g). This is believed to be the case for the clean {001} surface of silicon, as well as for overlayers of caesium and oxygen on Si{001}.[10]

We have recently studied the atomic arrangement in an overlayer of silicon on Mo{001}, which forms a 1×1 structure, and that of an overlayer of nitrogen on Mo{001}, which forms a $c(2 \times 2)$ structure. Both studies will be described in detail elsewhere [11, 12] but some results are presented below. Fig. 4(a) refers to the 1×1 overlayer of silicon on Mo{001} and displays comparison between 00 and $\bar{1}0$ spectra at two angles of incidence ($\theta = 8°$ and $\theta = 21°$). The model for which the theoretical curves were calculated is the one depicted schematically in fig. 1(a), with the Si atoms located in the four-fold symmetrical hollows formed by four adjacent Mo atoms on the surface, at a distance of 1.16 Å from the top plane of Mo atoms. The agreement between all spectra calculated for this model and their experimental counterparts is overall very satisfactory. Other models have also been tested, of course, in particular, the structure resulting from Si atoms adsorbed on the bridges across two adjacent Mo atoms, and that resulting from Si atoms adsorbed on the top of each surface Mo atom. The results indicate that while agreement may accidentally be found for some occasional beam, for most beams the correspondence between calculations and observations is very poor, so that the models tested must be discarded with the exception of the one, shown in fig. 1(a), that produces satisfactory correspondence for all beams measured experimentally.

The situation is much less clear for the $c(2 \times 2)$ structure of nitrogen on Mo{001}. In this case the location of the adsorbed atoms either in the four-fold symmetrical hollows, or in the bridge sites, or on the tops of the underlying substrate atoms (while keeping the periodicity required by a $c(2 \times 2)$ net) produces neither strong agreement nor strong disagreement with the experimental data.[12] For some beams (in particular, those chosen for display in fig. 4(b)) the model presented in fig. 1(b) produces somewhat better agreement with experiment than the other models tested. For other beams, however (not shown here), the correspondence between observations and calculations leaves much to be desired, so that the solution of this particular structural problem is still in doubt, although the simple model of fig. 1(b) (which was proved to be correct for other $c(2 \times 2)$ overlayer structures [1]) is somewhat favoured.

3. CRITIQUE

The results obtained from the studies discussed above, together with others reported in the literature, allow us to draw some conclusions at this stage of development of surface crystallography.

The work with clean unreconstructed metal surfaces is very satisfactory: no serious discrepancies are observed in the comparison between theory and experiment, although in some cases there is still room for improvement. The information that structure analysis provides concerns the interplanar spacing between the top and the second layer of atoms on the surface, the vibrational characteristics of the surface atoms, the inner-potential and the absorption component of electrons in the surface layer. The work with overlayer structures is not always as convincing. There are cases (the 1×1 overlayer of silicon on Mo{001} discussed above is one of them) for which the analysis and the results are as good and as convincing as those of clean metal surfaces.[1] But there are others (such as the $c(2 \times 2)$ structure of nitrogen on Mo{001} mentioned above) which appear puzzling because the agreement between calculations for a given model and experimental observations is very good for a few beams but rather unsatisfactory for a few others. It is difficult to say, in these cases, that the model is wrong, because it produces satisfactory results for a number of beams—it is worrying to think that a wrong model can produce such good agreement with experiment for several beams. But it is equally difficult to say that the model is right, because there are several beams for which the agreement with experiment is very unsatisfactory. The reasons for these discrepancies are not understood at present but it is probable that they are of experimental nature.

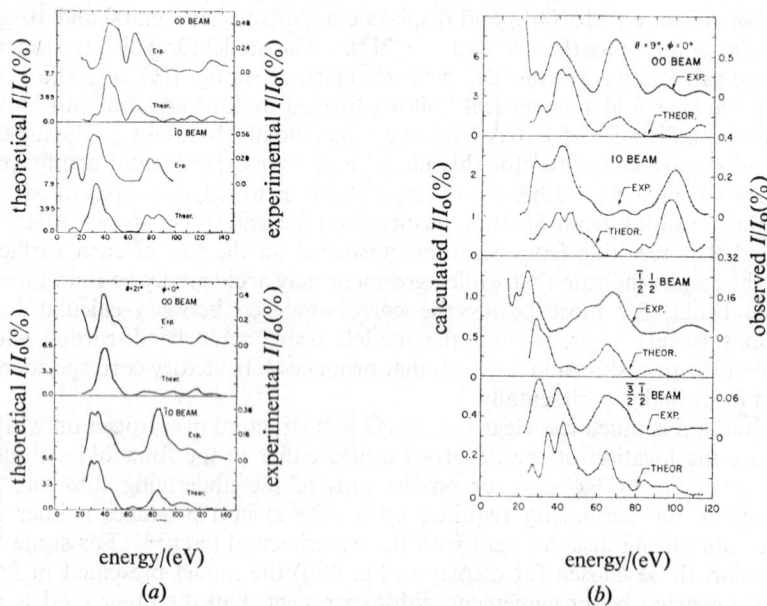

FIG. 4.—Observed and calculated LEED spectra for: (a) 1 × 1 structure of an overlayer of silicon on Mo{001}; (b) $c(2 \times 2)$ structure of an overlayer of nitrogen on Mo{001}.

The preparation of overlayer structures is very difficult to control. As long as the amount deposited on the substrate is less than that corresponding to monolayer coverage, multilayer formation is unlikely, but "defective" regions of the substrate surface may give rise to less well-ordered or even completely different phases. The observed LEED pattern would then result from a superposition of beams from all phases, ordered and disordered, present on the investigated portion of the surface. If the concentration of adsorbate is equivalent to, or somewhat larger than, that corresponding to one monolayer coverage then multilayer formation or partial diffusion into the substrate is possible over a substantial portion of the surface. In either case, some beams are likely to be affected more than others. In addition, present-day techniques for data collection are usually so slow that not all beams in any given large set are certain to refer exactly to the *same* surface structure. The photographic technique recently developed by Somorjai and coworkers [13] is a considerable improvement in this respect. Electronic systems that ought to make data collection trouble-free are likely to be introduced in the near future. Control and reproducibility of surface and overlayer-structure preparation are more difficult to achieve, but more attention will have to be devoted to these problems in the future than has been heretofore.

From the theoretical point of view, consideration will have to be given to the possible co-existence of different phases on the surface and to the calculation of their effects on the LEED pattern. This requirement makes it more and more desirable that computer programmes for dynamic calculations be made less demanding on computer time and core size than they are at present. Encouraging progress is being made in this direction in several quarters.[14-16]

The work discussed above has been made possible by partial support from the Air Force Office of Scientific Research, Air Force Systems Command, and the National Science Foundation; both are gratefully acknowledged.

[1] A recent review of surface crystallography, by J. A. Strozier, Jr., D. W. Jepsen and F. Jona, appears as chap. I in *Surface Physics of Crystalline Materials*, J. M. Blakely, ed. (Academic Press, New York, 1975).

[2] The (0001) surfaces of beryllium and zinc have been objects of earlier investigations (J. A. Strozier, Jr. and R. O. Jones, *Phys. Rev. Letters*, 1970, **25**, 516; 1971, **B3**, 3228; C. B. Duke and C. W. Tucker, Jr., *Phys. Rev. B*, 1971, **3**, 3561; F. Jona, J. A. Strozier, Jr., J. Kumar and R. O. Jones, *Phys. Rev. B*, 1972, **6**, 407; J. M. Baker and J. M. Blakely, *Surface Sci.*, 1972, **32**, 45), but the model calculations were made with approximations that seriously limited the accuracy of the results.

[3] D. W. Jepsen and P. M. Marcus, in *Computational Methods in Band Theory*, ed. P. M. Marcus, J. F. Janak and A. R. Williams (Plenum Press, New York, 1971), pp. 416-443.

[4] D. W. Jepsen, P. M. Marcus and F. Jona, *Phys. Rev. B*, 1972, **5**, 3933.

[5] J. E. Demuth, D. W. Jepsen and P. M. Marcus, *Phys. Rev. Letters*, 1973, **31**, 540; *Solid State Comm.*, 1973, **13**, 1311; *Phys. Rev. Letters*, 1974, **32**, 1182.

[6] A. Ignatiev, F. Jona, H. D. Shih, D. W. Jepsen and P. M. Marcus, submitted to *Phys. Rev.*

[7] H. D. Shih, F. Jona, D. W. Jepsen and P. M. Marcus, to be published.

[8] D. W. Jepsen, P. M. Marcus and F. Jona, *Phys. Rev. B*, 1973, **8**, 5523.

[9] A. Ignatiev, F. Jona, D. W. Jepsen and P. M. Marcus, *LEED 7 Seminar Notes* (San Diego, California, March 19-21, 1973), p. 74.

[10] B. Goldstein, *Surface Sci.*, 1973, **35**, 227.

[11] A. Ignatiev, F. Jona, D. W. Jepsen and P. M. Marcus, submitted to *Phys. Rev.*

[12] A. Ignatiev, F. Jona, D. W. Jepsen and P. M. Marcus, submitted to *Surface Sci.*

[13] P. C. Stair, T. J. Kaminska, L. L. Kesmodel and G. A. Somorjai, *Phys. Rev.*, in press.

[14] S. Y. Tong, *Bull. Amer. Phys. Soc.*, 1974, (II) **19**, 334.

[15] J. B. Pendry, private communication, 1974.

[16] S. W. Tam and J. A. Strozier, Jr., *Bull. Amer. Phys. Soc.*, 1974, (II) **19**, 333; *ibid.*, 1975, (II) **20**, 387.

Data Averaging and Pseudo-kinematical Approaches to the LEED Surface Structure Problem

By D. P. Woodruff

Physics Department, University of Warwick, Coventry CV4 7AL.

Received 4th March 1975

Some of the problems of constant momentum transfer averaging and Fourier transform approaches to the analysis of LEED data are investigated. Even if these techniques satisfactorily circumvent the multiple scattering problem, the complex nature of scattering factors in LEED leads to a phase problem not present in the commonly used analogue situation of X-ray diffraction. Model calculations show that this leads to a loss of information and hence sensitivity in the application of averaging to adsorbate structures, and to quite spurious results in the application of Fourier transforms to such structures.

While the technique of low energy electron diffraction (LEED) was " discovered " in 1927 [1] and has been worked on extensively both experimentally and theoretically, especially since about 1960, it is only recently that claims have been made for proper surface structure analysis using this technique.[2-5] The delay has been primarily associated with the difficulties created by the strong elastic scattering cross-sections the ion cores of a solid present to electrons in the LEED energy range, making multiple scattering an important process. Calculations of diffracted intensities to compare with experiment for structure determination must therefore be based on multiple scattering (" dynamical ") theories, and it is easy to show that a single scattering (" kinematical ") theory produces a very poor description of experimental results. It is now clear that the best dynamical calculations do permit structure determinations to be made for clean metal surfaces and simple adsorbates on these surfaces. Complex adsorbates, or covalently-bonded solids, still lie beyond the scope of present calculations. Moreover, even in the favourable cases, very long (and expensive) computer programmes are needed to determine a surface structure.

In view of these complexities, any method of circumventing the multiple scattering problem is extremely attractive and a number of proposals have been made to do this; none can be proved to be rigorously valid, but each has an essential plausibility which makes further study worthwhile. The three methods proposed so far are:

(i) energy averaging of experimental diffracted beam intensities,

(ii) constant momentum-transfer averaging of experimental intensities, and

(iii) Fourier transform methods and their developments.

The first of these methods, proposed by Tucker and Duke,[6] will not be discussed in detail in this paper. It is designed only to find the overlayer structure and does not attack the interesting problem of overlayer-substrate registry. There is conflicting evidence regarding its applicability for its intended objective.[7,8]

In order to assess the potential usefulness of these techniques for the analysis of specific structures, a brief description of the theoretical problem is necessary. Because these theories owe much, in their derivation, to the kinematical theory generally applicable to X-ray diffraction, it is useful to cast the theory in a similar form. Thus, the intensity of a particular diffracted beam is proportional to the square of the modulus of the geometrical structure factor,

$$F(hkl) = \sum_{n=1}^{N} f_n \exp [2\pi i (hx_n + ky_n + lz_n)],$$

where, in LEED, h and k denote the diffracted beam and l the momentum transfer perpendicular to the surface; l is a continous variable because of the relaxation of the third Laue condition due to the absence of translational symmetry perpendicular to the surface. Thus, while in X-ray diffraction the summation is over a unit cell, in LEED the summation is over a cell defined by the surface unit mesh and the surface normal (and is strictly infinite in the direction normal to the surface, although the electron damping ensures it need only be a few atom layers deep). The f_n are the scattering factors of the ion cores at sites (x_n, y_n, z_n). In X-ray diffraction the f_n are real and the phase or imaginary component of $F(hkl)$ is determined entirely by the positions (x_n, y_n, z_n) of the scatterers. In LEED the f_n are complex; moreover, in a proper formulation they are modified by the effects of multiple scattering in a complicated way with a dependence on incident beam direction, energy and scatterer site. The effects of electron damping may be included in the f_n or can be dealt with analytically to reduce the size of the " unit cell "; this separation will be assumed in the following discussion.

LEED data are generally presented in the form of plots of the intensity of a diffracted beam versus energy (I-E plots) and hence are probes in k-space. If the f_n were real, the phase of scattering amplitude from each ion core would be dictated entirely by its location, so that the position of intensity peaks would be *only* a function of the structure of the surface; the actual value of the intensity would, of course, still be a function of the way the f_n change with beam energy and direction.

The complex nature of the f_n thus gives rise to intensity peak shifts relative to the locations implied only by structure. Multiple scattering, by modifying both the modulus and phase of f_n, can thus introduce new peaks as well as shifting peaks. Moreover, even in the absence of multiple scattering, peak positions associated with interference between scattering from different atomic species (different complex f_n) will be a function of the f_n as well as the scatterer locations or structure. This effect will be referred to as the " phase problem ". The nature of this problem on the different averaging techniques will be assessed separately.

CONSTANT MOMENTUM TRANSFER AVERAGING

The constant momentum transfer averaging (CMTA) technique, proposed by Lagally et al.[9,10] is based on the idea that the effect of multiple scattering is to modulate the intensity-energy curves which would be obtained were LEED kinematical. Although we have already seen that shifts in single scattering peaks can also occur, it is certainly experimentally observed that most intense peaks occur in the vicinity of single scattering or " Bragg " peaks; this has been referred to by Tucker and Duke[6] as the " Bragg envelope " effect. Thus, as the location of single scattering peaks is (in the absence of phase differences in the f_n) only a function of the momentum transfer in the scattering (or the " scattering vector ") and multiple scattering effects are a function of other parameters, an averaging of many I-E plots for different angles of incidence and azimuth but constant momentum transfer might be expected to fortify single scattering effects and " wash-out " multiple scattering effects. Data from a number of clean metal surfaces have been subjected to this treatment; for f.c.c.

(111) surfaces (Ni,[10] Ag,[9] Cu,[11]) the resulting curves are extremely kinematical and lead to a structural conclusion that there is little surface layer expansion or contraction. Also studied are Al (100),[12] Cu (100), [11, 13] and W (110) [14, 15] surfaces where similar conclusions result but with a little more multiple scattering information remaining. In all these cases, however, all the f_n, in the absence of multiple scattering, are the same because the scatterer is an elemental solid, and hence the phase problem is not involved.

One of the most interesting problems in LEED, however, is the structure of adsorbate-covered surfaces. In this case the surface region contains two or more species of scatterer with different f_n even in a kinematical approximation. Thus, the location of intensity peaks associated with adsorbate-substrate scattering (giving the important information of adsorbate/substrate registry) is not only a function of the structure, but also of the different f_n. Moreover, the phase component of f_n is a function of energy and scattering angle so, in any two intensity-energy curves for the same beam at different incidence angle, different sets of f_n apply and peak locations will be at different momentum-transfer. Thus, when a momentum-transfer average is executed there will be a tendency for adsorbate/substrate kinematical peaks to wash out rather than be fortified. Of course, in any comparison of theoretical and experimental averages, the f_n will (in principle) be known, so that the final results should be properly comparable. However, the effect will clearly greatly *reduce the inherent sensitivity* of the technique relative to that shown in individual I-E plots or relative to averaged results from a reconstructed clean surface. While this is therefore a general statement of the problem it is impossible to assess its general importance quantitatively. However, some calculations of specific examples can be made to see if the effect can be important, at least in some cases. Model calculations were therefore performed for half a monolayer of an adsorbate sited at 4-fold coordination sites on a Cu(100) surface. The adsorbate/top-substrate layer spacing was set to 0.75 times the layer spacing of the substrate (1.805 Å). Fig. 1 shows the (00) beam intensity for angles of incidence of 8–33° for Cu, Ag and O adsorbates plotted against momentum transfer; the scale on this axis corresponds to the " Bragg order " for the substrate. Also shown are the averages of these curves. In the absence of phase effects all curves would be expected to show substrate peaks at integral Bragg orders, and adsorbate/substrate peaks at integral multiples of 1.33 (in particular, 2.67); they would also be similar for different angles of incidence apart from the effects of different scattering amplitudes (i.e., the modulus of f_n). As expected, the Cu on Cu calculations show this general behaviour although the large peak widths associated with the necessity to include the elastic scattering effects in the imaginary (damping) part of the potential [16] make the extra feature at 2.67 and the normal 2 peak wash each other out. Peak positions are, however, essentially independent of incidence angle. This is clearly not true for the foreign adsorbates. Peaks positions shift significantly with changing incidence angle so that the average shows structure which is some average of these shifts. If f_n is written in the form $Ae^{i\Delta}$ the two components, A (the modulus of f_n or " scattering strength ") and Δ, the phase component, can be plotted. Fig. 2 shows these components for the three species for three scattering angles of 164, 144 and 114°. These correspond to angles of incidence of 8, 18 and 33° for the (00) beam, and the results are plotted against the same parameter (Bragg order) used in fig. 1 so that the results can be compared. Clearly the phase problem is very real. The phase difference between different scatterers can be large and is a strong function of energy and scattering angle. These changes in phase difference will lead to a significant reduction in sensitivity of averaged data to adsorbate/substrate registry changes.

In view of this, it is interesting to consider the effect of inaccuracies in the description of the scattering properties of the different ionic species. It is possible that extensive averaging may make results more sensitive to these inaccuracies than in the normal, single curve matching used in dynamical theory approaches. To assess the extent of this effect, the problem of the $(\sqrt{2} \times \sqrt{2})R\,45°$ oxygen on Cu (100) structure, previously studied using the averaging technique,[11] has been re-assessed using a different set of scattering factors. The original analysis used values based on phase

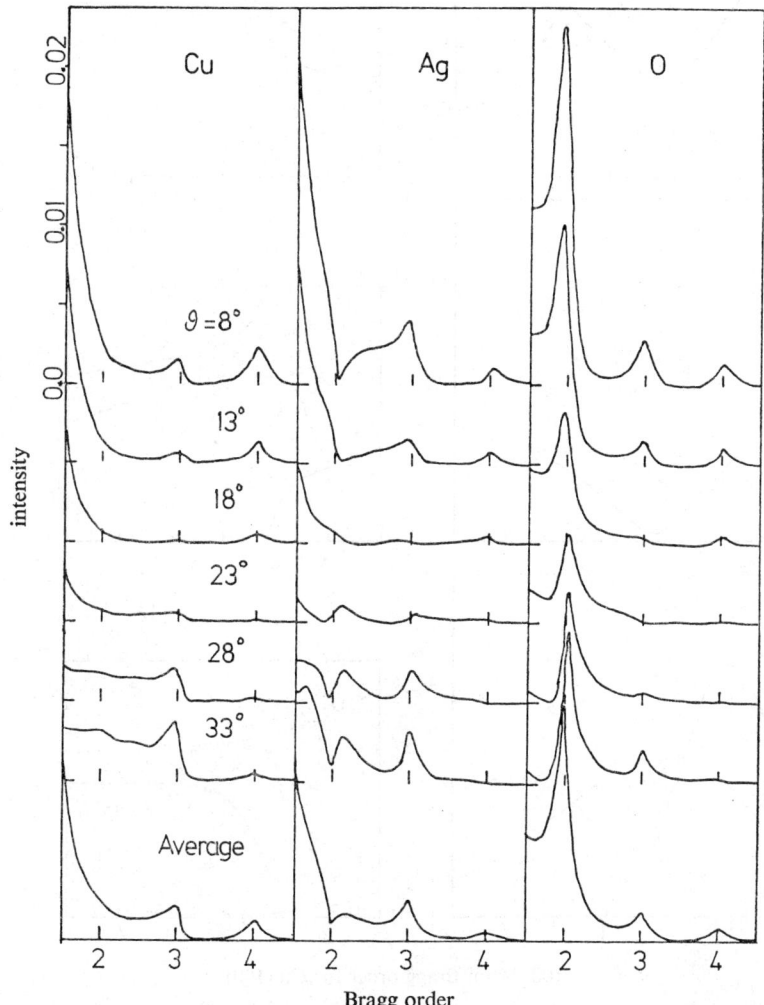

FIG. 1.—Calculated kinematical intensities for model structures of a planar monolayer of Cu, Ag or O on a Cu(100) surface at a spacing of 0.75 times the substrate layer spacing. θ is the angle of incidence and intensities are expresssed relative to the incident beam intensity.

shifts (in a partial wave expansion) calculated by Zimmer[17] using the scheme of Pendry.[18] These are the values used for calculations presented in this paper. Recent results in dynamical LEED calculations from Ni suggest that the method of calculation used for these phase shifts is not ideal [19,20] and an alternative set of phase shifts, based on a calculational scheme found best for Ni, have been used for Cu[21] and O.[22]

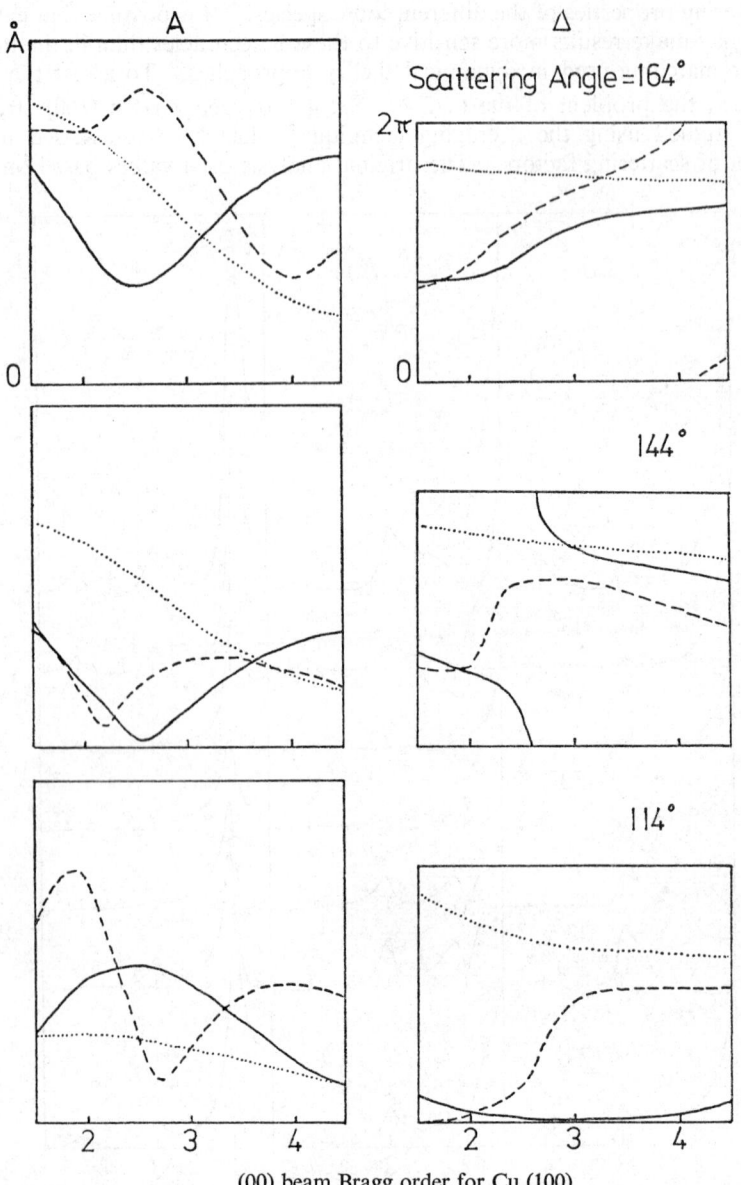

(00) beam Bragg order for Cu (100)

Fig. 2.—Modulus A and phase Δ of scattering factors for Cu (full line) Ag (dashed line) and O (dotted line) at three scattering angles equivalent of angles of incidence for an (00) beam of 8, 18 and 33°.

Using these phase shifts a new set of theoretical averages have been calculated for a range of trial structures; the general form of these results are very similar to those calculated originally. Fig. 3 and 4 allow a comparison of these results; in the original analysis, peak shifts were investigated for some peaks in (00), (10) and (11) beam data relative to the clean surface, and these shifts were expressed in momentum-transfer units as a function of a Bragg Peak order. Thus, a shift of 0.01 corresponds to a shift of 1 % of the spacing of the Bragg peaks. In fig. 3 a comparison of peak shifts for the " old " and " new " phase shifts is shown for a range of 4-fold coordinated atomic overlayer structures and fig. 4 shows a similar comparison for "reconstructed" structures with a top layer consisting of alternate Cu and O atoms. The area of the circles in the figures is proportional to the number of occurrences of a particular point.

The accuracy of shift measurement is certainly no better than $\pm$ 0.01 and in both cases 95 % of points occur within $\pm$ 0.02 of the lime of exact agreement; for recontructed surfaces, 90% lie within $\pm$0.01, for the atomic overlayer, 76 % are within $\pm$ 0.01. Thus the agreement is generally rather good and CMTA does seem relatively insensitive to imperfections in phase shift calculations; on the basis of dynamical theory experience with Ni [19, 20] the averaging technique is probably less sensitive to such changes than are dynamical theory calculations.

FOURIER TRANSFORM TECHNIQUES

Fourier transform techniques, and specifically the Patterson function, have proved extremely valuable in X-ray diffraction. They have a particular attraction because they provide a *direct* method of structure analysis rather than one relying on a comparison of experiment and theoretical calculations based on a range of possible trial structures. Interest in application of such methods to LEED was revived by a paper by Clarke, Mason and Tescari [23] who used a very simple modification of the Patterson function for LEED from adsorbates on Pt(100). Unfortunately, the apparent success of this work was misleading as the results were dominated by truncation effects and distorted by an unrealistic choice of inner ptoential.[24] A modified transform, the optical transform, largely overcomes the truncation effects as has been discussed by Buchholtz, Lagally and Webb [25] and Landman and Adams.[26] However, as has been previously pointed out,[24] the phase problem is very severe in Fourier transform methods because they implicitly assume the f_n to be real (or to have a constant phase factor). Thus, peak shifts in data due to differences in the phase of f_n for different scatterers are transformed into structural changes. This effect is clearly potentially serious, but can only be assessed quantitatively by model calculations. The extent to which Fourier transform techniques remove multiple scattering effects is also in some doubt. A careful study by Buchholtz et al.[25] has shown that experimental data taken at different angles of incidence from the same surface lead to appreciably different transforms. No attempt is made here to assess this further and, in order to investigate the extent of the phase problem, data are assumed to be kinematical; however, we include transforms performed on data subjected to CMTA as these may prove the only data sufficiently kinematical-like for the transform procedure to be valid, even in the absence of the phase problem. Of course, even if this proved necessary, the directness of the transform approach would still prove valuable; also, the averaging may reduce the phase problem, albeit at the expense of reduced sensitivity. Fig. 5 shows one dimensional " optical transforms " [25] ($|P(O, O, z)|$,[2] where P is the complex Patterson function) for a range of model structures of a planar monolayer of Cu, Ag or O on a Cu(100) surface with adsorbate/substrate layer spacings of c, $1.3c$ and $0.75c$, where c is the substrate interlayer spacing. Transforms are presented for a single kinematical theoretical (00) beam I-E plot (angle of incidence, $\theta = 8°$) and for similar data subjected to CMTA over a range $8° \leqslant \theta \leqslant 33°$. Not surprisingly, the best transform is for a Cu overlayer at a spacing of c (i.e., a clean surface); the single θ transforms shows some extra features but these are removed in the average transforms and are thus associated with fluctuations in $|f_n|$ with energy. Ag or O overlayers at c spacing produce considerable

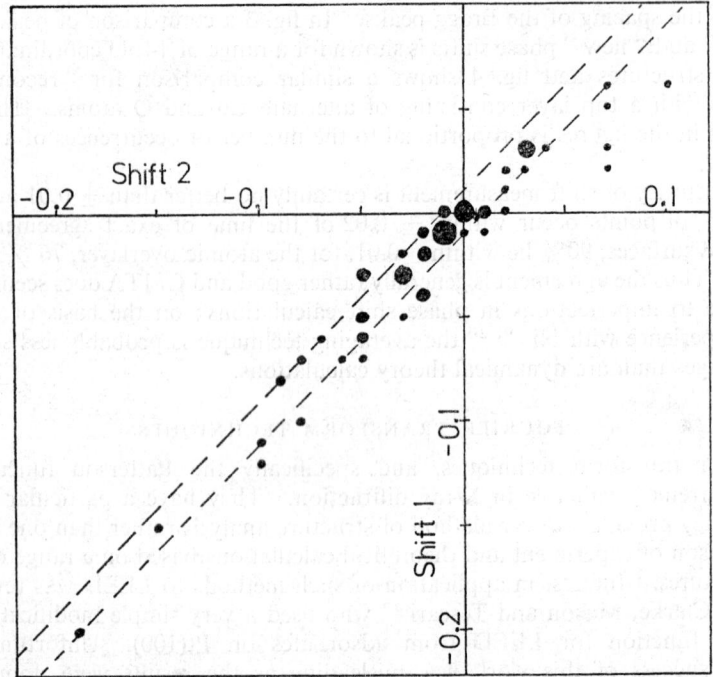

FIG. 3.—Correlation of peak shifts (expressed as a function of a Bragg order) for various 4-fold adsorbed ½-monolayer oxygen overlayers using original phase shifts [11, 17] to produce shift 1 and new values [21, 22] to produce shift 2. The dashed lines delineate the region in which the results agree to ±0.01. All structures are on a Cu(100) substrate.

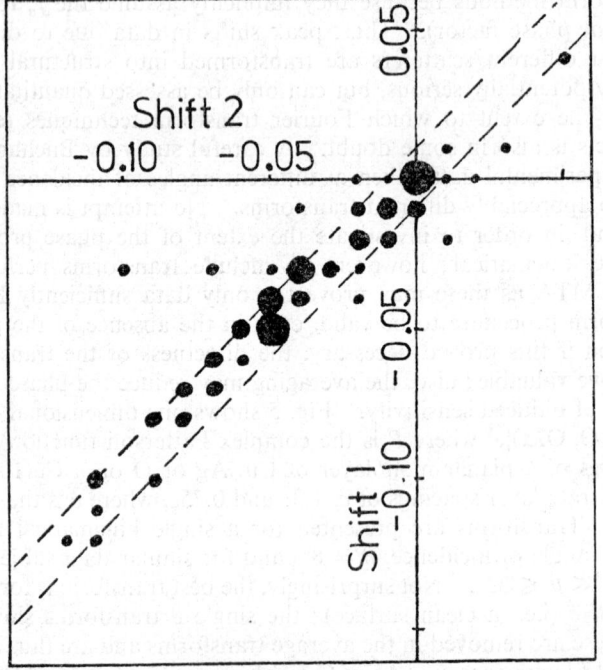

FIG. 4.—Similar plot to fig. 3 but for a range of reconstructed surface models.

weakening of the peaks expected for both adsorbate/substrate and substrate/substrate scattering interference, presumably due to the phase effect. Cu overlayers at other spacings give reasonable transforms (though with imprecisely located peaks, presumably associated with the limited data range of Bragg order 2 to $5\frac{1}{2}$), but foreign overlayers at these other spacings could barely be deemed to show any interpretable

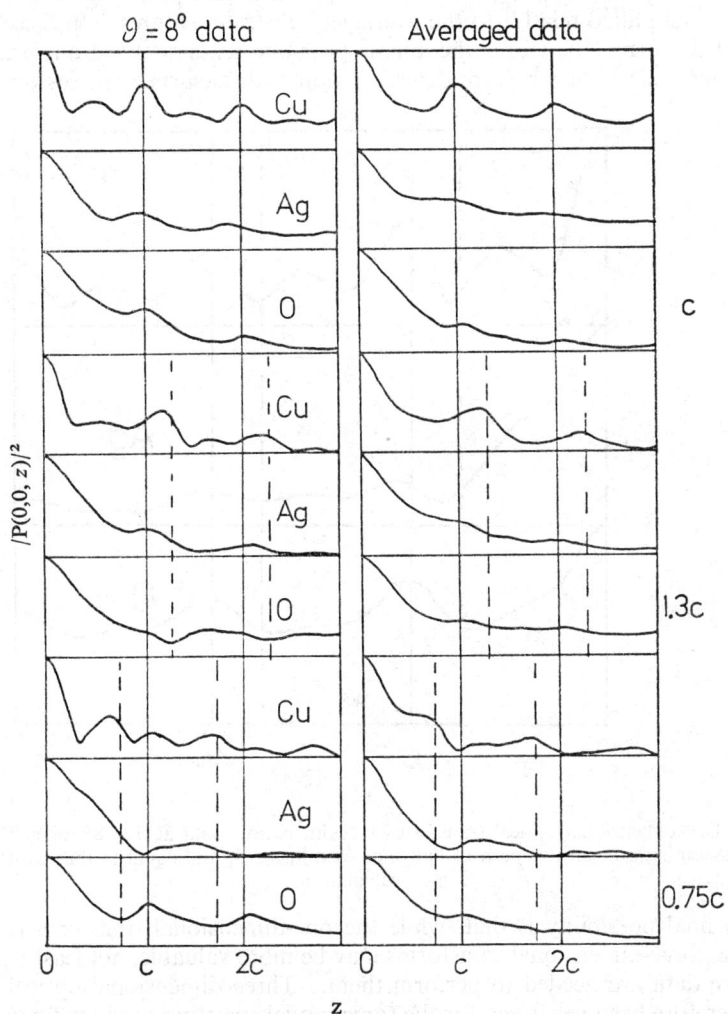

FIG. 5.—One-dimensional optical transforms of kinematical data for model structures of a planar monolayer of Cu, Ag or O at a spacing of c, $1.3\,c$ and $0.75\,c$ where c is the substrate layer spacing. The full lines at c and $2c$ show predicted substrate vectors; the dashed lines show the location of predicted adsorbate/substrate vectors. All data are calculated assuming a mean-free-path for electron damping appropriate to averaged data from Cu (100).[11]

adsorbate/substrate interference effects. The LEED data for these transforms are based on an imaginary component of the inner potential taken from experiments on averaging for Cu (100) and (111) data.[11] While this is certainly appropriate for averaged curves, it includes a contribution from elastic scattering [16] which may be deemed unsuitable for single θ curves. Fig. 6 shows the single θ ($\theta = 8°$) transforms of the 0.75c overlayer spacing now based on an imaginary potential half as large and thus roughly equal to the value used in dynamical theory calculations. These transforms therefore take a very favourable view of LEED data, i.e., that it is kinematical and that peak widths are as narrow as the effects of multiple scattering can make them. The transforms, however, are essentially the same as those in fig. 5 except that, as may be expected for a longer mean-free-path, the substrate/substrate peaks are amplified relative to the overlayer/substrate features. In these one-dimensional transforms therefore, the phase problem renders this transform technique essentially useless for structure determination with adsorbed species on the surface.

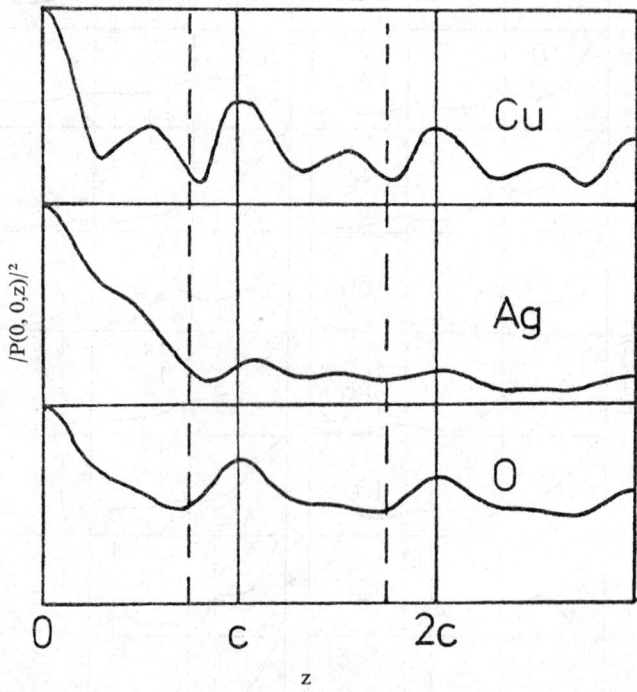

FIG. 6.—One-dimensional optical transforms for kinematical data at $\theta = 8°$ for a 0.75c overlayer spacing assuming a mean-free-path of half that used in fig. 5, and equal to that used in dynamical calculations.

One final possibility is that, while the one-dimensional transform is of little (if any) use, three-dimensional transforms may be more valuable, not least because more extensive data are needed to perform them. Three dimensional optical transforms have therefore been generated for the same model structure used for fig. 6, now using (00) (10) and (11) beam data over the same range of Bragg orders (2 to $5\frac{1}{2}$). Sections of transforms space were taken in 0.1c intervals from 0 to 2.2c perpendicular to the surface and for each peak shown in transform space the amplitude in the nearest section was noted. Peak locations are thus given to $\pm 0.05c$ and amplitudes will generally be lower than the true peak value. Table 1 shows the amplitudes and locations of all such peaks for these model structures, the amplitudes being normalised to a value of 100 at the origin of transform space. Locations in the plane parallel to the surface are given in terms of the primitive surface net vectors (along $\langle 110 \rangle$ directions and of length 2.55 Å). Note that the use of the primitive *surface* net means that the bulk crystal is effectively described by an *I*-tetragonal rather than *F*-cubic lattice

TABLE 1a.—PEAKS IN OPTICAL TRANSFORMS: SINGLE ANGLE OF INCIDENCE ($\theta = 8°$)

top layer spacing	c									1.3c									0.75c									
Adsorbate site	Cu			Ag			O			Cu			Ag			O			Cu			Ag			O			
z =	00	0½	½½	00	0½	½½	00	0½	½½	00	0½	½½	00	0½	½½	00	0½	½½	00	0½	½½	00	0½	½½	00	0½	½½	
0.0	100	22		100			100	10		100	22		100			100	19	6	100	17	13	100			100	20		
0.1																												
0.2			42		51							31		35									26				20	
0.3		17				13					15													27				
0.4								11			32									13						43		
0.5	41				48															21								
0.6																												
0.7		14	58		14	—		—			10	37		15	33		12			8	22		7	21				
0.8								13				□					10							□				
0.9	33			36													11					12						
1.0											7			13									7					
1.1		11		30																20	4			7				
1.2		28		31																								
1.3								18	6		—			—	15		19	5		12	6		10	2		22		
1.4	37	9		—	8															3								

TABLE 1b.—PEAKS IN OPTICAL TRANSFORMS: AVERAGED DATA

top layer spacing	c									1.3c									0.75c								
	Cu			Ag			O			Cu			Ag			O			Cu			Ag			O		
Adsorbate site z =	00	0½	½½	00	0½	½½	00	0½	½½	00	0½	½½	00	0½	½½	00	0½	½½	00	0½	½½	00	0½	½½	00	0½	½½
0.0	100	48	29	100	36	20	100		17	100	47	22	100		33	100	22	21	100		40	100		28	100	29	33
0.1																											
0.2																											
0.3																											
0.4																											
0.5																								28			37
0.6																										19	
0.7																											35
0.8								18																			
0.9								—						11													
1.0	29	51																									
1.1											—			13			31										
1.2				18	23							30 □		13 □							33 □			22			
1.3																		13 □									
1.4																											
1.5														15							7						
1.6																					8 □			2	27 □		9
1.7																											
1.8						31																					
1.9																											
2.0	33	18								21	7																
2.1								—										—									—

Expected substrate/substrate vectors are underlined   Expected adsorbate/substrate vectors denoted by boxes.

(although distances normal to the surface are measured in units of c rather than the height of the I-tetragonal unit cell which is $2c$). All peaks at $(0, \frac{1}{2}, z)$ are spurious and are associated with data truncation in taking only two non-specular beams. The result of this analysis, as may be seen in table 1, is similar to the one-dimensional transforms; various spurious peaks occur, Cu overlayers produce most of the expected peaks, but for foreign adsorbates *no* predicted adsorbate/substrate vector is correctly described in the transform. Some peaks, by virtue of their amplitude and occurrence in the correct surface net site, can probably be associated with adsorbate/substrate vectors, but all have incorrect z location. A reduction in imaginary component of the inner potential (cf. fig. (6)) does not change this picture.

Conclusions

The effect of the phase problem on the CMTA and Fourier transform approaches to LEED has been investigated. For CMTA, a marked reduction in sensitivity is observed relative to unaveraged LEED data (apart from that due to peak broadening [16]). The technique must therefore clearly be limited for the determination of foreign adsorbate structures. The effect on Fourier transforms is much more severe; not only the sensitivity but also the correctness of implied results are affected. Straight Fourier transforms are clearly of little, or no, value for adsorbate structure determination; no modification to the transform will alter this result, although it is conceivable that a deconvolution procedure, in which the known f_n are injected into the analysis, may be fruitful. However, the only method proposed so far [27] assumes dynamical effects are not important which is contrary to the findings of Buchholtz et al.[25]

[1] C. J. Davisson and L. H. Germer, *Phys. Rev.*, 1927, **30**, 705.
[2] S. Andersson and J. B. Pendry, *J. Phys. C, Solid State Phys.*, 1973, **6**, 601 but see also J. E. Demuth, D. W. Jepsen and P. M. Marcus, *J. Phys. C, Solid State Phys.*, 1975, **8**, L25.
[3] F. Forstmann, W. Berndt and P. Büttner, *Phys. Rev. Letters*, 1973, **30**, 17.
[4] C. B. Duke, N. O. Lipari, G. E. Laramore and J. B. Theeten, *Solid State Comm.*, 1973, **13**, 579.
[5] J. E. Demuth, D. W. Jepsen and P. M. Marcus, *Solid State Comm.*, 1973, **13**, 1311.
[6] C. W. Tucker and C. B. Duke, *Surface Sci.*, 1970, **23**, 411; *ibid*, 1972, **29**, 237.
[7] C. B. Duke and G. E. Laramore, *Surface Sci.*, 1972, **30**, 659.
[8] K. A. R. Mitchell, D. P. Woodruff and G. W. Vernon, *Surface Sci.*, 1974, **46**, 418.
[9] M. G. Lagally, T. C. Ngoc and M. B. Webb, *Phys. Rev. Letters*, 1971, **26**, 1557.
[10] M. G. Lagally, T. C. Ngoc and M. B. Webb, *J. Vacuum Sci. Techn.*, 1972, **9**, 546.
[11] L. McDonnell, D. P. Woodruff and K. A. R. Mitchell, *Surface Sci.*, 1974, **45**, 1.
[12] D. T. Quinto and W. D. Robertson, *Surface Sci.*, 1973, **34**, 501.
[13] J. M. Burkstrand, G. G. Kleiman and F. J. Arlinghaus, *Surface Sci.*, 1974, **46**, 43.
[14] M. G. Lagally, J. C. Buchholtz and G.-C. Wang, *J. Vacuum Sci., Techn.* to be published.
[15] J. C. Buchholtz, G. -C. Wang and M. G. Lagally, *Surface Sci.*, to be published.
[16] J. B. Pendry, *J. Phys. C, Solid State Phys.*, 1972, **5**, 2567.
[17] R. S. Zimmer, University of Warwick, private communication.
[18] J. B. Pendry, *J. Phys. C, Solid State Phys.*, 1971, **4**, 2514.
[19] J. E. Demuth, D. W. Jepsen and P. M. Marcus, *Phys Rev. Letters*, 1973, **31**, 540.
[20] J. B. Pendry, paper presented at meeting on "LEED and Surface Structure" (London, May 1974).
[21] J. F. Janak, A. R. Williams and V. L. Moruzzi, IBM Yorktown Heights, N.Y.
[22] As used in ref. (19): the author is grateful to Dr. J. E. Demuth for providing him with these phase shifts and those of Janak et al.[21]
[23] T. A. Clarke, R. Mason and M. Tescari, *Proc. Roy. Soc. A*, 1972, **331**, 321.
[24] D. P. Woodruff, K. A. R. Mitchell and L. McDonnell, *Surface Sci.*, 1974, **42**, 355.
[25] J. C. Buchholz, M. G. Lagally and M. B. Webb, *Surface Sci.*, 1974, **41**, 248.
[26] U. Landman and D. L. Adams, *J. Vac. Sci., Techn.* 1974, **11**, 195.
[27] D. L. Adams and U. Landman, *Phys. Rev. Letters*, 1974, **33**, 585 and paper presented at this meeting.

Fourier Transforms in Surface Structure Determination from LEED

The Transform-Deconvolution Method

By Uzi Landman

Institute for Fundamental Studies, Department of Physics and Astronomy,
University of Rochester, Rochester, New York 14627

Received 25th February, 1975

Fourier transforms of LEED intensities contain convolution products of functions of the interatomic vectors with data truncation, lattice vibration and potential windows. The dominant structural information in the transform originates from kinematical processes. Extraction of structural parameters of the surface, can be achieved upon a proper deconvolution of the windows from the transforms of the intensities.

Structural parameters are of fundamental importance for the development of theoretical understanding of the properties and behaviour of physical systems. In particular, when considering the surface region of materials one may expect structural changes at the surface compared to the bulk atomic arrangement, which may in turn be correlated to the electronic structure and reactivity of the surface.[1] The diffraction of low-energy electrons (LEED) from solid surfaces offers a method sensitive to the atomic arrangement at the surface (a few atomic layers in thickness), due to the strong attenuation[2] of the propagation of electrons in this energy range ($E \sim 0$ – 500 eV) inside the solid. The complete characterization of the structure requires the analysis of the diffraction patterns (angular distribution of the diffracted beams) which provide information about the two-dimensional symmetry in the direction parallel to the surface, and the much more complicated analysis of the intensity against incident energy curves of the various diffraction beams. One of the main themes in the study of LEED intensities has been the analysis via an *indirect approach* in which a detailed theoretical model of the diffraction process is utilized in computations of the intensities [3, 4] for *postulated* geometrical models of the surface which are then compared with experiment. The *intrinsic* model uncertainties originating from an incomplete description of the diffraction process, and the failure to perform a full variation over the geometrical and electronic parameters due to prohibitive expense in computational time and storage are the main disadvantages of the indirect methods. In a series of recent publications [5-8] we have discussed a *practical direct method* in which structure is determined via a direct analysis of the experimental intensities *without presumptions* about the structure. The main purpose of this paper is to discuss the analytical structure of the Fourier transform of dynamical electron diffraction intensities and to establish the structural content of the transformation. In addition, a brief review of our transform-deconvolution method is presented.

(2a) TRANSFORM CONSTRUCTION

Fourier transformations of intensities measured in diffraction and scattering experiments are useful in the extraction of structural information from experimental data.[5-8, 9-12] With no exception, the constructions of the transformations are based on single-scattering (kinematical) ideas. The strong interaction of low energy electrons with the ion-cores of the scattering material, results however, in the occurrence of higher order, multiple-scattering (dynamical) processes which reflects in the complicated nature of the diffracted intensity curves. The objective of the discussion in this section is to study multiple-scattering effects on transforms of LEED intensities. First, it is instructive to review the principles of the transform construction. In LEED, only the component of the electron momentum parallel to surface is conserved giving rise to diffracted beams characterized by discrete (hk) Miller indices of the two-dimensional surface net. The propagation of the electron normal to the surface is strongly attenuated and is characterized by the continuous variable s, where $2\pi s$ is the normal momentum transfer inside the solid for the incident and backscattered beam, i.e.,

$$2\pi s(\mathbf{g}_0) = \text{Re}[k_\perp(0) + k_\perp(\mathbf{g}_0)], \qquad (1)$$

where Re denotes the real part of $k_\perp(0)$ and $k_\perp(\mathbf{g}_0)$, the normal components of the momentum of the incident and $\mathbf{g}_0$-th backscattered beam, ($\mathbf{g}_0$ is a reciprocal vector of the two-dimensional net, for specular diffraction $\mathbf{g}_0 = 0$). Since only back-scattered electrons are recorded and due to the inner potential of the solid, $s > 0$.

The complex Fourier transform of the intensity is now defined [6] as

$$P(x, y, z) = \sum_{h,k=-\infty}^{\infty} \int_{-\infty}^{\infty} ds(\mathbf{g}_0) I_{hk}(s) \omega_B(s; s_1, s_2) \exp[-2\pi i(hx+ky)] \exp[-2\pi i s z], \qquad (2)$$

where $I_{hk}(s)$ is the intensity of the (hk) beam and $\omega_B(s; s_1, s_2)$ is a truncation function defined as

$$\omega_B(s; s_1, s_2) = 1 \text{ for } s_1 < s < s_2 \qquad (3)$$
$$= 0 \text{ otherwise.}$$

This function confines the integral in eqn (2) to the experimentally available momentum transfer range $[s_1, s_2]$. In structure determination it is useful to consider reduced forms of the three-dimensional transform defined in eqn (2): one-dimensional projections (line projections) and two-dimensional sections. The later are utilized in the determination of interlayer registries (as we have previously demonstrated [5, 8]), while the line projections are used in determination of interlayer spacings and will be the subject of our following discussion.

The one-dimensional projection is defined as [6]

$$P(z) = \int_{-\infty}^{\infty} ds \omega_B(s; s_1, s_2) I_{00}(s) \exp[-2\pi i s z] = W_B(z) * \int_{-\infty}^{\infty} ds I_{00}(s) \exp[-2\pi i s z], \qquad (4)$$

where W_B is the Fourier transform of ω_B; * denotes the convolution operation and I_{00} is the intensity of the specularly diffracted beam, and $s \equiv s(0)$ in eqn (1). As evident from the defining expression (eqn (4)), each real space point z of the transform, derives contributions from the scattered intensity over the entire available momentum transfer range (defined by the function ω_B), weighted by the corresponding phase factors. The Fourier transform, harmonically analyzes the intensity record; it identifies the periodic features (related by a fixed phase) and adds them coherently. The aperiodic features, on the other hand, are added with a variable phase, and hence are effectively averaged over by the transformation. The averaging aspect is demonstrated in fig. 1 where the transforms of measured intensities [7, 8] exhibiting pronounced multiple-scattering characteristics are compared with those of corresponding kinematically calculated intensities. The transforms of the intensities are remarkably similar despite the marked differences in the intensities.

(2b) MULTIPLE SCATTERING TRANSFORMS

A general expression for the diffracted intensity can be written in terms of the total scattering matrix **T** which satisfies the following integral equation,[13]

$$\mathbf{T}_v(\mathbf{k}, E) = \tau_v + \tau_v \sum \mathbf{G}^{vv'}(\mathbf{k}, E)\mathbf{T}_v(\mathbf{k}, E), \qquad (5)$$

where v is a layer index, with $v = 0$ the topmost layer defined by the xy plane (the semi-infinite solid filling the $z \geqslant 0$ half space). $\mathbf{G}^{vv'}$ is the *interplane propagator* matrix and τ_v is the total scattering matrix from plane v. The matrix $\mathbf{T}_v$ describes the

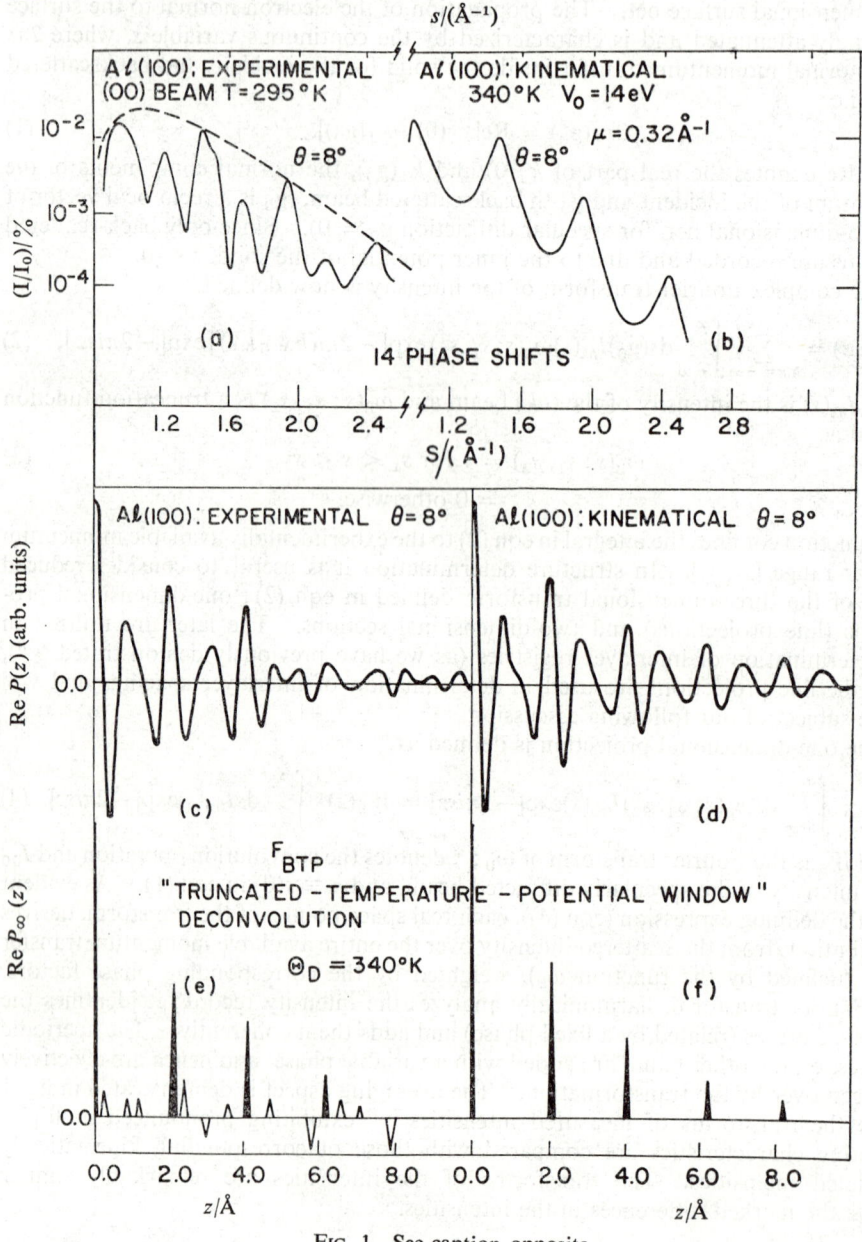

Fig. 1—See caption opposite.

amplitude of scattering processes in the v-th plane including those processes which have been propagated from plane v'. The scattering matrix τ_v satisfies a similar integral equaton

$$\tau_v = t_v[1 + G^{sp}t_v + G^{sp}t_v G^{sp}t_v + \cdots] \equiv t_v \beta, \tag{6}$$

where t_v is the individual ion-core scattering matrix and G^{sp} is the *intraplanar propagator matrix*. (When $\beta = 1$, scattering paths with consecutive scattering events occurring in the same plane are not allowed).

The diffracted intensity can now be written in terms of the Fourier transform of the total scattering matrix, in the spherical harmonics representation [13, 14]

$$T(\mathbf{k}_f, \mathbf{k}_i) = \frac{(8\pi^2)^2}{A} \left\{ \sum_{LL'} Y_L(\mathbf{k}_f) Y_{L'}^+(\mathbf{k}_i) \left(\sum_{v=0}^{\infty} \exp[i(\mathbf{k}_i - \mathbf{k}_f) \cdot \boldsymbol{\rho}_v][T_v(|\mathbf{k}_i|)]_{LL'} \right) \right\} \times \sum_{g_0} \delta(\mathbf{k}_{i\|} - \mathbf{k}_{f\|} + \mathbf{g}_0), \tag{7}$$

where $\mathbf{k}_f$ and $\mathbf{k}_i$ are the momenta of the scattered and incident electron respectively, with $\mathbf{k}_{f\|}$ and $\mathbf{k}_{i\|}$ their corresponding components parallel to the surface. $Y_L(\mathbf{k}) \equiv Y_{lm}(\mathbf{k})$ are spherical harmonics, and A is the area of the unit cell of the two-dimensional net. The vector between the origin of the surface plane ($z = 0$) and the origin in the v-th plane is denoted by $\boldsymbol{\rho}_v = \mathbf{d}_v + \mathbf{a}_v$ where $|\mathbf{d}_v| \equiv d_v$ is the normal distance between the surface plane and the v-th layer ($\mathbf{d}_0 = 0$), and $\mathbf{a}_v$ is a vector in the plane parallel to the xy plane, allowing for shifts in registry between the layers. The δ-function in the above equation, expresses the conservation of momenta parallel to the surface plane, and $\mathbf{g}_0$ is a reciprocal vector of the two-dimensional surface net. Working in a finite-number of-layers scheme the matrix eqn (5), can be solved by matrix inversion. An alternative way is to expand the T-matrix given in eqn (5) in a perturbation series with each term in the series describing a scattering path of the incident electron in the solid (for a discussion of the perturbation scheme and explicit expressions for a third-order perturbation expansion, see ref. (14)). The intensity is given as the modulus square of the amplitude ($\propto TT^+$, '+' denotes complex conjugation), thus it is sufficient for our purpose to examine the Fourier transform of the T matrix, since the transform of the intensity can be derived from it via the convolution theorem. To simplify our discussion we consider the following system: An electron incident normal to the surface of a solid which is composed of identical scatterers periodically arranged in planes (layers) with a uniform inter-layer distance (d). We consider only the specularly diffracted intensity, (I_{00}), since it is the only component of the diffracted intensity needed for evaluation of the line projection $P(z)$ eqn ((4)). The normal

---

FIG. 1(*a*).—Specular beam normalized intensity diffracted from the (100) face of Al at room temperature for azimuthal and polar angles of incidence $\phi = 45°$ and $\theta = 8°$ respectively (solid line). The dashed line describes the temperature renormalized potential scattering envelope (ω_{BPT} in ref. (8).) Fourteen phase-shifts derived from Snow's APW potential for Al (E. C. Snow, *Phys. Rev.*, 1967, **158**, 683), were used in the calculation, with $\Theta_D = 340$ K. Logarithmic intensity scale. 1(*b*). Kinematically calculated specular intensity profile from Al(100). Phase shifts and Debye temperature used as in (*a*) with an attenuation coefficient $\mu = 0.32$ Å^{-1}, an inner potential $V_0 = 14$ eV and uniform interlayer spacing $d = 2.05$ Å. Logarithmic intensity scale. 1(*c*)/1(*d*). The real parts of the line projections, $P(z)$, of the transforms of the experimental intensity shown in (*a*) and kinematical intensity shown in (*b*), respectively. Note the similarity between the transforms in contrast to the pronounced differences in the intensities. 1(*e*). Deconvolution ($P_\infty(z)$) of the experimental transform shown in (*c*), using the F_{BPT} window (the transform of the dashed curve in (*a*)). Peaks forming a weighted consistent vector set have been filled for clarity ($z = N2.05$ Å $\pm$ 0.05Å, $N = 0, 1, 2, 3$). Momentum transfer and coordinate grids of 0.02 Å^{-1} and 0.05 Å have been used in the calculation. The base width of the peaks is ± 0.05 Å. The deconvolution algorithm used is a modification of a relaxation method described by Southwell.[24] 1(*f*). Same as (*e*) for the transform (*d*) of the kinematical intensity (*b*).

incidence configuration simplifies our discussion due to the symmetry it introduces into the internal beams thresholds, (the analysis can however be extended easily to off-normal incidence configuration). After some algebraic manipulation the **T** matrix of the specular beam can be written as

$$\mathbf{T}(\mathbf{k}_f, \mathbf{k}_i) = \frac{(8\pi^2)^2}{A}[T^{(1)}(s)M_p(0,0) + T^{(1)}(s)Q(k_0,s)] \equiv \frac{(8\pi^2)^2}{A}(T_1 + T_{ms}) \quad (8)$$

with

$$T^{(1)} = \sum_{v=0}^{\infty} \alpha_1^v \exp[2\pi i s \, dv] \quad (9a)$$

and

$$M_p(\pm \mathbf{g}_1, \pm \mathbf{g}_2) = \frac{1}{4\pi} \sum_{l=0}^{\bar{l}-1} t_l(k_0)(2l+1) P_l(\cos \Omega). \quad (9b)$$

Here s is the normal momentum transfer given by eqn (1) with $\mathbf{g}_0 = 0$, $\alpha_1 = \exp(-2\mu d)$ where μ is the attenuation coefficient, l is the number of phase shifts considered, $k_0 = [(2m/\hbar^2)(E+V_0)]^{\frac{1}{2}}$, with V_0 the inner potential of the solid and E the external energy of incidence. Ω is the scattering angle which is defined by the propagation vectors $\mathbf{k}_\pm(\mathbf{g}_i)$ ($i = 1, 2$) given by

$$\mathbf{k}_\pm(\mathbf{g}) = \mathbf{k}_{i\parallel} + \mathbf{g} \pm k_\perp(\mathbf{g})\mathbf{e}_\perp, \quad (10)$$

where $\mathbf{e}_\perp$ is a unit vector in the $+z$ direction and the $+$ sign is for the case $(\mathbf{d}_v - \mathbf{d}_{v'})_\perp > 0$ and the $-$ sign for the case $(\mathbf{d}_v - \mathbf{d}_{v'})_\perp < 0$. The directional scattering matrix M_p describes the potential scattering in a partial wave expansion [15, 16] with the l-th partial wave component t_l, given by

$$t_l(k_0) = -\left(\frac{\hbar^2}{2m}\right)[(\exp(2i\delta_l(k_0))-1)/2ik_0]. \quad (11)$$

The first term in eqn (8) describes the contribution to the scattered amplitude from single-scattering processes, the second term includes contributions from higher-order processes. From the above, the multiple-scattering contribution can be viewed as a modulation of the single-scattering geometrical factor $T^{(1)}$ by a function Q, representing dynamical effects.

The Fourier transform of the single scattering contribution can be written as

$$F[T_1] = F[T^{(1)}] * F[M_p(0,0)] = [\sum_{v=0}^{\infty} \alpha_1^v \delta(z-vd)] * F[M_p(0,0)] \quad (12)$$

which exhibits an attenuated "δ-function" structural information convoluted with the transform of the potential-scattering factor. The Fourier transform of the multiple-scattering contribution, can be written as

$$F[T_{ms}] = F[T^{(1)}] * Y(z) \quad (13)$$

where

$$Y(z) = F[Q]. \quad (14)$$

Finally the expression for the Fourier transform of the total scattering matrix **T** (eqn (8)) can be written as a convolution of the structural function with a function L which contains the transform of the ion-core potential-scattering factor for single specular reflection $[M_p(0,0)]$, and multiple scattering effects

$$F[T] = \left[\sum_{v=0}^{\infty} \alpha^v \delta(z-vd)\right] * L(z) \quad (15a)$$

$$L(z) = F[M_p(0,0)] + Y(z). \quad (15b)$$

In the kinematical approximation $Y(z) = 0$ and
$$L_1(z) = F[M_p(0, 0)]. \tag{15c}$$

Substituting $L_1(z)$ for $L(z)$ in eqn (15a) yields the expression for the transform of the single-scattering scattering matrix (see eqn (12)).

To study the effects of multiple-scattering we evaluate the function $Y(z)$. Due to the complexity [14] of the function Q defined in eqn (8), we consider, in the following, scattering paths composed of double-scattering events occurring at different layers (i.e., $\beta = 1$ in eqn 6). (The extension to higher order processes can also be evaluated [17]). The function Q_{DD} corresponding to these double-scattering processes can be written as [18]

$$Q_{DD}(s) = 2 \sum_g M_p(0, \mathbf{g}) M_p(\mathbf{g}, 0) \sum_{v=0}^{\infty} \alpha_2^v \cos(\mathbf{g} \cdot \mathbf{a}v) I(s; \mathbf{g}, v), \tag{16}$$

where the layer attenuation coefficient is

$$\alpha_2 = \exp\left[-\mu d\left(1 + \frac{1}{\cos \Omega}\right)\right] \tag{17}$$

and

$$I(s; \mathbf{g}, v) = [(\pi s)^2 - g^2]^{-\frac{1}{2}} \exp\{i[\pi s + \sqrt{(\pi s)^2 - g^2}]dv\}. \tag{18}$$

The transform of the function Q_{DD} given by eqn (16) can be written as

$$Y_{DD}(z) = 2 \sum_g F[M_p(0, g) M_p(g, 0)] * \{\sum_{v=0}^{\infty} \alpha_2^v \cos(\mathbf{g} \cdot \mathbf{a}v) \kappa(z; \mathbf{g}, v)\} \tag{19}$$

where

$$\kappa(z; \mathbf{g}, v) = \int_c ds \exp(-2\pi i s z) I(s; \mathbf{g}, v). \tag{20}$$

The first term in the convolution is the Fourier transform of the functions associated with the potential scattering from the individual ion cores, while the second term is an attenuated transform of the function I(eqn (18)) related to the structural parameters (layer spacing d and layer shift $\mathbf{a}$). The integral in eqn (20) is performed along the contour c to be specified in the following. Substituting eqn (18) into the expression for κ(eqn (20)) we get

$$\kappa(z; \mathbf{g}, v) = \frac{1}{\pi}\int_c dx[(x^2 - g^2)]^{-\frac{1}{2}} \exp\{i[\sqrt{x^2 - g^2}\, dv - x(2z - dv)]\}. \tag{21}$$

The branch points of the integral correspond to thresholds for excitation of non-specular internal beams. The path of integration relative to these branch points is chosen by envoking the causality condition for double-scattering, i.e., $z \geqslant d$. For $dv > z(v = 1, 2, \ldots)$ we choose a contour closed in the upper half-plane. Since the integrand does not have any singularities in the upper half-plane the value of the integral is 0. For $dv < z$ a contour which includes all the branch points is used.[17] After change of variables the integral can be transformed into the integral representation of the zero-order Hankel function of the first kind, $H_0^{(1)}(x)$[19]. The integral κ (eqn (21)) can now be written as

$$\kappa(z; \mathbf{g}, v) \propto H_0^{(1)}[2g\sqrt{z(z-vd)}]\, \mathscr{H}(z - dv), \tag{22}$$

$\mathscr{H}$ is the Heaviside function.[20] For values of the coordinate z equal to multiples of the layer spacing ($z = vd, v = 1, 2, \ldots$) the integral κ is a constant independent of the geometrical variables. We note that also for the case of specular propagation between the two scattering events ($g = 0$) the value of κ reduces to a constant independent of the geometrical variables.

The main conclusion of our analysis is that the Fourier transformation of the kinematical scattering intensity yields geometrical "δ-function" information convoluted with the effects of date truncation (W_B in eqn (1)) and scattering potential (which can be re-normalized to include vibronic effects). The inclusion of higher-order scattering processes merely adds a monotonically decreasing modulation to the, kinematical in origin, structural function (first term in the convolutions in eqn (12) and (15a)). Due to the reciprocal dependence on the momentum of the elastic scattering vertices from individual ion-cores (eqn (11) and (9b)), the strong attenuation of the propagation of electrons in the energy range of interest inside the solid (eqn (17)) and the properties of the integral κ (eqn (21)), the modulation due to multiple-scattering of the transform of the single scattering process, is rather weak. Thus, transforms of electron diffraction intensities constructed *from kinematical considerations* can be used, upon a proper deconvolution [5-8] of truncation, potential and vibronic effects, for the extraction of structural information.

(3) TRANSFORM-DECONVOLUTION

We conclude our presentation by a brief discussion of the deconvolution step in our method. Since the intensity is proportional to the squared modulus of the scattering matrix ($I \propto TT^+$) we may derive the expression for the transform of the kinematical intensity by employing the convolution theorem

$$F[I_{00}^{\text{kin.}}] \propto F[T_1] * F[T_1^+]. \tag{23}$$

Substituting eqn (12) for the transform of T_1 and the corresponding equation for the transform of its complex conjugate, T_1^+, in eqn (23) and using the result in eqn (4) we get the following expression for $P(z)$ [8]

$$P(z) = F_{\text{BTP}}(z) * G(z), \tag{24}$$

where

$$G(z) = (1 - \alpha_1^2)^{-1} \sum_{v=-\infty}^{\infty} \alpha^{|v|} \delta(z + vd) \tag{25}$$

and

$$F_{\text{BTP}}(z) = \int_{-\infty}^{\infty} ds \, \omega_B(s; s_1, s_2) \omega_T(s; \Theta_D) \omega_p(s) \exp[-2\pi i s z] \tag{26}$$

where

$$\omega_p(s) = M_p(0, 0) M_p^+(0, 0) \tag{27}$$

with M_p defined by eqn (9b). Here we have factored the vibrational effect which is described by the Debye-Waller factor [21] $\omega_T(s; \Theta_D)$, with Θ_D the Debye temperature. Note that the function $G(z)$ in eqn (24) and (25) corresponds to the uniform clean substrate case, similar functions can be written for non-uniform structures (see eqn (4c) in ref. (8)). In the analysis of overlayer systems, the difference in scattering characteristics of atoms has to be included, and a substrate subtraction method to deal with this situation has been proposed by us.[6] A few comments about the deconvolution of the truncated potential-scattering and temperature window function, which is necessary to achieve an unambiguous structural determination, are now in place. From the defining eqn (4) we observe that eqn (24) does not have a mathematical unique solution for (Gz). This is because an infinite number of arbitrary functions can be constructed identical to $I_{00}(s)$ in eqn (4) over the range of s defined by the boxcar function ω_B, but different from $I_{00}(s)$ outside this range. The problem is "ill-posed", and a *physically* meaningful solution can only be achieved if sufficient *a priori* information about the desired solution is incorporated into a numerically stable algorithm, thus allowing to discriminate between physical and non-physical solutions.[22] As is observed from eqn (25) and similar expressions for non-uniform systems,[6, 8] the required answer consists of sets of "delta-functions" which occur periodically in z and decay in amplitude with increasing z. Coupled with the above, a solution satisfying the mentioned criteria was achieved by us via a relaxation deconvolution scheme. The deconvolution prodecure consists of: (i) a sequential deconvolution, and a feed-back mechanism whereby the result of a particular step is used to refine previous steps, (ii) mathematical convergence criteria

$$\varepsilon = \sum_z |P(z) - P_\infty(z)^* F_{BTP}(z)| / \sum_z |P(z)|,$$

where ε is the error indicator, $P(z)$ is the complex transform of the intensity and $P_\infty(z)$ is the real deconvolution solution, (iii) a search procedure, which identifies and assigns weights to vector sets, (iv) evaluation of the deconvolution result in terms of a signal-to-noise estimator constructed by a linear combination of (ii) and (iii). The Debye temperature serves as an adjustable parameter via a variation to minimize the value of the estimator in (iv).[25]

Typical results of the deconvolution for the transforms of the intensity curves shown in figs. 1*a* and 1*b* are presented in figs. 1*e* and 1*f*. As is evident from these results a perfect deconvolution is achieved for the kinematically calculated case, fig. 1*f*. The final deconvolution result for the experimental intensities shows (fig. 1*e*) a significant enhancement of features in the transform which are structural in origin (the vector set, indicated by filled triangles of base width ± 0.05 Å, at 2.05 NÅ, $N = 0, 1, 2, \ldots$) and a reduction of the non-structural features (aperiodic and small value, open triangles). As we have demonstrated elsewhere,[6-8] this result is obtained consistently for diffraction intensities recorded for various experimental configurations.

The main purpose of this study was to establish the dominant role of the single-scattering transform in the analysis of LEED intensities via Fourier transformations. As we have shown in § (2), the effect of multiple-scattering is to modulate only weakly the transform based on kinematical ideas.[23] The extraction of structural information from the transform requires a deconvolution of data truncation, scattering potential and vibronic effects which has been achieved via a stable relaxation-deconvolution algorithm. It is worthwhile noting that the method is practical (the analysis of an intensity curve spanning ~ 300 eV with a momentum trnasfer grid of 0.02 Å^{-1} and a coordinate grid of 0.05A, including a Θ_D optimization, takes less than 2 mi (corresponding to ~ 1 μs cycle time) and requires less than 10 K 32 bit words of storage). In addition, speed and storage requirements are merely slightly affected by the complexity of the system.

In the study of complex systems, such as in the presence of overlayers and/or reconstruction, two-dimensional projections of the transforms of the diffracted intensities are of importance. The use of the deconvoluted line projections (and the substrate-subtraction method [6] for overlayers) in conjunction with the two-dimensional sections of the transform could provide a way for a complete characterization of the surface structure. We also note that whereas in the *indirect*, model calculation approach, small changes in the description of the non-structural parts of the problem (scattering amplitude and number of phase shifts used, inner potential value, Debye temperature, the description of the attenuation, etc.) cause significant changes in the results, they do not have such a crucial effect on the results derived via the *direct* transform-deconvolution method, i.e., small changes in the window function affect the "noise" level in the deconvolution but do not destroy the overall result. In general, it is advisable to use in the analysis normalized intensities recorded over extended energy range for several non-equivalent diffraction beams and various experimental configurations. This reduces the effects of data truncation and those due to multiple-scattering, and allows consistent checks of the results.

Part of this work was done at Xerox, Webster Research Center, Webster, New York 14580. I thank my colleague Dr. D. L. Adams for many helpful discussions. The author is indebted to Professor E. W. Montroll for his help and encouragement at the final stages of the work.

[1] T. N. Rhodin and D. L. Adams, *Adsorption of Gases on Solids* in *Treatise on Solid State Chemistry*, Vol. 6, ed. N. B. Hannay, (Plenum, N.Y. 1975).
[2] C. J. Davisson and L. H. Germer, *Phys. Rev.*, 1927, **30**, 705.
[3] *Physics Today*, 1974, **27**, 17 and references cited therein.
[4] J. B. Pendry, *Low Energy Electron Diffraction*, (Academic Press, London, 1974).
[5] U. Landman and D. L. Adams, *J. Vac. Sci. Techn.*, 1974, **11**, 195.
[6] D. L. Adams and U. Landman, *Phys. Rev. Letters*, 1974, **33**, 585.
[7] D. L. Adams, U. Landman and J. C. Hamilton, *J. Vac. Tech.*, 1975, **12**, 260.
[8] U. Landman and D. L. Adams, *Surface Sci.*, 1975, **51**, 149.
[9] R. Hoseman and S. N. Bagchi, *Direct Analysis of Diffraction by Matter*, (North-Holland, Amsterdam, 1962).
[10] M. J. Burger, *Vector Space* (Wiley, New York, 1954).
[11] D. E. Sayers, E. A. Stern and F. W. Lytle, *Phys. Rev. Letters*, 1971, **27**, 1204.
[12] E. A. Stern, *Phys. Rev. B*, 1974, **10**, 3027,
[13] J. L. Beeby, *J. Phys. C*, 1968, **1**, 82.
[14] S. Y. Tong, T. N. Rhodin and R. H. Tait, *Phys. Rev. B*, 1973, **8**, 421.
[15] L. J. Schiff, *Quantum Mechanics*, (McGraw-Hill, N.Y., 1955), sec. 19.
[16] D. W. Jepsen, P. M. Marcus and F. Jona, *Phys. Rev. B*, 1972, **5**, 3933. In § 5, these authors show that the Debye averaged rigid-lattice elastic scattering vertex, can be expressed in terms of effective complex phase shifts.
[17] In extending the analysis to more general dynamical processes (for example, $\beta \neq 1$) a complete knowledge of the analytical properties of the scattering matrix is useful. This has been the subject of extensive studies by E. G. McRae and collaborators (see J. I. Gersten and E. G. McRae, *Surface Sci.*, 1972, **29**, 483; *Surface Sci.*, 1974, **42**, 413; *Z. Naturforsch.*, 1972, **27a**, 373). Further details and numerical examples will be published in a forthcoming article by this author.
[18] This expression can be obtained from eqn (17) in ref. (14), with the assumption of a uniform, energy-independent attenuation of the electron propagation inside the solid.
[19] A. Erdelyi et al., *Tables of Integral Transforms*, (McGraw-Hill, New York, 1954), vol. I.
[20] N. W. McLachlan, *Complex Variable Theory and Transform Calculus*, (Cambridge University Press, 1963).
[21] C. Kittel, *Quantum Theory of Solids*, (John Wiley, New York, 1963), chap. 19.
[22] For a discussion of a similar "ill-posed" problem, see H. D. Hagstrum and G. E. Becker, *Phys. Rev. B*, 1971, **4**, 4187.
[23] Fourier transforms applied to the study of EXAFS [11, 12] (extended X-ray absorption fine structure) may be analyzed in a similar manner. One can view this problem as a "spherical LEED" system where both the electron source and a phase sensitive detector are embedded in the solid, (see C. A. Ashley, SSRP report No. 74/01, Stanford University, March 1974).
[24] R. V. Southwell, *Relaxation Methods in Theoretical Physics*, (Oxford University Press, Oxford, England, 1946).
[25] The attenuation coefficient on the other hand is not an input parameter of the method. An estimate of it's value is provided by the decay of structural maxima in the final deconvolution result. In addition, the inner potential V_0 can also be treated on a variable parameter.

GENERAL DISCUSSION

Dr. U. Landman (*Rochester*) said: In the MXSα method, the radius of the model spheres and boundary conditions on them can have a marked effect on the results of calculations. The need for " charge addition " may result from truncation effects, i.e., neglect of certain interatomic overlap integrals. Could one substitute charge addition by a modification of sphere radii, boundary conditions and inclusion of non-spherical components?

Dr. U. Landman (*Rochester*) said: The basic point that Dr. D. L. Adams and I would make in our discussion of Woodruff's paper is that of his criticisms of the use of Fourier transform methods for the analysis of LEED intensities some are inapplicable and the others have been already met in the case of our transform-deconvolution method. We should like to clarify two main issues: (1) The distinction between analyses based on simple Fourier transforms of LEED intensities,[1] including the so-called optical-transform method,[2] and the transform-deconvolution approach which we have originated.[3-6] (2) The " phase-problem "[7] and the consequences of multiple-scattering in methods involving Fourier transformations. Concerning the first issue we should note that we are in substantial agreement with Woodruff's criticism of the simple Fourier transform and optical-transform approaches[8] and, in fact, made similar criticism in our first paper on the subject.[3] We note, however, that in his paper Woodruff attributes to us the conclusion that the optical-transform method " largely overcomes the truncation effects ". Since Woodruff *et al.* made a similar attribution in an earlier paper,[8] we should like to take this opportunity to correct this misunderstanding of both our findings and the method we advocate. As we have discussed previously,[3] the optical transform method alleviates the problem of data truncation at the expense of quite unacceptable loss of resolution. The problem of data truncation is solved in our transform-deconvolution method by a deconvolution procedure, yielding excellent resolution. Contrary to the understanding of Woodruff *et al.* we have not employed the optical transform approach, rather, we have *deconvoluted* the *complex transforms* of *individual* intensity curves.

With respect to Woodruff's analysis of calculated LEED intensities by an optical-transform we have no comment other than to agree that the completely unsatisfactory results are a consequence of data truncation and the fact that this method does not take into account the different scattering factors of adsorbate and substrate atoms nor their individual dependence upon electron energy. We emphasise again that these factors are dealt with in the transform-deconvolution method. Finally, we note that the nomenclature $P(0,0,Z)$ used by Woodruff in this paper and previously, and also by Mason and co-workers, is at variance with the conventional X-ray notation. For the one-dimensional transform obtained by Fourier transformation of the specular beam intensity, following Burger[9] we advocate the use of the notation $P(Z)$. In the

[1] T. A. Clarke, R. Mason and M. Tescari, *Proc. Roy. Soc. A*, 1972, **331**, 321.
[2] J. C. Buchholz, M. G. Lagally and M. B. Webb, *Surface Sci.*, 1974, **41**, 248.
[3] U. Landman and D. L. Adams, *J. Vac. Sci. Tech.*, 1974, **11**, 195.
[4] D. L. Adams and U. Landman, *Phys. Rev. Letters*, 1974, **33**, 585.
[5] U. Landman and D. L. Adams, *Surface Sci.*, 1975, **51**, 149 and D. L. Adams, U. Landman and J. C. Hamilton, *J. Vac. Sci. Tech.*, 1975, **12**, 260.
[6] U. Landman, paper at this Discussion.
[7] D. P. Woodruff, paper at this Discussion.
[8] D. P. Woodruff, K. A. R. Mitchell and L. McDonnell, *Surface Sci.*, 1974, **42**, 355.
[9] M. J. Burger, *Vector Space* (Wiley, New York, 1954).

ideal kinematical case of unit scatterers and in the absence of data truncation, $P(Z)$ represents the *projection* of *all* the interatomic vectors on the surface normal (Z) direction. In X-ray studies, use of the function $P(0,0,Z)$ represents a one-dimensional line *section* through vector space, containing vectors between atoms at different levels of Z but having the *same* X, Y coordinates.

We turn now to the more important concerns of the atomic scattering factors (referred to as the " phase-problem " by Woodruff—somewhat misleading in view of the more general use of this term in analysis of structures via diffraction experiments) and the effects of multiple-scattering. As discussed by Woodruff, the use of simple Fourier transform methods is subject not only to problems of data truncation but also to problems associated with the energy dependence of the complex atomic scattering factors and the phase difference resulting from scattering from atoms of different scattering factors in the case of overlayer systems. We are encouraged to note that Woodruff believes [1] that a deconvolution procedure in which the atomic scattering factors and their energy dependence are taken into account may be fruitful, since we have described such a procedure.[2, 3] In fact our deconvolution procedure and the substrate-subtraction method were designed to deal with these problems. Finally we note that we have shown both analytically (see following paper in this volume) and by analysis of experimental data [2-4] that the effects of multiple-scattering do not invalidate the transform-deconvolution method. Contrary to Woodruff's conclusion, the results of Buchholz et al.[5] do not necessarily indicate the importance of dynamical effects. It is entirely conceivable that the appreciably different transforms obtained by these authors for different angles of incidence result mainly from changes in the atomic scattering factor with angle of incidence, which is not accounted for in their analysis.

We hope that the above remarks would clarify some of the problems and remove some of the misconceptions of the use of transforms in the analysis of LEED data, and would help further constructive investigations into their use in surface structure determination.

Dr. D. P. Woodruff (*Warwick*) (*communicated*): Landman appears to be most concerned that his work with Adams be properly referenced, a state of affairs which I sympathise with, but wonder why he takes issue with me as his comments largely restate my own references to their work. In case there is any doubt about this, however, I should say firstly that their early paper is largely concerned with the optical transform method, an approach which undoubtedly does remove much of the truncation effect associated with simple approaches even if Landman does not wish to be associated with this view. However, I believe I have made it clear that Landman and Adams are primarily concerned with the deconvolution technique which is designed, in its simplest form, to remove truncation errors (without the side effects of the optical transform alone), and in its extended form to try to circumvent what I have referred to as the " phase problem ". (We have already made clear the relationship of this problem to the " phase problem " in X-ray diffraction in an earlier paper.[6]) Had these authors demonstrated that this technique achieves its aim in its extended

---

[1] D. P. Woodruff, paper at this Discussion.
[2] D. L. Adams and U. Landman, *Phys. Rev. Letters*, 1974, **33**, 585.
[3] U. Landman and D. L. Adams, *Surface Sci.*, 1975, **51**, 149; and D. L. Adams, U. Landman and J. C. Hamilton, *J. Vac. Sci. Tech.*, 1975, **12**, 260.
[4] U. Landman and D. L. Adams, *J. Vac. Sci. Tech.*, 1974, **11**, 195.
[5] J. C. Buchholz, M. G. Lagally and M. B. Webb, *Surface Sci.*, 1974, **41**, 248.
[6] D. P. Woodruff, K. A. R. Mitchell and L. McDonnell, *Surface Sci.*, 1974, **42**, 355.

form, my paper, in so far as it is concerned with discussion and quantifying the problems underlying the transform approaches, would be largely redundant. However, despite several publications on the method which imply that its success had been demonstrated, these authors have so far failed to present a single relevant test of the technique as applied to adsorbate structures (when the phase problem is most important). I will reserve my detailed comments on this method and on Landman's current paper until after the presentation of this paper.

Dr. D. P. Woodruff (*Warwick*) said: Landman has stated that in earlier publications [1, 4] with Adams he has discussed a "practical direct method" for structure determination via LEED. As the word "practical" seems to me to imply not only that the method is computationally possible (perhaps even easy) but also that the method is successful (i.e., it leads to valid structure determinations from experimental LEED data) I feel the contribution of this work deserves some assessment. In discussing the Fourier transform-deconvolution procedures it is important to draw a distinction between studies of clean surfaces (particularly clean, unreconstructed surfaces) and adsorbate-surface structures. For clean surface studies the deconvolution procedure porposed by Adams and Landman [2, 3] is simply a method of removing the truncation effects inherent in the transform technique [5] when applied to LEED and, less importantly, correcting for the variation in effective scattering amplitude with energy across the truncated range of data used. Unlike the optical transform [1, 6] approaches, this technique appears to remove truncation effects without destroying resolution in the resulting transform. In the paper in this Discussion Landman has analysed a very special and simplified model of a clean surface to try to demonstrate that multiple scattering does not seriously influence the resulting transform. More significantly, I think, their earlier papers have shown that such transform-deconvolution procedures, applied to experimental data from a small number of clean surfaces, does indeed provide results not obviously seriously disturbed by the known presence of multiple scattering effects in this data. However, the clean unreconstructed surface is a very simple problem; all the surface layers are providing identical single scattering interferences and it is well known to LEED experimentalists that data from such systems, while strongly affected by multiple scattering, shows clear single scattering Bragg peaks (or at least Bragg "envelopes" [7]) which would allow a reasonably confident analysis of the structure. The success of the transform-deconvolution procedure in these cases cannot, therefore, be regarded as profound.

The adsorbate-surface problem, on the other hand, is not only more interesting as a valuable application of LEED, but is also much more difficult. In this case the phase problem [2] exists and a deconvolution procedure to remove truncation effects in no way corrects for this problem. Instead, Adams and Landman [2] have proposed a method to overcome the phase problem; this method is arrived at by considering the exact nature of the intensities in a single scattering approximation and devising a deconvolution procedure which is applied to a difference function between transforms from the clean surface data and the adsorbate surface data; included in this

[1] U. Landman and D. L. Adams, *J. Vac. Sci. Techn.*, 1974, **11**, 195.
[2] D. L. Adams and U. Landman, *Phys. Rev. Letters*, 1974, **33**, 585.
[3] D. L. Adams, U. Landman and J. C. Hamilton, *J. Vac. Sci. Techn.*, 1975, **12**, 260.
[4] U. Landman and D. L. Adams, *Surface Sci.*, 1975, **51**, 149.
[5] D. P. Woodruff, K. A. R. Mitchell and L. McDonnell, *Surface Sci.*, 1974, **42**, 355.
[6] J. C. Bucholtz, M. G. Lagally and M. B. Webb, *Surface Sci.*, 1974, **41**, 248.
[7] C. W. Tucker and C. B. Duke, *Surface Sci.*, 1970, **23**, 411; 1972, **29**, 237.
[8] D. P. Woodruff, paper in this Discussion.

deconvolution are functions which are derived from the calculated scattering phase shifts of adsorbate and substrate scatterers. Thus they have taken specific account of the phase problem, but have applied it only to data calculated from kinematical theory for Na on Al. The fact that their deconvolution produces the known structure used in the calculation demonstrates that their deconvolution scheme is self-consistent and can correctly deconvolute a *known convolution*. Evidently, until this procedure has been applied to real experimental data (or, at the least, calculations based on dynamical theory), it is impossible to assess quantitatively the effect of multiple scattering on this procedure.

It is clear, however, that there are reasons for believing that multiple scattering will be far more troublesome in this case. Adsorption generally produces rather small changes in the intensity-energy spectra of LEED beams from a surface. Changes in these spectra between clean surface and adsorbate surface are very much less than between clean surface and single scattering theory calculations. Thus, in the data themselves, multiple scattering produces much larger changes than adsorption. For this reason, it seems likely that any difference spectrum will be dominated by changes in multiple scattering effects and that a deconvolution based on single scattering theory will not be successful.

Of course, the final assessment must come from quantitative tests of the procedure. In view of the many systems which have been studied by LEED using dynamical theory calculations [1] it is unfortunate that this test has not been applied to some of these experimental data from systems which are now regarded as having " known " structures.

Dr. U. Landman (*Rochester*) (*communicated*): The transform-deconvolution method that we have proposed [2-5] is intended as a *direct and practical* method of analysis of LEED intensities for the determination of surface structures. Since the prohibitive computational problems associated with model calculations of LEED intensities are well recognized the motivation for the construction of a practical method does not need further elaboration. As for the methodology employed in the analysis, a clear distinction between *indirect* and *direct* methods should be made. In the indirect approaches (model calculations) structures are postulated, followed by a variation of the model-parameters (structural and non-structural) with the idea of achieving agreement with experimental observations, the basis of comparison being a tested, objective criterion. However, in practice only a restricted variation is performed and " agreement " is stated on the basis of subjective criteria. These limitations and others have led occasionally to erroneous structural assignments, discrepancies and much controversy.[6] In the direct approach on the other hand, structure is an output rather than an input, with an associated objective criterion of " success ". Therefore, in our studies of LEED data from clean surfaces (Ni(100),[8] Al(100),[3-5] Al(111),[7,8] Cu(100) [8]) using the transform-deconvolution method, *no apriori assumptions* concerning the occurrence of relaxation or reconstruction have

[1] e.g., a review of many of these is given in P. M. Marcus, J. E. Demuth and D. W. Jepsen Proceedings of Interdisciplinary Surface Science, Warwick, March 1975; to be published in *Surface Sci*.
[2] U. Landman and D. L. Adams, *J. Vac. Sci. Tech.*, 1974, **11**, 195.
[3] D. L. Adams and U. Landman, *Phys. Rev. Letters*, 1974, **33**, 585.
[4] D. L. Adams, U. Landman and J. C. Hamilton, *J. Vac. Sci. Tech.*, 1975, **12**, 260.
[5] U. Landman and D. L. Adams, *Surface Sci.*, 1975, **51**, 149.
[6] T. N. Rhodin and D. S. Y. Tong, *Physics Today*, 1975 October, p. 23.
[7] U. Landman and D. L. Adams, *Elec. Y Fis. Applic.*, 1974, **17**, 144.
[8] D. L. Adams and U. Landman, to be published.

been made. In fact, our analytical procedure *allows* and *searches for any and all* structural periodicities in the transforms of the intensity data to the limit of our stated resolution. Thus, the above studies provide *general* valuable information about the method and its applicability.

As we have discussed, a deconvolution of a (truncated potential, temperature) envelope from the transforms of the experimental data is essential in order to extract accurately and unambiguously structural information from the transforms. Truncation of LEED data is not only a function of the productivity of the experimentalist, but is mainly a consequence of the nature of the scattering processes and their dependence on the electron's momentum. Thus, the elastic scattering vertex (see eqn (9b), (11)) falls off for high values of the incident momentum and the Debye–Waller factor is exponentially decreasing as the momentum of the electron is increasing. As a result, it is the combined effect of all these factors, all of which contribute to non-structural features in the transforms, which has to be considered. Similar problems (though in general in less acute form) have been encountered in various other fields, like : X-ray diffraction,[1, 2] electron scattering,[3] neutron diffraction[2] and EXAFS,[4] and solutions to these hard problems have been proposed with variable degrees of success. Having experimented with various deconvolution techniques we have concluded that our relaxation-deconvolution algorithm provides the best solution to our problem. The main advantages being the stability of the mathematical procedure, the minimal loss of resolution and the reliance on objective criteria in achieving convergence to a *physical* solution.

In discussing the effects of multiple-scattering on the transforms of LEED intensities, we have chosen in the present paper a simple model of the scattering system. However, despite its simplicity the model contains the germane factors which would appear in more complex situations, and the conclusions reached on the basis of this model apply to more general cases with no major modifications.

We turn now to Woodruff's remarks concerning the analysis of scattering from overlayer systems. We do not share his feelings on the role of multiple scattering and the general structure of LEED intensities from overlayer systems, nor do we see the " reasons for believing that multiple scattering will be far more troublesome in this case ". Careful examination of LEED data for a large number of systems reveals that upon adsorption, marked changes in relative peak intensities and position of maxima occur. In a transformation of the intensity record *both these features* play an important role. The *substrate-subtraction* method that we have proposed[5] for the analysis of overlayer systems, consists of a prescribed subtraction of the transform of the intensity scattered by a clean substrate from the transform of the intensity measured after adsorption. The resulting residual transform is then subjected to a deconvolution of functions constructed from the atomic scattering factors of the substrate and overlayer atoms. It should be noted, that by subtracting the substrate-transform, contributions from multiple scattering processes which involve the substrate alone have been partly removed. This procedure enhances features due to the overlayer and allows for improved identification. Our application of the method to calculated intensities from overlayer systems is intended to test our ideas and to

[1] B. E. Warren, *X-ray diffraction*, (Addison-Wesley, Reading Mass., 1964), chap. 10.
[2] A. C. Wright, in *Advances in Structure Research by Diffraction Methods*, vol. 5, ed. W. Hoppe and R. Mason, (Pergamon, Oxford, 1974).
[3] D. B. Dove, in *Physics of Thin Films*, vol. 7, ed. G. Hass, M. H. Franscombe and R. W. Hoffman (Academic, New York, 1973).
[4] E. A. Stern, D. E. Sayers and F. W. Lytle, *Phys. Rev.*, 1975, **B11**, 4836.
[5] D. L. Adams and U. Landman, *Phys. Rev. Letters*, 1974, **33**, 585.

demonstrate that given high quality intensity data and reliable atomic scattering factors, overlayer structures could be determined. Although the substrate-subtraction method has not been fully tested yet for the analysis of experimental data (work in progress will be reported in due course), based on our experience with the application of the transform-deconvolution method to clean surface systems and for the above reasons, we expect it to be of great use in overlayer structural studies.

In view of the complexity of the problems involved in the analysis of LEED data for surface structure determination, we have taken a careful stepwise approach in the formulation, testing and application of our method. We believe that by adopting this approach we would avoid the unfortunate occurrences of discrepancies and incorrect conclusions which have occasionally plagued and cast doubt on the value of LEED in surface structure studies,[1] and would provide a sound basis for the application of the method.

Prof. E. A. Stern (*University of Washington*) said: It does not appear to be widely recognized that the Fourier transform method of Landman also contains a model calculation. There appears to be a tendency to polarize the methods of determining the structure from LEED measurements into two competing camps, the Fourier transform method and the model calculation method as described by Jona. Woodruff has correctly pointed out that a pure Fourier transform method has phase problems which would give spurious structural features.

However, the method of Landman described here is not a pure Fourier transform method but a hybrid of the Fourier transform method and the model calculation method. This hybrid method combines the best features of both methods and has decided advantages over each alone. The criticisms of both methods have been outlined by Jona and Woodruff. The model calculation, in spite of its outstanding successes, has the important fault that Jona mentioned, namely no objective criteria to determine whether satisfactory agreement exists between the calculation and the measurements. Jona showed a case where good agreement between calculations and measurements occurs among several beam directions yet poor agreement occurs among other beam directions. In this case there is no way to decide how close the assumed structure is to the real one and how to modify this assumed structure to obtain better agreement. This result also makes one uneasy about other model calculations that appear to agree with measurements. What if a new beam is measured? Is there any guarantee that the agreement between the model calculation and the measurements will continue?

The hybrid Fourier transform method of Landman does have the important advantage of having an objective criterion for determining agreement between calculation and measurements. In addition, it has the advantage of not requiring an assumption of the structure, but of determining it from the measurements. The Fourier transform method separates the structural and non-structural terms, at least when multiple scattering effects are small. Landman calculates the non-structural terms and then the Fourier transform determines the structure. If the non-structural terms are correctly calculated then the Fourier transform will consist of a series of δ-functions in the $\mathrm{Re}P_\infty(Z)$ plot of Landman locating the planes of atoms near the surface. Any errors in the non-structural terms will appear as additional features in the $\mathrm{Re}P_\infty(Z)$ plots. This noise can be compared with the δ-function signal for an objective criterion of the accuracy of the fit.

[1] T. N. Rhodin and D. S. Y. Tong, *Physics Today*, 1975, October, p. 23.

Inelastic Low-energy Electron Scattering from Solid Surfaces and Adsorbed Species

The Observation of Adsorbate Surface Resonances

By Roy F. Willis

Surface Physics Group, Space Sciences Department,
European Space Research Organization, Noordwijk, Holland

Received 5th March, 1975

Results are presented for angle-resolved secondary-electron emission normal to a W(110) surface for the special case of " band gap emission ", which arises when the energy of the emitted electron corresponds to an energy gap in the final states' bands located along a principal symmetry direction. Under such conditions, surface enhanced emission occurs, and scattering out of adsorbate-induced (i.e., " extrinsic ") surface states is resolved. Spectra are presented for the specific cases of O_2 and CO adsorption on W(110) which, together with parallel LEED studies, show that such effects only appear in the presence of *ordered* adsorbate monolayers, and are extremely sensitive to the surface unit cell dimensions of the adsorbate–substrate superstructure. The results are interpreted in terms of *adsorbate surface resonances* associated with excited two-dimensional surface Bloch states produced by the periodicity of the adsorbate layers.

Inelastic electron–electron scattering processes are responsible for the " cascade " distribution of secondary electrons observed at low kinetic energies, $E_k \leqslant 50$ eV, in the energy distribution of low energy electrons " back-scattered " into vacuum from solids bombarded by an incident " primary " electron beam. Fine spectral structure, superimposed on the smoothly varying cascade background curve, has been shown [1] to be due to scattering out of excited Bloch-states located above the vacuum level, E_v; the secondary-electron emission (SEE) spectral intensity relates to the density of unfilled or " final " states profile, $D(E)$, just as XUV-photoemission provides the density of filled or " initial " states.[2]

Angle-resolved measurements using small-angle selective analyzers limit scattering out of final states in **k**-space to a small volume of the Brillouin zone and, as recent studies have shown,[3] this results in increased spectral definition of density of final-states structure related to the region of **k**-space sampled. A special case occurs for emission along a direction normal to a low-index crystal face, corresponding to small or vanishing momentum parallel to the surface, $\mathbf{K}_\parallel \to 0$.[4] Energy-dependent refraction effects are negligible and the energy distribution curve relates directly to the density of those final states which are located along a single symmetry line, corresponding to the low-index crystal face under investigation. Scattering from final states distributed about general points throughout the Brillouin zone is excluded, and the one-dimensional density of final-states profile, which is what is essentially observed, is sufficiently detailed that a critical comparison with the calculated energy band structure is possible.

In this paper, results are presented for the interesting case of angle-resolved secondary electron emission normal to a single-crystal tungsten surface, such that the final state Bloch wave $\psi(\mathbf{K})$, with crystal momentum, **K**, and energy E relative to the

Fermi level, corresponds to a band-gap for all $K_\perp$ in the crystal. Under these conditions, it has been postulated [5] that the *surface* contribution (e.g., emission from excited " intrinsic " surface states or " extrinsic " adsorbate-induced surface states) becomes greatly enhanced relative to the usual *bulk* contribution. A first hint of support for such a view comes from the fact that in the presence of adsorbed mono-layers, additional SEE fine spectral structure is observed at energies which lie within the final-states band gap of W(110), which extends about 6 to 11 eV above the Fermi energy ($0.8 < E_k < 5.4$ eV).[4] Spectra are presented for the specific cases of O_2 and CO adsorption on W(110) and show that such effects appear only in the presence of *ordered* adsorbate layers. The results are interpreted in terms of scattering from *adsorbate surface resonances* associated with Bloch-like excited surface states. Such effects manifest themselves as a direct consequence of the lattice periodicity of the adsorbate superstructure and the special circumstances associated with " band gap secondary-electron emission ".

EXPERIMENTAL

The experimental arrangement is shown schematically in fig. 1. The SEE spectra were measured with a 130°-sector electrostatic analyzer with an energy resolution of 0.5 % equivalent to an angular resolution of less than 0.1°, as defined by the entrance slit of the

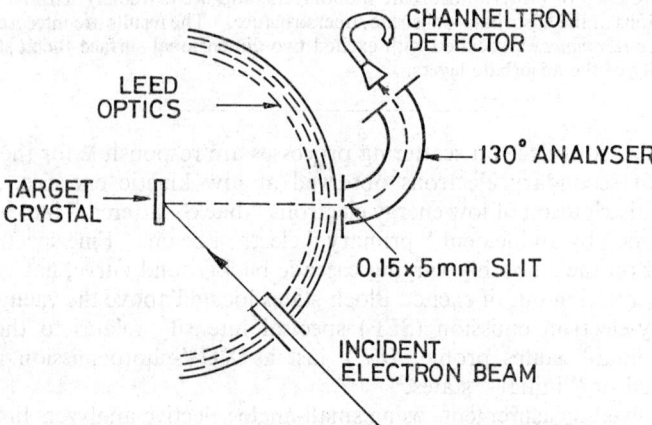

FIG. 1.—Schematic diagram of the experimental system.

analyzer. Tungsten single crystals were cut to expose the low-index planes to an accuracy within 3° and, after a standard cleaning procedure, were oriented within 1°.[6] The samples' surfaces were regularly checked *in situ* for cleanliness and structure by a LEED-Auger display system, the first grid of which was grounded in order to provide a drift region free from electrostatic fields.[7] Ambient magnetic fields were reduced below 10 mG by enclosing the scattering and analyzing region completely with μ-metal shells. The electron beam was incident at an arbitrary angle of 45° to the crystal surface normal and had a beam-energy spread of around 0.3 eV at a beam energy, $E_p \simeq 100$ eV. Adsorption experiments have been performed using gases of research-grade purity under conditions similar to those described for previous photoemission measurements.[8] Scanning times were such that exposure to background pressure was less than 1 % of the exposure to the adsorbant gas. Ordered adsorbate layer structures were determined from the characteristic LEED patterns.

RESULTS

The calculated energy band structure of tungsten [9] exhibits a gap in the final states of varying width and energetic position along the principal symmetry directions (shaded region, fig. 2(a)). Angle-resolved energy distribution spectra of secondary

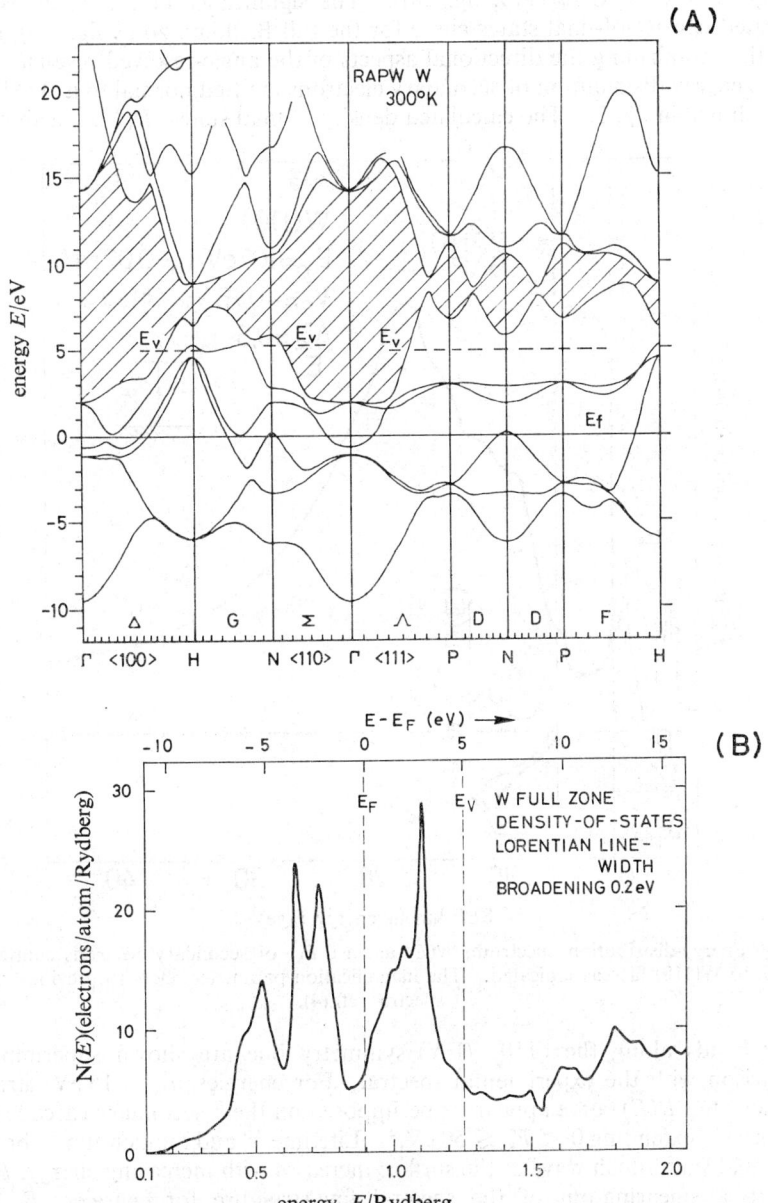

FIG. 2.—Band structure of tungsten along symmetry lines (A) and density of states for the full zone (B) as calculated by a relativistic augmented-plane-wave-method at 300 K (Christensen, ref. (9)). The work functions of each low-index face are indicated by the positions of the respective vacuum levels, E_v, relative to the Fermi level, $E_F = 0$ eV in (A).

electrons emitted normal to the clean low-index (100), (110) and (111) faces of tungsten crystals have been published [4] and show clearly resolved intensity-minima, extending from $2.5 < E_k < 4.5$ eV, $0.8 < E_k < 5.4$ eV and $3.2 < E_k < 4.4$ eV, respectively, for the three faces, i.e., the widths of the minima correspond to minima in the density of final states associated with the energy gaps located along the symmetry directions, $\Delta\langle 100\rangle$; $\Sigma\langle 110\rangle$ and $\Lambda\langle 111\rangle$, fig. 2(a). The significance of this result is that the integrated density-of-final states curve for the full Brillouin zone, fig. 2(b), shows no gaps, thus confirming the directional aspects of the angle-resolved results.

The energy distribution of secondary electrons emitted normal to a W(110) crystal face is shown in fig. 3. The calculated density of final states, $D(E)$, together with the

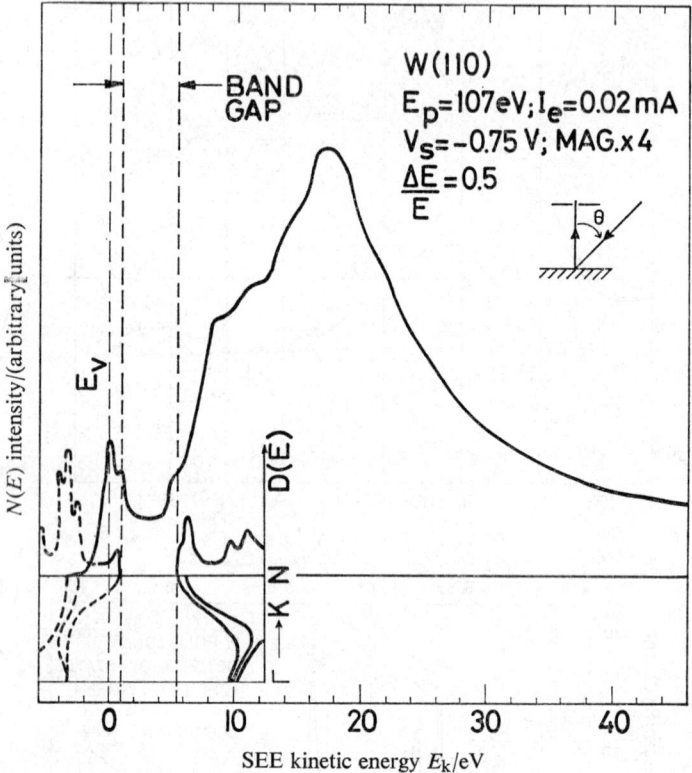

FIG. 3.—Energy-distribution spectrum, $N(E)$ against E_k, of secondary electrons emitted normal, $K_\| = 0$, to W(110) face as indicated. The magnification parameter refers to previously published spectra, ref. (4).

energy bands along the $\langle 110\rangle$ (ΓN) symmetry line are shown superimposed for comparison with the experimental spectra. For energies, $E_k \lesssim 15$ eV, structure in the calculated $D(E)$ curve appears superimposed on the " secondary cascade distribution curve ", extending $0 < E_k \lesssim 50$ eV.[4] Lifetime [10] and momentum [11] broadening of the final state Bloch wave at the surface increases with increasing energy, E_k, which leads to a smearing-out of the spectral fine-structure for energies, $E_k \gtrsim 10$ eV. However, the envelope of the calculated $D(E)$ fine structure continues to show good agreement with the broad features appearing in the SEE spectra.[4] The large peak at $E_k = E_v = 0$ eV is spurious in that it arises from scattering effects in the LEED-optics grids at the entrance to the analyzer, fig. 1.

The band gap, appearing at $0.8 < E_k < 5.4$ eV, is clearly resolved in the W(110) SEE spectrum, fig. 3. After exposure to 1.0 L of oxygen at 300 K (fig. 4), the SEE spectrum shows little change compared with that for the clean crystal, except for a slight loss of definition of the $D(E)$ fine-structure at the extremes of the gap. With progressively increasing exposure up to 10 L of O_2, additional SEE spectral features evolve in the band gap, i.e., at $E_k \simeq 1.0$ and 2.0 eV after a gas dosage of 10 L, fig. 4.

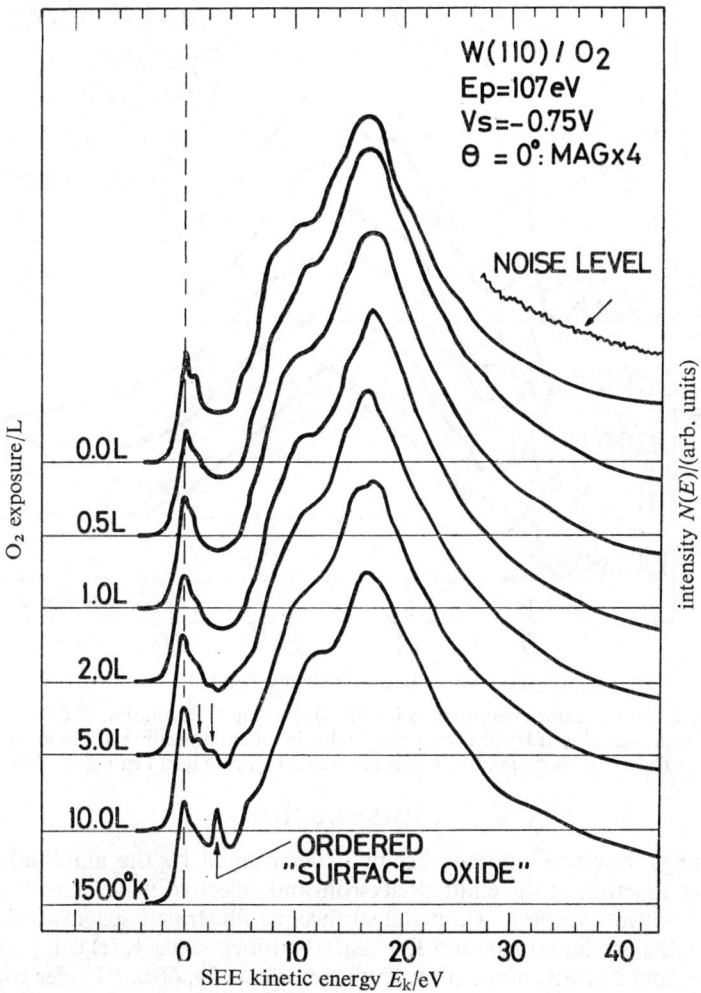

FIG. 4.—SEE spectra for emission normal to W(110) surface showing the evolution of adsorbate-induced scattering features in the energy gap region, $0.8 < E_k < 5.4$ eV, with increasing exposure to O_2 at 300 K and after subsequent annealing at 1500 K. (1 L $\equiv 10^{-6}$ Torr s). The spectra are displaced vertically for clarity.

This exposure corresponds to a half-monolayer coverage and the formation of a $C(2 \times 1)$ overlayer of atomic oxygen.[12] Subsequent flashing to approximately 1500 K gave rise to a single peak centered about $E_k \simeq 3.0$ eV of increased intensity. In contrast, exposures of up to 10 L of CO at 300 K produced no such well-defined spectral features in the energy band gap of W(110), fig. 5. Subsequent annealing to 1100 K, however, produced " adsorbate-induced emission features " at $E_k \simeq 1.5$ and

3.5 eV. Further heating to progressively higher temperatures resulted in first, the decay of the 1.5 eV peak after flashing to 1300 K and then, a single weak peak at $E_k \sim 2.5$ eV after flashing to 1500 K. Heating for a prolonged period at 2500 K was required to reproduce the original SEE spectrum of the clean surface, fig. 3, after exposure to both O_2 and CO.

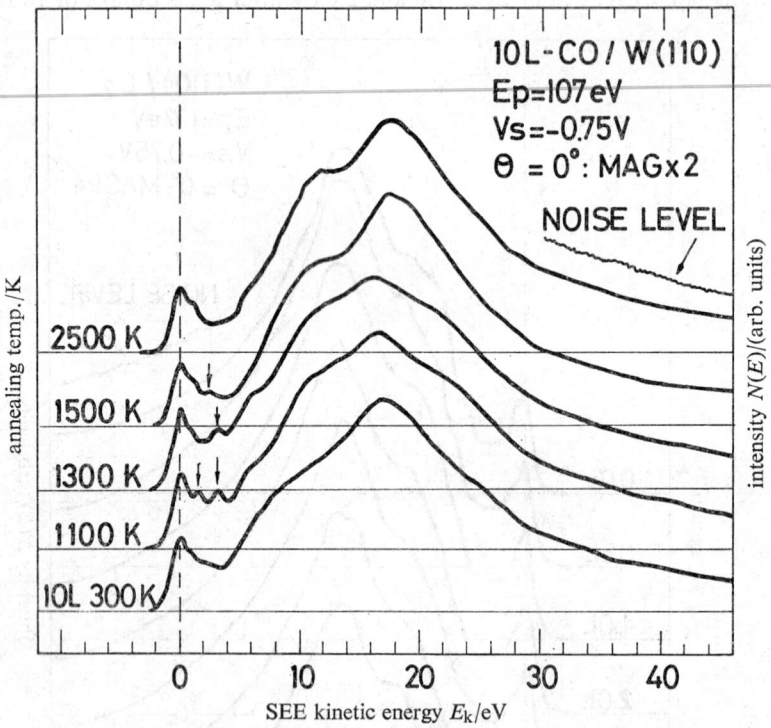

FIG. 5.—SEE spectra (same conditions as for fig. 4) showing the evolution of CO adsorbate-induced emission features produced by 10 L exposure and subsequent annealing to the temperatures shown. The clean spectrum, fig. 3, was recovered by prolonged heating at 2500 K.

DISCUSSION

The angle-resolved emitted current is determined by the amplitude of the final state wave function of the emitted electron and selective wave matching with plane wave components, $\exp(i\mathbf{p}\cdot\mathbf{r})$, at the surface, as illustrated schematically in fig. 6.[13] Providing the crystal momentum $\mathbf{K}$ is real, the Bloch wave $\psi_\mathbf{K}(\mathbf{r})$ can propagate inside the crystal and be transmitted through the interface, fig. 6(a). Under conditions such that the energy of the final state wave function corresponds to a band gap for all $\mathbf{K}_\perp$ in the crystal, $\psi_\mathbf{K}(\mathbf{r})$ comprises only surface evanescent waves, the amplitudes of which decay exponentially into the solid, fig. 6(b). Solutions of the Schrödinger equation are of the form:[14]

$$\psi_\mathbf{K}(\mathbf{r}) \simeq \exp(-z/\lambda) \cos(\pi z/a + \delta) \qquad (1)$$

such that only wave functions within a small distance of the surface, decay length $\approx \lambda(E)_{\text{gap}}$, contribute to the emitted current. For emission normal to the W(110) surface and final state energies corresponding to the band gap, fig. 3, Feibelman and Eastman[5] have estimated, $\lambda(E)_{\text{gap}} \approx 3$ Å. They conclude, therefore, that the *surface contribution* to the emitted current can be significantly enhanced, e.g., emission from

" intrinsic " or " extrinsic " (i.e., adsorbate) surface states, fig. 6(c). The observation of " adsorbate-induced " SEE spectral features located at energies within the W(110) band gap, fig. 4 and 5, are a first hint of support for such a view.[15]

Matching wave functions on a surface *plane* is more involved than is implied by the one-dimensional argument, fig. 6. A consequence of the lattice periodicity at the surface is that the surface states become two-dimensional Bloch functions extending over the whole surface.[16] The nature of such states is illustrated schematically in fig. 7 for a simple cubic surface with lattice constant, a. The coordinates are chosen

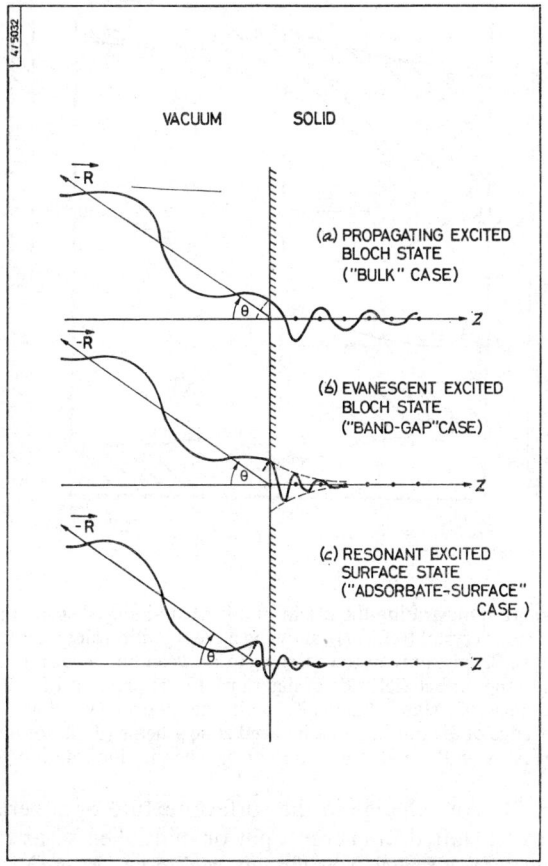

Fig. 6.—Schamatic diagram illustrating the matching at the solid-vacuum interface of the final state wave function for (a) a propagating excited state, (b) a rapidly decaying surface evanescent wave, corresponding to band gap SEE, and (c) and adsorbate-induced " extrinsic " surface state or " adsorbate surface resonance " ($E > E_{vac}$).

such that the xy-plane lies in the surface and the z-axis is normal. *Bands* of surface states can arise as a consequence of a periodic atomic arrangement. Considering a cross-section along the x-axis in the surface plane, $K_y = 0$, for each value of K_x, one surface-state and many bulk states (due to many possible K_z) exist (shaded regions, B, in fig. 7). Surface states, S, can exist *only* in the gap between the bulk states. Their energies, $E_s(K_x)$, vary with K_x (and K_y) to form *bands* of states in the surface. If the states are plotted on an energy scale without consideration of the wave vector K_x (left edge of fig. 7) the *bands* of the bulk, $E_n(K)$, and surface states, $E_s(K)$ can overlap.

The situation on the formation of a surface superstructure is shown on the right edge, fig. 7. Any increase in the dimensions of the surface unit cell leads to a decrease of the surface Brillouin zone, as shown in fig. 7 for a (2×2) superstructure. The surface state band is split at the new zone boundary (dashed line) to form two sub-bands. In the experimental arrangement, fig. 1, where only scattering out of states close to the centre of the surface Brillouin zone are sampled ($K \simeq 0$), surface states appear in the band gap at the positions encircled A and A', and at energies indicated

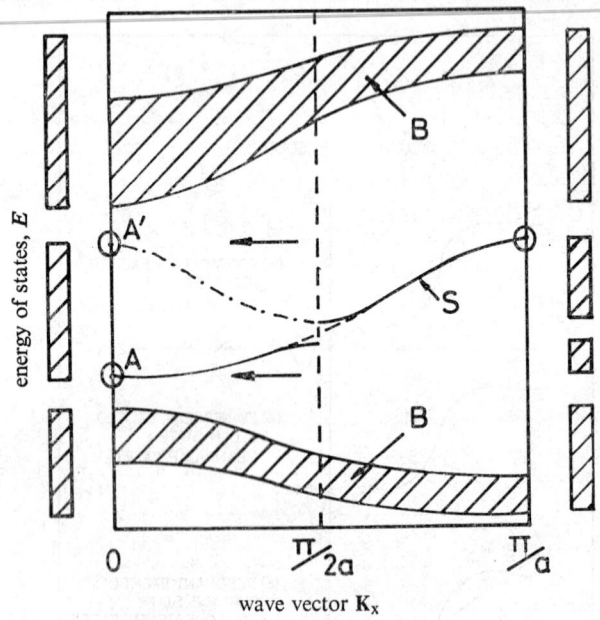

FIG. 7.—Schematic diagram illustrating the origin of adsorbate-induced surface states in k-space for a (100) face of a single cubic crystal (ref. (17)), showing a cross-section along the x-axis (parallel to the surface). Surface states, S, exist only in the energy gap between bulk states, shaded regions, B, and form a band of states in the surface (left side of diagram). In the presence of a (2×2) superstructure, the original single surface state band is split at the new zone boundary (dashed vertical line and right edge of diagram). In a reduced zone scheme (dash-dot line), this creates multiple states A, A' in the bulk gap region at the energies indicated for $K_x \Rightarrow 0$.

by the arrows, fig. 7. Any change in the surface texture or superlattice periodicity causes the bands to be shifted in energy, split or multiplied so as to create complex changes among the states appearing in the gap region (A, A' etc.).

While the above discussion follows the general ideas existing for surface states on clean surfaces,[17] the SEE results, fig. 4 and 5 indicate strongly that the argument is applicable to *ordered* adsorbate overlayers. The observation of the development LEED patterns from the W(110) surface exposed to O_2 and CO served to endorse this conclusion. For O_2 adsorption, the development of a bright $O(2 \times 1)$ pattern at half-monolayer coverage (10 L at 300 K), was paralleled by the evolution of the adsorbate-induced SEE spectral features, fig. 4. Similarly, for CO adsorption, fig. 5, which has a disordered structure at room temperature,[18] SEE spectral fine-structure appeared in the gap only after heat treatment and the formation of an ordered overlayer, the LEED pattern being at its most intense at 1100 K. Annealing the exposed W(110) surface by flashing to higher temperatures produces complex superstructures,[12, 18] which give rise to " band-gap-emission features " which split, move in energy and change in intensity, in a manner which one might expect from the scheme outlined in fig. 7.[19]

CONCLUSIONS

The angle-resolved secondary-electron-emission spectroscopy described above provides a powerful technique for investigating excited electron states corresponding to the unfilled region of the energy band structure above the vacuum level.[3, 4] An important special case is that of " band gap emission ", which arises when the energy of the emitted electron corresponds to an energy gap in the final states' bands along the specific symmetry line corresponding to the crystal surface normal direction. The excited states are evanescent Bloch functions under these circumstances, leading to enhanced surface sensitivity. This view is endorsed by the observation of adsorbate-induced SEE spectral features in the band gap region. Such features appear, however, only after a thermal treatment that gives rise to an *ordered* adsorbate layer. The results for adsorbed monolayers of O_2 and CO are indicative of scattering out of excited *adsorbate surface resonances* [15] produced by the two-dimensional periodicity of the substrate–adsorbate overlayer. The angle-resolved SEE-spectra distinguish clearly between both " bulk " and " surface " electronic excitations, and are sensitive to the surface unit cell dimensions of the adsorbate–substrate superstructure. The technique should prove useful, therefore, for studying such effects as superlattice formation, adsorbate reordering and order–disorder surface phase transitions, thus complementing surface structural studies using other techniques, such as LEED.

I am grateful to M. R. Barnes for his technical assistance with the measurements, Dr. B. Feuerbacher for useful discussions and Prof. J. M. Blakely of Cornell University for kindly supplying the tungsten crystal.

[1] Here we refer specifically to spectral features which appear as discrete maxima and minima in SEE *energy-distribution* curves (R. F. Willis, B. Feuerbacher and B. Fitton, *Phys. Letters A*, 1971, **34**, 231; *Phys. Rev. B*, 1971, **4**, 2441); not to be confused with weaker structure resolved in *derivative* spectra associated with Auger relaxation, temperature-dependent diffraction processes etc.

[2] See references in *Band Structure Spectroscopy of Metals and Alloys*, ed. D. J. Fabian and L. M. Watson (Academic Press, London, 1973).

[3] R. F. Willis, *Proc. IV Int. Conf. VUV Radiation Physics*, ed. E. E. Koch, R. Haensel and C. Kunz (Pergamon-Vieweg, 1974).

[4] R. F. Willis, *Phys. Rev. Letters*, 1975, **34**, 670; directional *photo-emission* results have been published, B. Feuerbacher and B. Fitton, *Phys. Rev. Letters*, 1973, **30**, 923.

[5] P. J. Feibelman and D. E. Eastman, *Phys. Rev. B*, 1974, **10**, 4932.

[6] The SEE spectra were insensitive to variations in polar, θ, or azimuthal angle, ϕ, up to 5° off normal direction to the crystal face; see ref. (3).

[7] Those features appearing at $E_k \lesssim 10$ eV in the SEE spectra were found to be extremely sensitive to electrostatic patch-field effects in the vacuum chamber; these effects could be balanced by careful adjustment of a small negative potential, $V_s \simeq -0.75$ eV, on the target crystal.

[8] B. Feuerbacher and M. R. Adriaens, *Surface Sci.*, 1974, **45**, 553.

[9] N. E. Christensen and B. Feuerbacher, *Phys. Rev. B*, 1974, **10**, 2349; N. E. Christensen, unpublished results.

[10] E. O. Kane, *J. Phys. Soc. Japan*, 1966, **21**, suppl., 37.

[11] E. A. Stern, *Phys. Rev.* 1967, **162**, 565; also ref. (5).

[12] L. H. Germer and J. W. May, *Surface Sci.*, 1966, **4**, 452.

[13] The wave function is identical to that which one would employ to describe low-energy-electron-diffraction (LEED) intensity measurements, except that it is the components of $\psi_K(\mathbf{r})$ that are *transmitted* out of the crystal rather than components *reflected* back into vacuum, as in LEED. The role of *Umklapp* or " multiple-scattering " processes has been described, R. F. Willis and B. Feuerbacher, *Surface Sci.*, 1975, to be published.

[14] V. Heine, *Proc. Roy. Soc. A*, 1972, **331**, 307.

[15] The electron can never be completely trapped in a surface state for energies above the vacuum level since it can always match into a plane wave in vacuum. The formal expression is that the surface state is a " resonance " to distinguish it from the usual " bound " surface states, which occur for $E < E_{vac}$, and require that δ is positive in eqn (1); ref. (14).

[16] The probability amplitude of an electron in such a surface state decreases exponentially inside and outside the vacuum-solid interface, i.e., it is " localized " at the interface, but " non-localized " in that the electron can be found anywhere in the surface plane.

[17] M. Henzler, *Surface Sci.*, 1971, **25**, 650.

[18] J. W. May and L. H. Germer, *J. Chem. Phys.*, 1966, **44**, 2895.

[19] The LEED observations were essentially those of ref. (12) and (18). The superstructures associated with the complex patterns observed after high-temperature treatment are the subject of controversy (see, e.g., C. Kohrt and R. Gomer, *Surface Sci.*, 1971 **24**, 77; J. C. Tracy and J. M. Blakely, *Surface Sci.*, 1969, **15**, 257), making more detailed arguments difficult at the present time. Further experiments are in progress, R. F. Willis, to be published.

Electronic Excitations in Adsorbed Alkali Metal Layers

By S. Andersson and U. Jostell

Department of Physics, Chalmers University of Technology,
Fack, S-402 20 Göteborg 5, Sweden

Received 17th March, 1975

Adsorption of Na and K on Ni(100) and Na on Ni(100) $c(2\times2)$S was investigated by electron energy-loss spectroscopy in conjunction with measurements of work function changes and lateral adatom distributions. The electronic excitations observed were found to be characterized by their dependence on the surface density and charging state at the adatoms. Tentative models concerning the nature of the electronic excitations are discussed.

For the electron structure of alkali atoms adsorbed on metal surfaces, one distinguishes two limiting cases, namely, the single-atom adsorption limit with properties expressed in terms of the adatom valence electron level,[1-3] and the monolayer with properties approaching those of the corresponding alkali metal.[4] This is unlike, e.g., the adsorption of chalcogens on nickel where UPS [5] and INS [6] clearly show a molecular-like localized bond picture throughout the range of coverage. For alkali atom adsorption most of our knowledge about the electron structure originates from experiments recording work function changes and desorption energies. Although these experimental observables are related to the electron structure in an implicit way, they do reveal some characteristic features of alkali atom adsorption as the charge transfer and its variation with the surface density of alkali atoms. The observed behaviour can be visualized in an atomistic picture by the virtual level model.[1-3] Owing to the interaction with the substrate the alkali valence electron level is shifted and broadened into a resonance. The adatom density of states extends below the Fermi level and gives rise to a non-zero expectation value q^- of the number of electrons on the adatom. As the density of adatoms is increased the electrostatic interaction among them shifts the virtual level downwards with respect to the Fermi level, hence increasing q^-. Evidently, the Fermi level probes the resonance and a measurement of the charging would give information about the density of states distribution although over a limited energy range. This has been elucidated by several authors.[1,7-8] The uncertainties in the predicted properties of the virtual valence level are, however, rather large. This is not surprising since the work function is a measure of the electrostatic potential far outside the surface and it is not in a simple way related to the local density of states at a particular adatom.

There have been a few experimental reports on the observation of electronic excitations in adsorbed alkali atom layers. The excitations have been observed in photoemission [9] and electron energy loss experiments [10-11] and are believed to be of a collective nature, i.e., plasma excitations. Recently, we have reported about an investigation of the electron structure of Na and K adsorbed on Ni(100).[12-13] Electronic excitations from occupied core and valence states to unoccupied valence states and a collective excitation were investigated by electron energy loss spectroscopy. These results are summarized here and complemented with new information in particular concerning the adsorption of Na on Ni(100) $c(2\times 2)$S surface structure. The experiments have been performed in conjunction with measurements of two-dimensional adatom distributions and work function changes.

EXPERIMENTAL

The experimental technique and procedure is briefly summarized here; a more detailed account was given previously.[13] Na and K were deposited from breakseal glass ampoules containing high-purity alkali metal (>99.97 %). Deposition rates were in the range 1 monolayer per 1-25 min controlled by the source temperature. The Ni substrate temperature was, in general, 25°C ($-100°$C in some cases to be specified below). Prior to the alkali depositions, clean Ni surface conditions were established by argon ion bombardment (0.5 μA, 300 eV A$^+$ ions, 10-20 h) at a specimen temperature of about 350°C. The final cleaning was a brief heating to 900°C. The $c(2\times 2)$S structure was formed by decomposition of H$_2$S on the clean Ni surface kept at 250°C. The exposures were around 5×10^{-5} Torr s. The $c(2\times 2)$S was annealed to perfect surface order at this temperature after evacuating the H$_2$S gas. The alkali depositions were carried out at ambient pressures in the 10^{-12}-10^{-11} Torr range and the following experiments were performed at different alkali atom coverages.

(i) The lateral 2D distribution of the adsorbed atoms was determined by LEED. Tentative structure determinations based on intensity analysis have been achieved for some of the 2D-ordered structures (in particular for $c(2\times 2)$Na, $c(2\times 2)$S$-p(2\times 2)$Na and $c(2\times 2)$S$-c(2\times 2)$Na).[14]

(ii) Work function changes were determined by the retarding field method. The accuracy is better than 0.1 eV and can be checked via the low-energy cut off in the electron energy loss spectra ((iii) below).

(iii) Electron energy loss spectra were obtained by using the retarding system of the LEED optics. The retarding grids were modulated by a 0.3 V peak-to-peak sinusoidal signal and the energy resolution was about 0.5 eV at electron energies 5-30 eV. The energy loss scale is calibrated to better than 0.1 eV. Angular resolved energy loss spectra were measured by means of a PM tube (EMI 9789QB) reading the modulated signal from the fluorescent screen of the LEED optics. The detector angle was about 0.8° as seen from the crystal. The energy resolution was somewhat decreased (~ 0.7 eV) because of the high potential on the screen.

RESULTS AND DISCUSSION

ADATOM DISTRIBUTIONS

An important observable in these experiments is the lateral distribution of Na and K at various surface densities and this is deduced from the LEED patterns. The density is expressed in terms of coverages θ defined such that $\theta = 1$ corresponds to the density of atoms in the reference Ni(100) surface which has ideal lateral periodicity with a unit mesh edge $a = 2.49$ Å.

On the clean Ni(100) surface, Na and K distribute *uniformly* [15] at coverages $\theta < 0.5$ for Na and $\theta < 0.29$ for K. The relatively well-defined mean separation R of adatoms is determined from the LEED pattern and is used to deduce θ. The absolute accuracy is about 15 % (the relative is considerably better) depending on the assumed symmetry (square-hexagonal). This is, however, better than the accuracy of the vapour rate determinations (≈ 50 %). At $\theta = 0.5$ Na forms an ordered $c(2\times 2)$ structure and at $\theta = 0.29$, K forms a coincidence site hexagonal structure (see fig. 1). The ultimate monolayer coverages at 25°C are $\theta = 0.5$ for Na and $\theta = 0.38$ for K. Further deposition gives rise to second-layer formation which is disordered for K but an ordered hexagonal "duolayer" structure for Na. From LEED intensity measurement we find the Na-Na layer spacing to be about

3.0 Å, i.e., approximately the spacing between the densest Na planes in Na metal. Continuing the Na deposition results in the growth of an expitaxial Na(110) film.

Na was also adsorbed on the Ni(100) $c(2 \times 2)$S surface structure. At 25°C Na distributes approximately uniformly to $\theta = 0.25$ where an ordered $p(2 \times 2)$Na structure forms. Further deposition results in island growth of a $c(2 \times 2)$Na structure which is complete at $\theta = 0.50$ (see fig. 1). Keeping the Ni(100) $c(2 \times 2)$S specimen at $-100°C$ results in a relatively uniform Na distribution until formation of the ordered $c(2 \times 2)$Na structure at $\theta = 0.50$. These differences in the lateral Na distribution at two substrate temperatures are crucial in order to understand the changes in the electron energy loss spectra reported below.

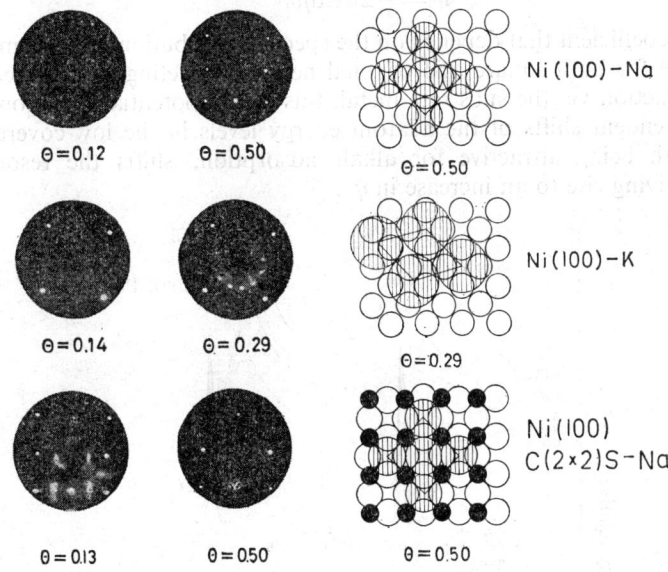

FIG. 1.—Diffraction patterns at various Na and K coverages θ. Tentative models for the ordered structures are sketched.

ELECTRONIC EXCITATIONS

In this section we discuss the experimental electron energy loss spectra and the work function data and explore three types of electronic excitation:

(i) excitations from occupied to unoccupied valence states.

(ii) excitation from a core state to unoccupied valence states.

(iii) a collective excitation in the denser alkali layers.

Obviously (i) and (ii) have implications with respect to the atomistic picture and this will be briefly outlined here. Within the virtual level model the originally sharp alkali valence level is shifted and broadened into a resonance owing to the interaction with the substrate. The adatom density of states extends below the Fermi level and gives rise to a non-zero expectation value q^- of the number of electrons on the adatom, with a resulting positive charging $q = 1 - q^-$. At zero temperature the effective electron charge on the adatom is given by

$$q^- = \int_{-\infty}^{E_F} A(E) \, dE \qquad (1)$$

where E_F is the Fermi energy and $A(E)$ is the adatom density of states distribution. In a classical picture the charged adatom and its image constitutes the fundamental dipole of the surface dipole layer, giving rise to a change in the work function

$$\Delta\phi = -4\pi\mu n, \quad (2)$$
$$\mu = qd, \quad (3)$$

where μ is the dipole moment, d is the dipole length and n the surface density of adatoms. This is valid far outside the surface as compared to the mean separation of the adatoms, while at the position at a particular adatom the electrostatic potential due to the surrounding dipoles is given by

$$\Phi = -2Ae\,d\mu n^{\frac{3}{2}} \quad (4)$$

where A is a coefficient that depends on the specific distribution of adatoms and takes the value ≈ 9 for both square and trigonal nets. Neglecting any indirect adatom–adatom interaction via the substrate metal, this dipole potential is the only cause of coverage dependent shifts of the adatom energy levels in the low-coverage regime. The potential, being attractive for alkali adsorption, shifts the resonance level downwards giving rise to an increase in q^-.

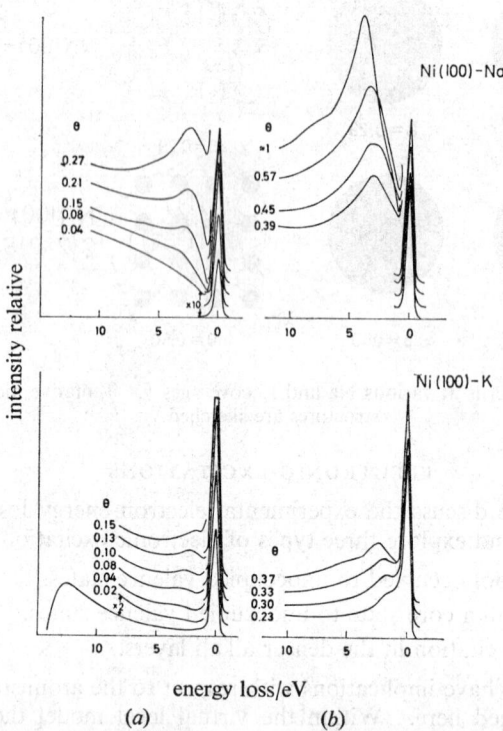

FIG. 2.—Experimental energy loss spectra for Ni(100)–Na and Ni(100)–K at various coverages θ. Primary electron energies were 32 eV (Na) and 18 eV (K) relative to the Ni Fermi level.

Accompanying redistributions of charge among the valence states, there will be changes in the intra-atomic Coulomb repulsions that manifest themselves as so-called chemical shifts of inner atomic levels. The energy to excite an adatom core electron to unoccupied valence states reflects the energy differences between initial and final state configurations. The difference in excitation energy relative to that of the free atom will primarily be related to the valence level occupancy q^- through the intraatomic Coulomb repulsions in the initial as well as final states. A crude estimate of this difference would be

$$\Delta E_c - \Delta E_c^\circ = (1-q^-)(U_{c-ns} - U_{ns-ns}) \tag{5}$$

where ΔE_c is the excitation energy for the charged atom and ΔE_c° that for the free atom. The Coulomb repulsion among core electrons and valence electrons, U_{c-ns}, as well as among the valence electrons, U_{ns-ns}, will, however, be different from the free atom values due to different valence charge distributions and due to screening by the metal substrate.

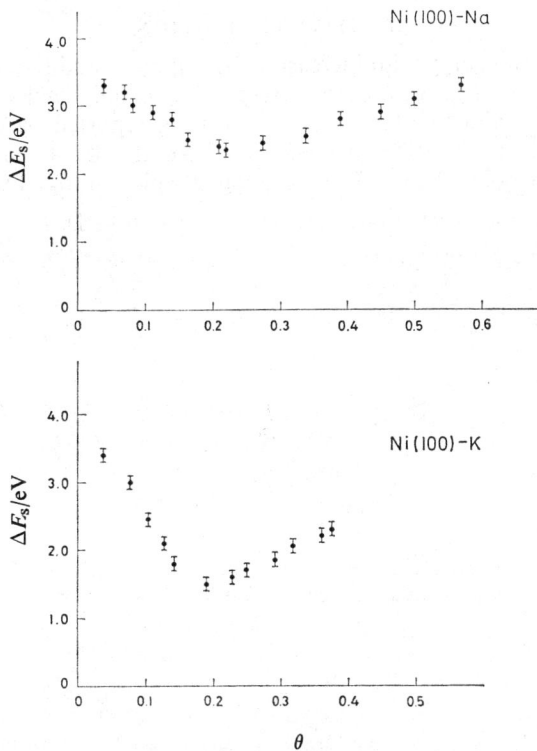

FIG. 3.—Experimental energy losses ΔE_s against coverage θ, for the Ni(100)-Na and Ni(100)-K adsorbate systems.

Concerning collective excitations in monolayer alkali films (iii), several theoretical treatments have been presented in the literature.[16-19] Thus the continuum model treated by Gadzuk [18] and the "box model" treated by Newns [19] yield essentially the same nature and dispersion of the mode. For small momentum Q, the dispersion is linear and the frequency is in the continuum model given by

$$\omega(Q) \approx \omega_p(1-\tfrac{1}{2}Ql) \tag{6}$$

$$\omega_p = [4\pi n e^2/ml]^{\frac{1}{2}} \tag{7}$$

where l is the film thickness, m and e are the effective electron mass and charge respectively.

Fig. 2 shows the experimental electron energy loss spectra, $I_s'(E)$, obtained for various coverages of Na and K on Ni(100). These spectra were recorded for normal incidence of the primary electron beam and at primary energies 32 eV (Na) and 18 eV (K). For these scattering conditions the clean Ni(100) spectra are relatively structureless. As a consequence of the alkali atom adsorption a new loss develops, the loss energy being dependent on the alkali atom density. The observed loss energies are plotted against θ in fig. 3. At the lowest coverages, a weak loss develops that decreases in energy from about 3.4 to 2.3 eV for Na in the range $0 < \theta < 0.25$ and from about 3.5 eV to 1.5 eV for K in the range $0 < \theta < 0.2$. As the coverage is further increased the loss energy increases and the intensity increases appreciably. These spectra finally turn into the plasmon spectra for thick alkali films.

COLLECTIVE EXCITATION

The increase in loss energy with increasing θ (for large θ) and the conversion of the spectra into the plasmon spectra for the corresponding alkali metal strongly suggest a collective excitation. Similar observations have been reported by others [10,11] and we have discussed this excitation in some detail for the K on Ni system.[12] We explore some further points here. From the log–log plots in fig. 4 we get

$$\Delta E_s \approx 4.2\, \theta^{0.4} \propto n^{0.4} \quad \text{for Na } (0.25 < \theta < 1),$$
$$\Delta E_s \approx 5.1\, \theta^{0.8} \propto n^{0.8} \quad \text{for K } (0.20 < \theta < 0.38),$$

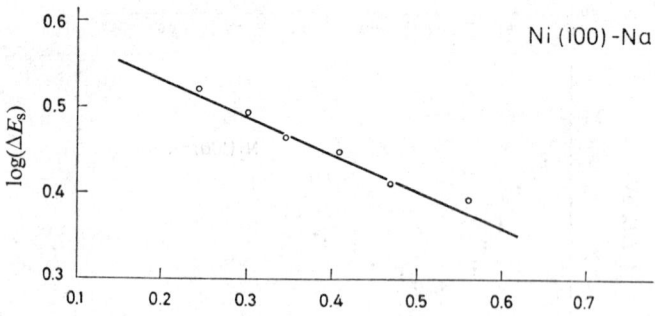

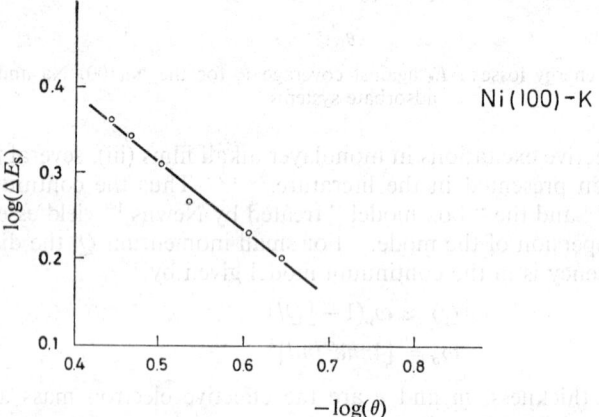

FIG. 4.—Logarithmic plots of experimental energy losses, $\log \Delta E_s$ against $\log \theta$, for Ni(100)–Na and Ni(100)–K.

i.e., the density dependence is close to $n^{\frac{1}{2}}$ (eqn (7)) for Na. Table 1 shows the experimental loss energies for the dense monolayers and the Na " duolayer " together with the $\hbar\omega_p$ values calculated from eqn (7) assuming free electron values for m and e. The l values used are the *n.n.* distances in the respective metals (monolayer case) and $l+3.0$ Å for the Na " duolayer " (3.0 Å Na–Na layer spacing, see above). The

TABLE 1.—COMPARISON OF EXPERIMENTAL ENERGY LOSS VALUES ΔE_s AND CALCULATED PLASMON ENERGIES $\hbar\omega_p$ FOR Na AND K LAYERS ON Ni(100). THE LAYER THICKNESSES l ARE THE *n.n.* DISTANCES IN Na AND K METAL FOR THE DENSE MONOLAYERS, AND THE *n.n.* DISTANCE PLUS THE Na LAYER SEPARATION 3.0 Å FOR THE Na DUOLAYER

	adsorbate coverage	ΔE_s(eV)	l(Å)	$\hbar\omega_p$(eV)
Na	$\theta = 0.50$ dense monolayer	3.1	3.71	5.5
Na	$\theta \approx 1$ duolayer	3.9	6.7	6.2
K	$\theta = 0.38$ dense monolayer	2.3	4.62	4.3

experimental ΔE_s values are considerably lower than $\hbar\omega_p$. Previously,[12] we argued about deviations in the factor e^2/ml (lower value of e because of charge transfer, uncertainties in l, etc.) but this seems unlikely in the light of the Na data. We would expect an error in e^2/ml to be smaller for the duo-layer than for the mono-layer but this is not the case. Since our analyzer integrates over Q we also argued about dispersion which according to eqn (6) will lower $\hbar\omega$ with a lower limit $\hbar\omega_p/\sqrt{2}$.[18]

Though angular-resolved experiments are hazardous to interpret in the case of back scattering because of the convolution of inelastic scattering and elastic scattering[20] we performed a series of such experiments on the $c(2\times2)$Na and the Na duolayer systems. Fig. 5 shows angular resolved, $I'(E)$ for the Na duolayer. The

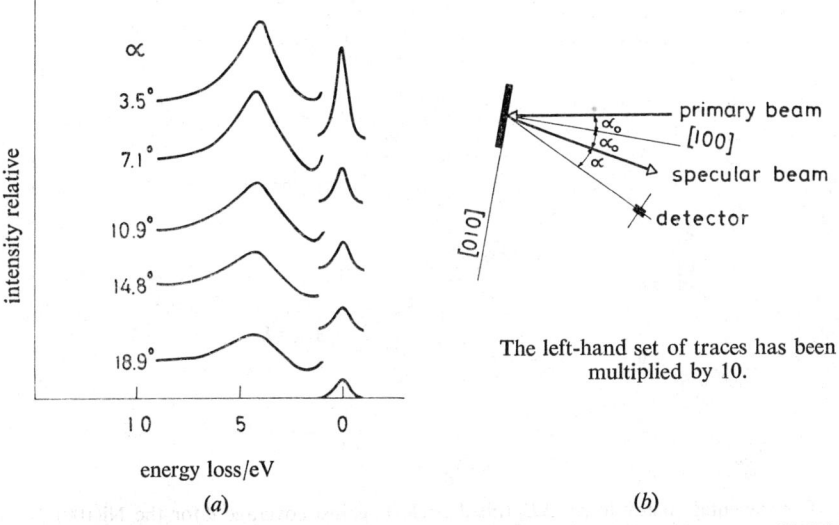

FIG. 5.—(a) Angular resolved energy loss spectra for the Ni(100)–Na duolayer system. (b) Schematic drawing of the experimental geometry. Detector opening $\approx 0.8°$.

detector is moved away from the specular, 00, beam in the plane of the surface normal and the [010] surface direction. The primary electron beam was incident in this plane at an angle $\alpha_0 = 10°$ relative to the surface normal. The energy loss peak around 4 eV is found to increase slightly in energy from about 3.9 to 4.3 eV and decrease slowly in intensity as the detector moves away from the specular beam. Similar observations were made for the $c(2 \times 2)$Na structure and were in that case almost independent of the diffraction conditions and whether the detector or the specimen was moved. Thus, the angular-resolved experiments indicates a weak positive dispersion rather than the negative in eqn (6). If this observation is correct, then the discrepancy of ΔE_s and $\hbar\omega_p$ can not be due to dispersion. It is interesting to note that the maximum energy 4,3 eV ($\alpha \approx 25°$, i.e., $Q \approx 0.5$ Å with respect to the 00 beam) is not far from the calculated $\hbar\omega_p/\sqrt{2} = 6.2/\sqrt{2} = 4.4$ eV. It seems as if the model (eqn (6), (7)) does not correctly describe the mode we observe.

VALENCE ELECTRON EXCITATION

The loss that is observed at low Na and K coverages and decreases in energy with increasing alkali atom density can hardly be interpreted in terms of a plasmon-like excitation and we have previously explored an interpretation in terms of a possible excitation from occupied to unoccupied valence states. The observed energies ΔE_s are plotted on an extended θ-scale in fig. 6. The solid curves are calculated assuming

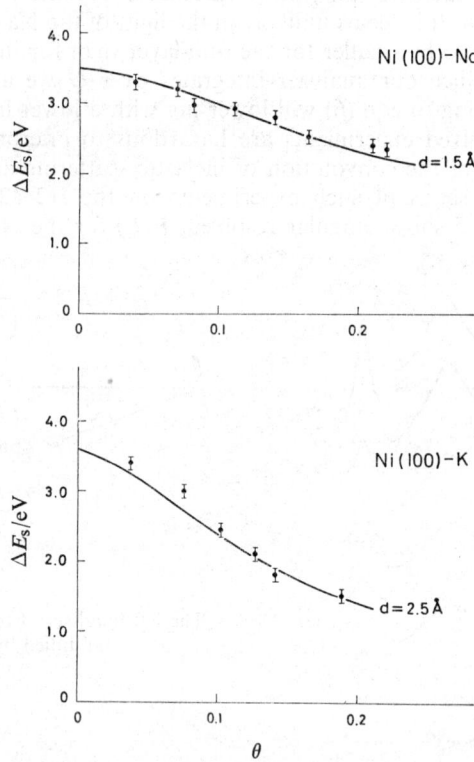

Fig. 6.—Experimental energy losses ΔE_s (filled circles) against coverage θ for the Ni(100)–Na and Ni(100)–K systems. The solid curves denoted by d were calculated from the electrostatic potential Φ as described in the text.

the decrease in excitation energy to be related to a decrease in splitting of the initial and final states caused by the change in the local electrostatic potential Φ (eqn (4)). Thus the excitation energy would vary as

$$\Delta E_s = \Delta E_{so} + \Phi_{if}, \tag{8}$$

where ΔE_{so} is the loss energy at $\theta = 0$ and Φ_{if} is the difference in Φ between the two states. From eqn (2) and (4) we have

$$\Phi_{if} = (A/2\pi)edn^{\frac{3}{2}}\Delta\phi. \qquad (9)$$

The curves in fig. 6 were then obtained from eqn (8) and (9) by using the experimental $\Delta\phi$ values of fig. 7. The d-values found to give the fit in fig. 6 are 1.5 and 2.5 Å for

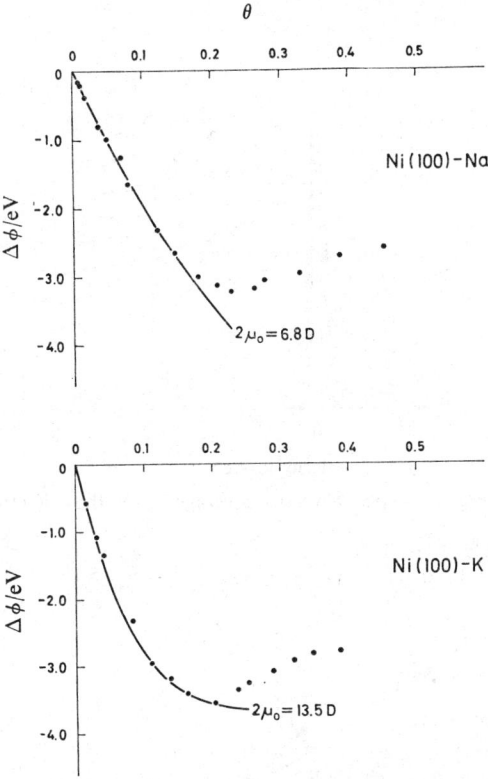

FIG. 7.—The change in work function $\Delta\phi$ against coverage θ, for Ni(100)–Na and Ni(100)–K. Filled circles are experimental values and the solid curves denoted by $2\mu_0$ were calculated from the point dipole depolarization formula.

Na and K respectively. We argued previously [13] that the excitation was a charge transfer transition from a bonding state to a state localized at the adatom and found the d-values to describe reasonable bond geometries and found it plausible that the initial state in the excitation via the bonding, involves the d-states of Ni.

When Na is adsorbed on the Ni(100) $c(2\times2)$S surface at $-100°$C, we observe a similar excitation in the low coverage regime. The energy loss spectra are shown in fig. 8. The spectrum for Ni(100) $c(2\times2)$S exhibits a loss peak around 6.5 eV which plausibly is due to an excitation on the sulphur atom, from a valence state observed ≈ 5 eV below E_F by INS,[6] to a final state involving sulphur valence orbitals overlapping Ni d-orbitals. Na adsorption gives rise to a new loss peak that decreases in energy. The loss energy, ΔE_s versus θ is shown in fig. 9. There are a few interesting items to note about fig. 9. First, at low coverages ($\theta \leqslant 0.1$), ΔE_s is about 0.3–0.5 eV larger than for Na on clean Ni(100). This increase can approximately be accounted for by the observed increase in work function of ≈ 0.4 eV related to the $c(2\times2)$S structure.

Thus the final state on the Na adatom is raised by the value of Φ appropriate for the adatom site. The geometry of the Ni(100) $c(2\times 2)$S–Na system is, at present, only tentative but preliminary LEED intensity calculations [14] for a $c(2\times 2)$Na support the structure model in fig. 1 and indicate the S layer to be about 1.3 Å [21] and the Na layer about 2.6 Å above the centre of the last Ni layer.

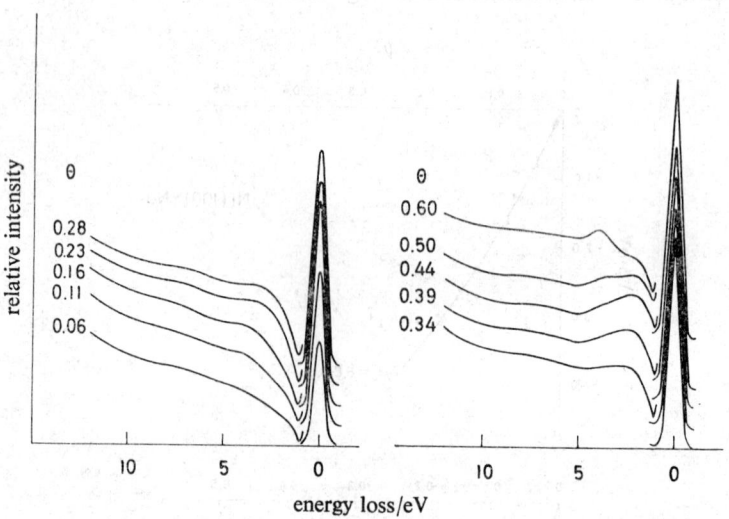

FIG. 8.—Experimental energy loss spectra versus coverage θ, for the Ni(100) $c(2\times 2)$S–Na system.

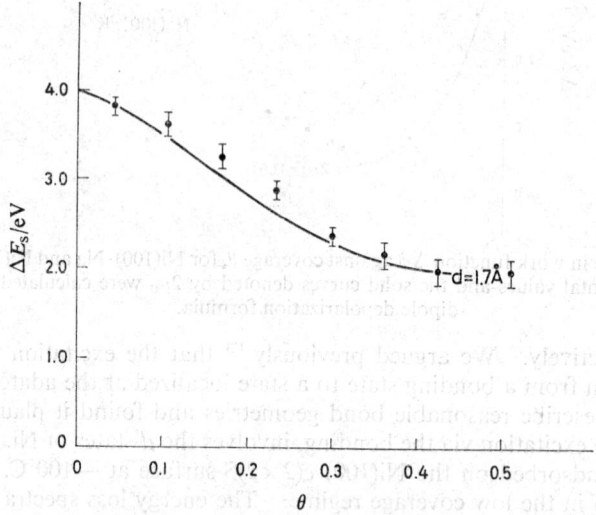

FIG. 9.—Experimental energy loss ΔE_s against coverage θ, for the Ni(100) $c(2\times 2)$S–Na adsorbate system where the solid curve denoted by d was calculated from the electrostatic potential Φ as described in the text.

Furthermore, only a weak loss peak that increases in energy with increasing θ is observed. It seems to merge out of the low-energy peak for $\theta > 0.4$ and is observed at $\Delta E_s \approx 3.8$ eV for $\theta = 0.5$. Instead, the low-energy loss is observed for all mono-layer coverages ($\theta \leqslant 0.5$) and decreases in energy to about 1.9 eV. The solid curve

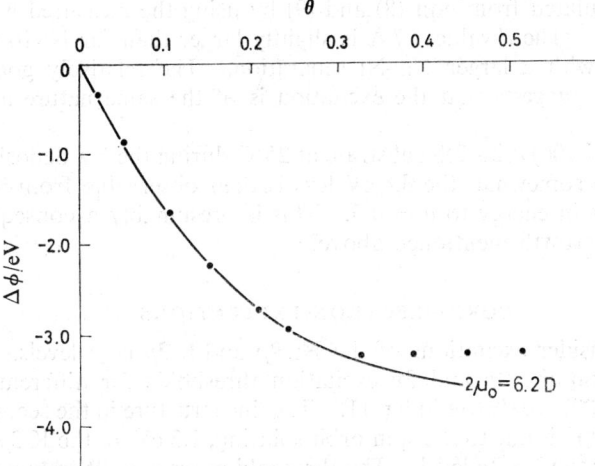

FIG. 10.—The change in work function $\Delta\phi$ versus coverage θ for the Ni(100) $c(2\times 2)$S–Na system measured relative to Ni(100) $c(2\times 2)$S.

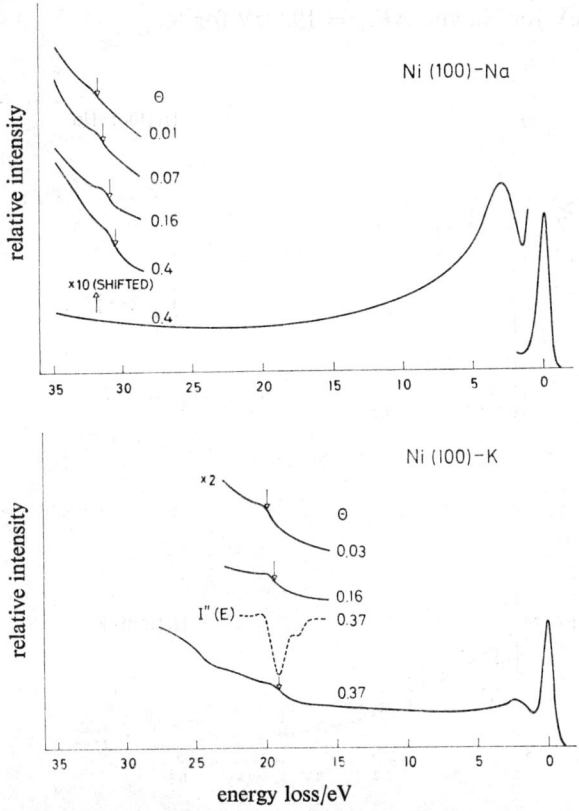

FIG. 11.—Experimental energy loss spectra around the Na $2p$ and K $3p$ core level excitation thresholds. The spectra recorded at various coverages θ are amplified and shifted along the ordinate (intensity). $I''(E)$ (at $\theta = 0.37$ for K) is a second derivative spectrum. The primary electron beam energies were 72 eV (Na) and 42 eV (K) relative to the Ni Fermi level.

in fig. 9 was calculated from eqn (8) and (9) by using the measured work function changes in fig. 10. The d-value 1.7 Å is slightly larger than for Ni(100)–Na. This seems consistent with a larger Na–Ni separation. The relatively good fit to the experimental data suggests that the excitation is of the same nature as for Na on Ni(100).

Keeping the Ni(100) $c(2 \times 2)$S substrate at 25°C during the Na deposition changes the spectral shape somewhat, the 3.8 eV loss is then observable from $\theta > 0.25$ and relatively constant in energy to $\theta = 0.5$. This is presumably a consequence of the $c(2 \times 2)$Na island growth mentioned above.

CORE-ELECTRON EXCITATIONS

We finally consider excitations of the Na $2p$ and K $3p$ core levels. The energy loss spectra around the $2p$ and $3p$ excitation thresholds for different Na and K coverages on Ni(100) are shown in fig. 11. The fine structure in the second derivative spectrum $I''(E)$ for K is due to the spin-orbit splitting, 1.3 eV, of the K $3p$ level. This is not resolved for the Na $2p$ level. The threshold energies shift to lower energies as θ increases. The observations are summarized in fig. 12 where it is also evident that the threshold energies approach the thick film values 30.7 and 19.2 eV already at the dense monolayer stage. If we extrapolate the data to $\theta = 0$ (dashed curves) we get $\Delta E_{2p} = 31.9$ eV for Na and $\Delta E_{3p} = 19.9$ eV for K.

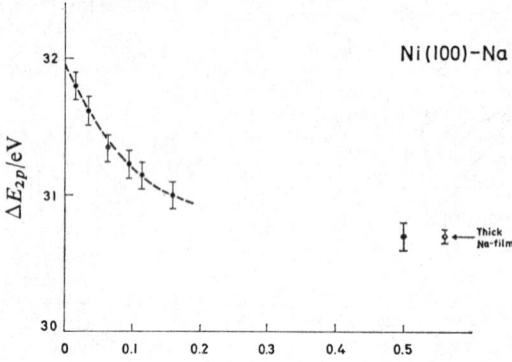

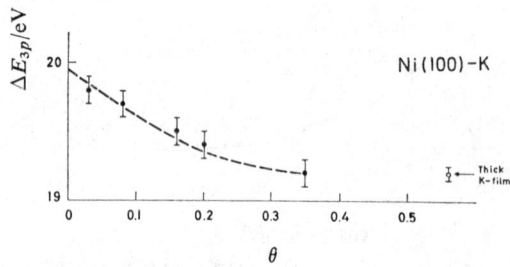

Fig. 12.—Experimental energy losses ΔE_{2p} and ΔE_{3p} (Na $2p$ and K $3p$ core level excitation thresholds) versus coverage θ. The dashed curves are fitted through the experimental points. The points denoted " thick film " are inserted for comparison.

In accordance with the model described above, we account for these shifts in terms of the charging state q^-. This is most readily done for Na where we can rely on soft X-ray absorption (SXA) data for the free Na atom [22] and Na metal.[23] The free atom transition $2p^63s-2p^53s^2$ is reported to occur at 30.8 eV which is very close to the absorption edge at 30.7 eV of Na metal (in accordance with ΔE_{2p} for a thick Na film above). This small difference in excitation energy in going to the solid state is plausibly due to compensating effects in the initial and final states. The experimental difference in binding energy of the $3s$ electron in the free atom configurations $2p^63s$ and $2p^53s^2$ amounts to 2.6 eV. We take this to be the difference in Coulomb attraction to the core hole and repulsion among the valence electrons. The value

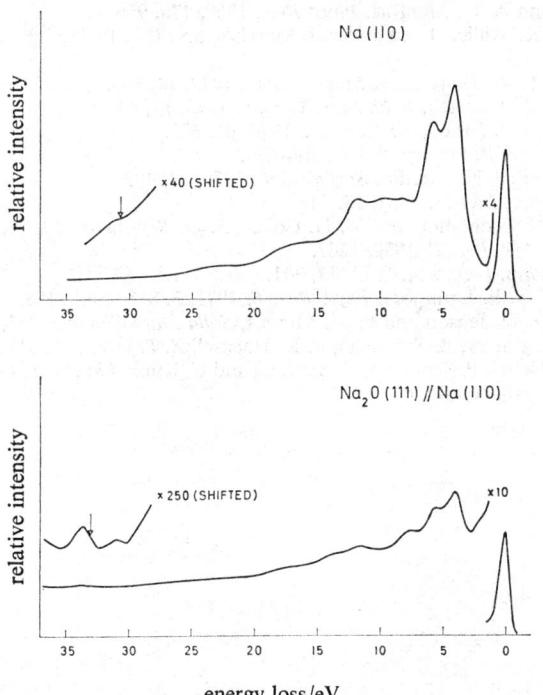

FIG. 13.—Experimental energy loss spectra for a clean Na(110) surface and a Na(110) surface exposed to 2×10^{-5} Torr s of oxygen (results in growth of Na$_2$O(111)//Na(110)). The primary electron energy was 72 eV relative to the Fermi level.

2.6 eV is, in fact, close to the shift, 2.4 eV, we observe for Na$_2$O. The Na $2p$ threshold is observed at 33.1 eV (see fig. 13) i.e., 2.4 eV higher than for Na metal. Hartree–Fock calculations [24] give $U_{2p-3s}-U_{3s-3s} = 7.8-5.9 = 1.9$ eV, i.e., somewhat lower than the experimental difference 2.6 eV. From our experimental shift 1.2 eV, and the value 2.6 eV, we deduce from eqn (5)

$$(1-q^-) = 1.2/2.6 = 0.46.$$

This figure can be compared with that which we derive from the extrapolated $\theta = 0$ dipole moment $2\mu_0 = 6.8$ Debye using $d = 1.5$ Å (found above) as the dipole length.

$$q_0 = (1-q^-) = 0.47.$$

The agreement between these two independent charge estimates lends some support for our simple models.

The authors are indepted to O. Gunnarsson, H. Hjelmberg, B. Kasemo, I. Lindgren and S. Lundqvist for helpful and stimulating discussions. Financial support from the Swedish Natural Science Research Council is gratefully acknowledged.

[1] A. J. Bennet and L. M. Falicov, *Phys. Rev.*, 1966, **151**, 512.
[2] J. W. Gadzuk, *Surface Sci.*, 1967, **6**, 159
[3] D. M. Newns, *Phys. Rev.*, 1969, **178**, 1123.
[4] N. D. Lang, *Phys. Rev. B*, 1971, **4**, 4234.
[5] D. E. Eastman and J. K. Cashion, *Phys. Rev. Letters*, 1971, **27**, 1520.
[6] G. E. Becker and H. D. Hagstrum, *Surface Sci.*, 1972, **30**, 505.
[7] J. W. Gadzuk, *Phys. Rev.*, 1967, **154**, 662.
[8] J. P. Muscat and D. M. Newns, *J. Phys. C*, 1974, **15**, 2630.
[9] T. A. Callcott and A. U. MacRae, *Phys. Rev.*, 1969, **178**, 966.
[10] A. U. MacRae, K. Müller, J. J. Lander, J. Morrison and J. C. Phillips, *Phys. Rev. Letters*, 1969, **22**, 1048.
[11] S. Thomas and T. W. Haas, *Solid State Comm.*, 1972, **11**, 193.
[12] S. Andersson and U. Jostell, *Solid State Comm.*, 1973, **13**, 833.
[13] S. Andersson and U. Jostell, *Surface Sci.*, 1974, **46**, 625.
[14] S. Andersson and J. B. Pendry, to be published.
[15] R. L. Gerlach and T. N. Rhodin, *Surface Sci.*, 1970, **19**, 403.
[16] F. Stern, *Phys. Rev. Letters*, 1967, **18**, 546.
[17] K. L. Ngai, E. N. Economou and M. H. Cohen, *Phys. Rev. Letters*, 1970, **24**, 61.
[18] J. W. Gadzuk, *Phys. Rev. B*, 1970, 1267.
[19] D. M. Newns, *Phys. Letters A*, 1972, **39**, 341.
[20] C. B. Duke and G. E. Laramore, *Phys. Rev. B*, 1971, **3**, 3183 and 3198.
[21] J. E. Demuth, D. W. Jepson and P. M. Marcus, *Solid State Comm.*, 1973, **13**, 1311.
[22] H.-W. Wolff, K. Radler, B. Sonntag and R. Haensel, *Z. Physik*, 1972, **257**, 353.
[23] R. Haensel, G. Keitel, P. Schreiber, B. Sonntag and C. Kunz, *Phys. Rev. Letters*, 1969, **23**, 528.
[24] J. B. Mann, Los Alamos Report LA 3690.

Beam Effects in AES revealed by XPS

By J. P. Coad*, M. Gettings and J. C. Rivière

A.E.R.E., Harwell, Didcot, Oxon.

Received 26th February, 1975

The XPS and AES analyses of virgin surfaces are frequently in disagreement; for example, the carbon and oxygen levels appear much larger by XPS. By rastering an electron beam over the area analysed by XPS it is shown that the XPS analysis changes to become more comparable to that by AES. The initial disagreement is thus due to the interaction of the electron beam used in AES with the surface. SIMS has been used as well as XPS in a preliminary study of the nature of the beam interaction with the surface.

Among the many papers on Auger electron spectroscopy (AES) that have appeared over the last few years, there have been several that have commented on the effect of the electron beam on the surface. The electron beam has been shown to affect radically the interaction of gases with surfaces; for example, the mechanism of the interaction of CO (and other gases) with the Si(111) surface has been studied and shown to be altered by electron irradiation.[1-4] Carbon monoxide has also been shown to decompose on Ni[5] and Pt[6] in an electron beam, whilst Barrie and Brundle[7] have shown that for CO on W some beam effects can be observed even with a defocussed beam of 0.5 μA (a primary beam flux at the surface about two orders of magnitude lower than that normally employed). There have also been reports of surfaces which are unstable in an electron beam—the alkali halides dissociate rapidly[8,9] and surface oxides may be reduced.[10,11] Localised diffusion of elements to, or from, the irradiated area has also been observed.[11-13] In addition, an electron beam can desorb material from many virgin surfaces (electron-induced desorption (EID) or electron-stimulated desorption (ESD)) and this process can sometimes be followed by AES.[14,15] The beam-induced desorption greatly alters the secondary emission coefficient,[15] an effect that can be seen readily in a visual display based on secondary emission yield,[16] when contrast develops between the irradiated and unirradiated areas. It is also common experience that when an electron beam is moved across many surfaces there is a burst of desorbed gas.

In many cases, therefore, the analysis obtained by AES does not give an accurate view of the " as-received " surface of a material because of beam-induced effects which can occur either during the course of examination, or have occurred almost instantaneously in the electron beam. In this laboratory, all the different types of beam interaction effects mentioned above have been observed, and some of these observations are presented here to augment the volume of work already published. Furthermore, results are included on the analysis of surfaces by both AES and X-ray photoelectron spectroscopy (XPS), and some preliminary results on the combination of AES, XPS and secondary ion mass spectroscopy (SIMS). Since XPS uses soft X-rays as the means of excitation, the *primary* electron beam effects of AES are absent. XPS produces *secondary* electrons at the surface as does AES, and they may also cause desorption and diffusion effects, but the density of such electrons present in XPS is orders of magnitude lower than in AES under normal operating conditions.

As the sampling depths for the two techniques are different, and quantification procedures are based on assumption of uniform elemental distribution with depth, some differences may be expected between the XPS and AES analyses of a surface. In general, the mean escape depths for XPS are greater than those of AES by a factor of about two, the latter being about 0.7 nm.[17] Thus the contribution from a monolayer of contaminant on a specimen surface might be 20 % of the AES analysis, but only 10 % of the XPS analysis; concentrations which are uniform to a depth exceeding the escape depths, however, give the same result in each analysis.

Despite these disparities between XPS and AES analyses, it would be instructive to note what differences exist in the surface analyses obtained by the two techniques, and to see whether the XPS analysis of a surface changes as a result of a period of electron bombardment.

EXPERIMENTAL

Two different spectrometer systems were used during the course of the work to be described. In the first (VG) system the XPS analysis was performed using a Vacuum Generators ESCA 3, whilst Auger analyses were obtained with a hemi-cylindrical mirror analyser designed at A.E.R.E., Harwell.[18] The specimen could be moved readily within the chamber from one analyser to the other by means of a long-travel manipulator. The second (PHI) system contained a cylindrical mirror analyser by Physical Electronics Inc (PHI) for both AES and XPS,[19] and facilities for SIMS (Balzers High Vacuum). It required only a simple rotation of a manipulator to move the specimen from the AES/XPS analysis position, to the point of examination by SIMS. The electron beam used for AES could be rastered over an area of the surface of about 3×3 mm at 3 keV and variations in the secondary electron yield were synchronised with the sweep generator to give an image of the specimen on an oscilloscope.[16] Auger spectra were recorded from single points, however, with a spot size of about 30 μm. Specimens were examined in the PHI system without baking at a pressure of less than 10^{-7} Pa.

It is important to note the differences in the area analysed by XPS in each instrument. The ESCA 3 analyser accepts electrons from an area on the specimen of about 10 mm $\times$ 2 mm, whilst XPS using the PHI analyser is restricted to an area of 1 to 2 mm². Thus, using the PHI system, it was possible to raster the electron beam over an area greater than that contributing to the XPS analysis (and since the area analysed by SIMS was comparable to the rastered area, it was useful to perform SIMS analysis as well in the PHI system). It was not as simple to ensure uniform bombardment over the large area analysed in the VG system, however, especially since large specimen movements were involved.

The specimens to be described in this paper are listed in table 1 with the system and techniques used. All specimens tabulated showed effects ascribable to electron beams.

TABLE 1.—SPECIMENS DESCRIBED IN THIS PAPER, TOGETHER WITH THE SYSTEM (SEE TEXT) AND TECHNIQUES USED

specimen type	analysis system	techniques reported
chrome steels	VG	XPS, AES, ion profiling
stainless steels	VG	XPS, AES, ion profiling
17 % Cr steel	PHI	XPS, electron irradiation, SIMS
nickel	PHI	XPS, AES, electron irradiation, SIMS
platinum electrode	PHI	XPS, AES, electron irradiation
glass	PHI	XPS, electron irradiation
brass	PHI	XPS, AES, electron irradiation
oxide on steel	PHI	XPS, AES, electron irradiation, ion profiling

RESULTS

XPS and AES analyses as a function of the amount of surface removed by ion bombardment have been obtained from many chrome and stainless steels, using the

VG system. The details of such profiling experiments on a few of the chrome steels have been described elsewhere.[20] In those experiments quite good agreement was obtained between the concentrations of Fe, Cr, C and O as determined by the two techniques, *except* from measurements taken from the virgin surface. It was interesting to note, too, that the XPS profiles showed comparable depth resolution with those using AES despite the greater sampling depth of XPS. Agreement between the profiles for Fe, Cr, C and O from a number of stainless steels was also obtained [13] (again, *except* in the initial condition). The profiles for Ni (in type 316 stainless steel) did not agree, however, as can be seen in fig. 1; XPS clearly showed a depletion of Ni at the surface whilst AES did not. Also shown in fig. 1 is the Mo profile obtained by XPS; it was not possible to construct a similar profile for AES, however, since the major Auger peaks for Mo coincide with peaks from S, Cl and Ar, all of which were present. The Mo profile has been included since the significant depletion of Mo at the surface confirms the AES results of Lumsden and Staehle,[21] while conflicting with those of Barnes et al.[22]

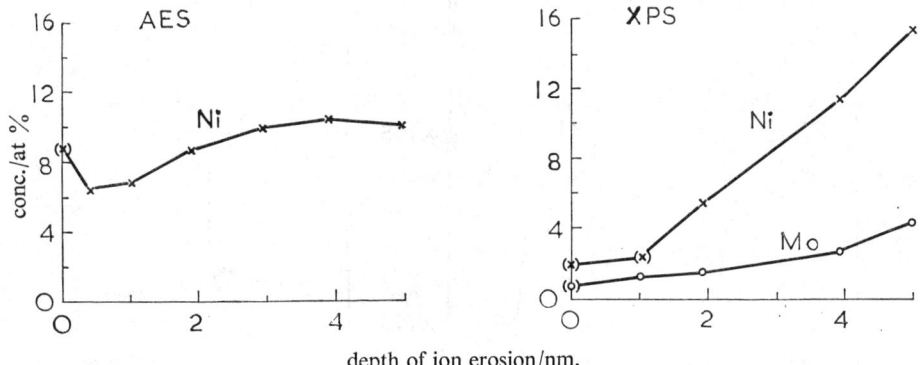

FIG. 1.—Concentration profiles for Ni (AES and XPS) and Mo (XPS only) as a function of the approximate depth of material removed from the surface by argon ion bombardment. The concentrations are the atomic percentages of the total amount of Fe, Cr, Ni and Mo detected.

Mention was made above of the disagreement between AES and XPS when applied to virgin oxide surfaces. XPS analyses of un-eroded air formed oxides on steel surfaces normally reveal approximately twice as much carbon and oxygen (relative to the amount of metal) as do AES analyses, despite the fact that a layer of contamination should contribute *less* to the XPS analysis. In XPS, the oxygen peak from the virgin surface is usually resolvable into two components, and as the peak of lower kinetic energy (higher binding energy) decreases rapidly on ion bombardment it is assumed to be associated with surface contamination. Better agreement is found between the initial XPS and AES oxygen levels when only the XPS peak at higher kinetic energy (lower binding energy) is considered (although this XPS concentration value is still greater than that from AES).

Because of the above discrepancy the oxygen peak from a 17 % Cr steel was studied before and after electron irradiation of the analysed region, in the PHI system. Complete XPS spectra (from 450 to 1500 eV) are shown in fig. 2, with details of the oxygen peak, before and after electron irradiation of the analysed area. An electron beam (30 μA at 3 keV) was rastered over an area of 3 × 3 mm for 2400 s, corresponding to an irradiation dose of 10 μA into any 30 μm diam. region (the normal AES analysis area) for a total time of about 1 s. There were some obvious differences in the XPS spectra before and after irradiation; the heights of the oxygen and carbon peaks clearly fell relative to the iron peaks and the oxygen peak shape changed, in that the lower kinetic energy component (not as well resolved with the PHI CMA as it was with ESCA 3) decreased markedly. When allowance was made for the change in width of the oxygen peak, the amounts of both C and O relative to Fe fell by a factor of 2.4 on electron beam irradiation. From the remarks in the previous paragraph it can be seen that this factor is enough to give good agreement between the AES and the modified XPS analyses *after* irradiation (with due allowance for the difference due to escape depths). An interesting side effect was observed during this experiment in that although the specimen was large compared to the analysed area, no other region was found after electron bombardment where XPS showed as much contamination as that in trace A of fig. 2. The implication is that there were some changes to the surface up to about 10 mm from the irradiated area.

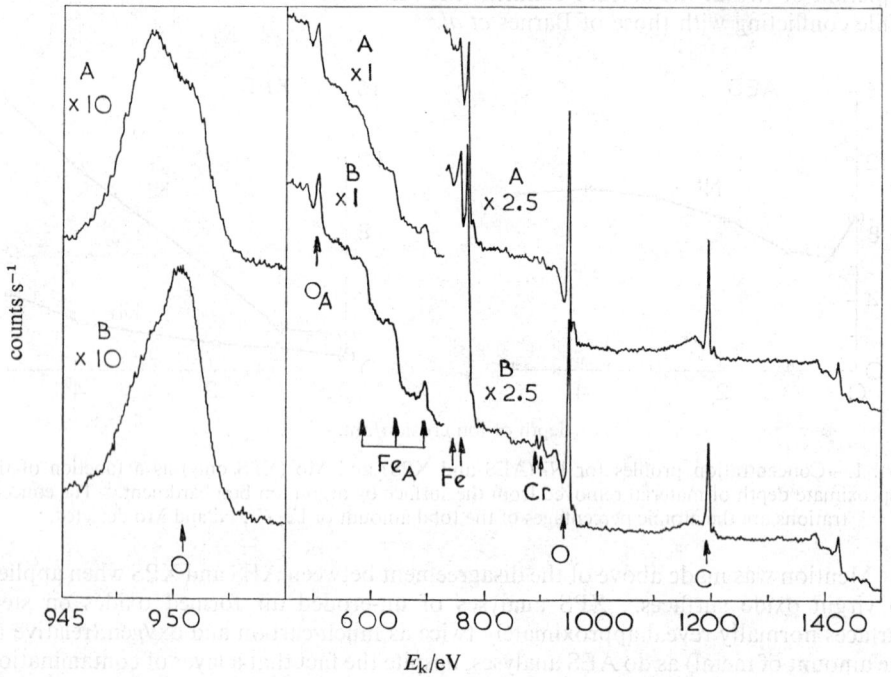

Fig. 2.—XPS spectra from a 17 % Cr steel surface before (traces labelled A) and after (B) electron irradiation. On the right, spectra from 425 to 1500 eV recorded at 100 eV pass energy; on the left, details of the oxygen peak recorded at 50 eV pass energy. The suffix A denotes an Auger peak. Al K_α radiation 1486.6 eV.

A more exhaustive SIMS/XPS experiment was performed using a large piece of Ni foil (15 × 50 mm). Towards one end of the foil (area 1) the techniques were employed in the following order: XPS, electron irradiation (including AES analysis), XPS, SIMS, XPS. Such a sequence determined (i) the XPS spectra before and after electron irradiation, (ii) the AES analysis for comparison with XPS, (iii) the SIMS spectra after irradiation (with a careful check kept on the amount of ion bombardment used at each point in the spectra), and (iv) the change in the XPS spectrum due to the ion bombardment required for SIMS. The electron irradiation conditions were 40 μA at 3 keV beam energy for 1800 s over an area of 3 × 3 mm, with 15 μA at

4 keV for 900 s from a 30 μm spot at the centre of the analysed area to obtain the Auger spectrum shown in fig. 3. The XPS spectra before and after the irradiation are shown in fig. 4; the most notable changes detected were an increase in Ni and a decrease in O peaks on irradiation. There was also a slight increase in N on irradiation. The XPS spectra before and after SIMS were indistinguishable under the Ar ion beam conditions used (5×10^{-10} A at 3 keV for 2700 s).

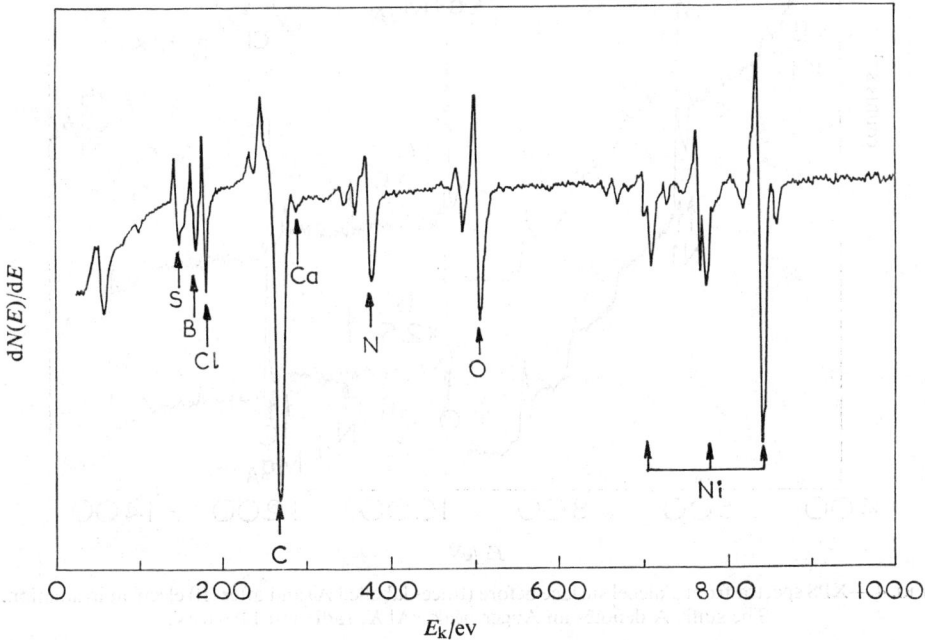

FIG. 3.—The Auger spectrum from a nickel surface, using an electron beam of 15 μA at 4 keV.

Positive and negative SIMS spectra from a fresh region of the specimen (area 2) were obtained for comparison with those from area 1 (after irradiation); this new unirradiated area was first checked with XPS which proved that contamination levels were the same as those originally on area 1. Care was taken to maintain the same SIMS conditions. The positive and negative spectra from each area are shown in fig. 5 and 6, respectively. In the positive spectra (fig. 5) peaks due to boron (masses 10 and 11), carbon (mass 12) and the peak at mass 24 (C_2 or magnesium) remained unchanged with irradiation while hydrogen (mass 1) fell by about 50 %. Nickel (masses 58 and 60) also decreased, a result which contrasts with the behaviour of the Ni peaks in XPS. The ion yields from the alkali metals Na and K, to which SIMS is very sensitive, doubled after irradiation, whilst the Ca level also increased slightly. Although the carbon peaks remained unchanged, the peaks at masses 13, 14, 15, 25, 26 and 29 which could all be attributed to hydrocarbon groups, showed significant reductions in intensity after irradiation.

In the negative spectra (fig. 6) the hydrogen peak again showed a large decrease with irradiation as did those due to oxygen (mass 16) and the OH radical (mass 17). The sulphur (mass 32) and chlorine (masses 35 and 37) peaks were unaffected by irradiation. Unlike their corresponding peaks in the positive spectrum, the carbon peaks C_1^- and C_2^- showed some increase with irradiation.

No differences were detected using XPS analysis on the fresh area before and after

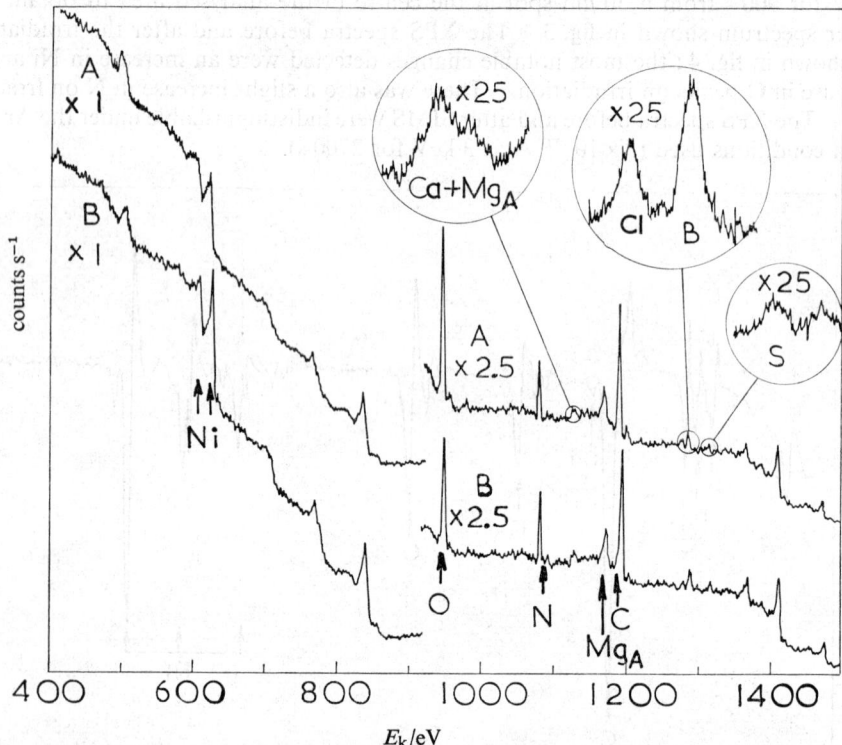

Fig. 4.—XPS spectra from a nickel surface before (traces labelled A) and after (B) electron irradiation. The suffix A denotes an Auger peak. Al K_α radiation 1486.6 eV.

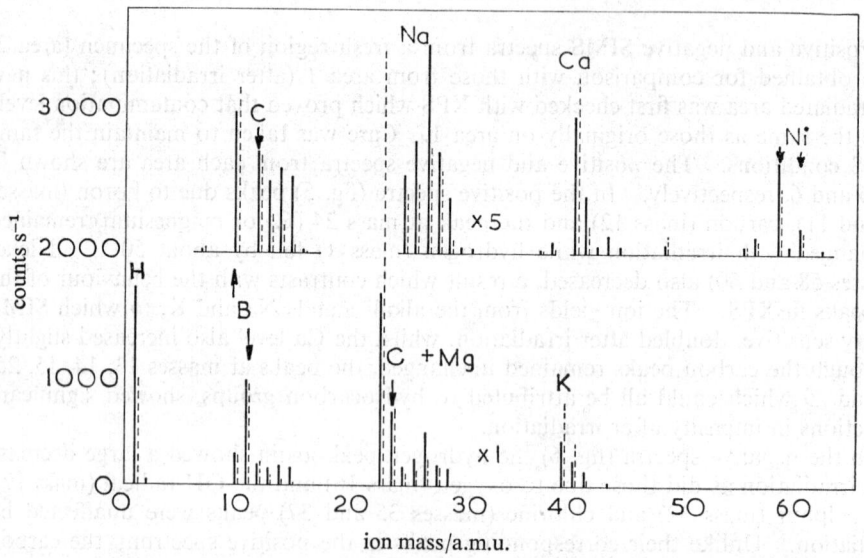

Fig. 5.—Positive SIMS spectra from a nickel surface. The solid and dashed lines correspond to conditions before (area 2) and after (area 1) electron irradiation, respectively.

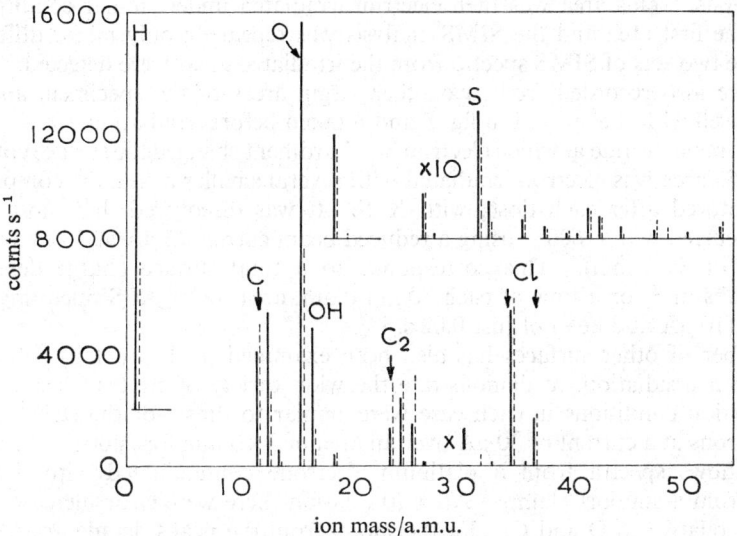

FIG. 6.—Negative SIMS spectra from a nickel surface. The solid and dashed lines correspond to conditions before (area 2) and after (area 1) electron irradiation, respectively.

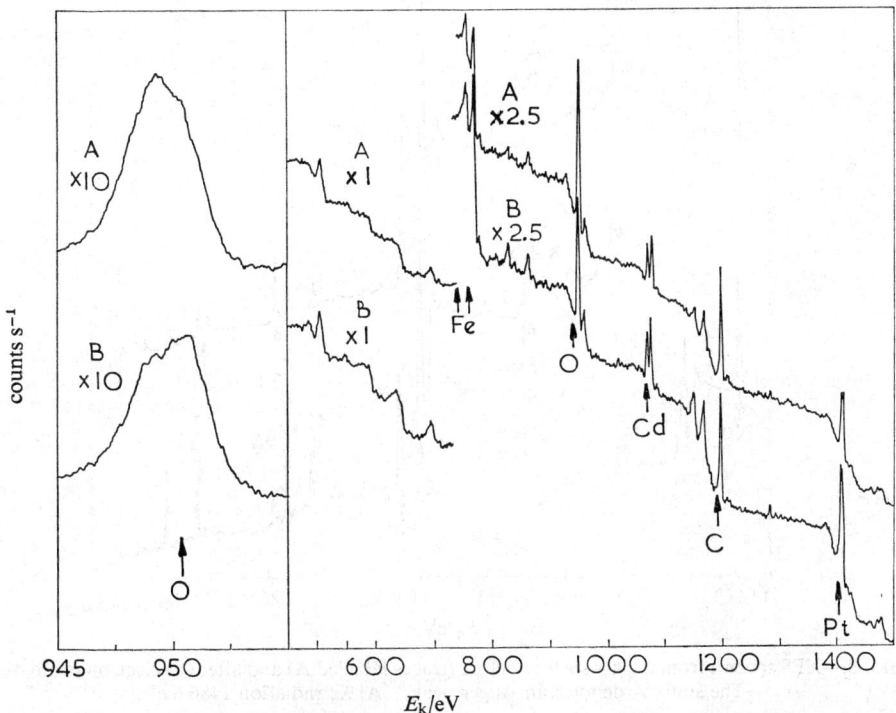

FIG. 7.—XPS spectra from a contaminated Pt electrode before (traces labelled A) and after (B) electron irradiation. On the right, complete spectra from 450 to 1500 eV recorded with 100 eV pass energy; on the left, detailed scans of the oxygen peak recorded at 50 eV pass energy. Al K_α radiation 1486.6 eV.

SIMS analysis. This area was then electron irradiated under similar conditions to those for the first area and the SIMS analysis was repeated; only minor differences between the two sets of SIMS spectra from the irradiated areas were detected. SIMS spectra were also recorded from two other virgin areas of the specimen, and they correlated well with the spectra in fig. 5 and 6 taken before irradiation.

To determine the rate at which electron bombardment changed the surface composition, a virgin area was electron irradiated with several smaller doses, the composition being monitored after each dose, with XPS. It was discovered that most of the changes occurred within 600 s, using a reduced beam current (5 μA at 3 keV over the same area of 3×3 mm). This corresponds to a total surface charge density of 240 coulombs m^{-2}, or a time at each 30 μm diam. area under AES operating beam conditions (10 μA at 3 keV) of just 0.02 s.

A number of other surfaces has also been examined (in less detail) before and after electron irradiation, to demonstrate the wide variety of electron beam effects. The irradiation conditions in each case were similar to those for the steel, namely, 3 keV electrons at a current of 30 μA over an area of 3×3 mm for 2400 s. Fig. 7, for example, shows spectra from a platinum electrode contaminated with iron and cadmium from a support clamp. After irradiation there were clear increases in Pt, Cd and Fe relative to O and C. Furthermore, from the peaks detailed on the left-hand-side of the figure, the O peak shape can be seen to have changed in a manner analogous to that shown in fig. 2 for a steel surface.

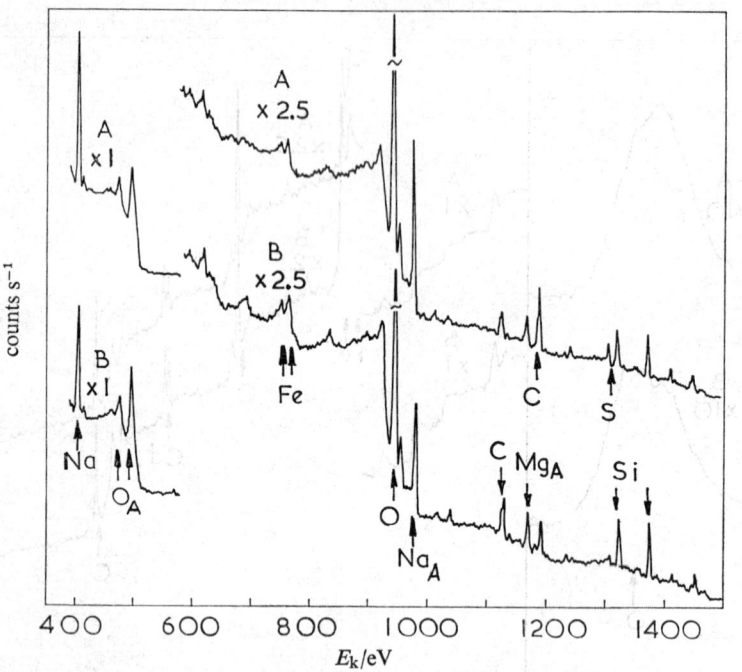

FIG. 8.—XPS spectra from a glass surface before (traces labelled A) and after (B) electron irradiation. The suffix A denotes an Auger peak. Al K_α radiation 1486.6 eV.

In another example the XPS spectra before and after irradiation of a glass are displayed in fig. 8 (parts of the oxygen peaks have been omitted for clarity). In this case the Na, C and S decreased on irradiation, whilst the Ca, Si, K and B increased.

It is likely that some of these changes were caused either by beam-assisted segregation or by decomposition of the glass rather than desorption of surface contamination.

Segregation effects were also evident on a brass sample treated with HF. The initial XPS analysis showed the concentration of Zn to be a factor of five greater than that of Cu at the surface. AES, however, gave a Cu level much greater than that of Zn, and this was verified in the XPS spectrum recorded *after* electron irradiation (which also showed a large reduction in the oxygen level).

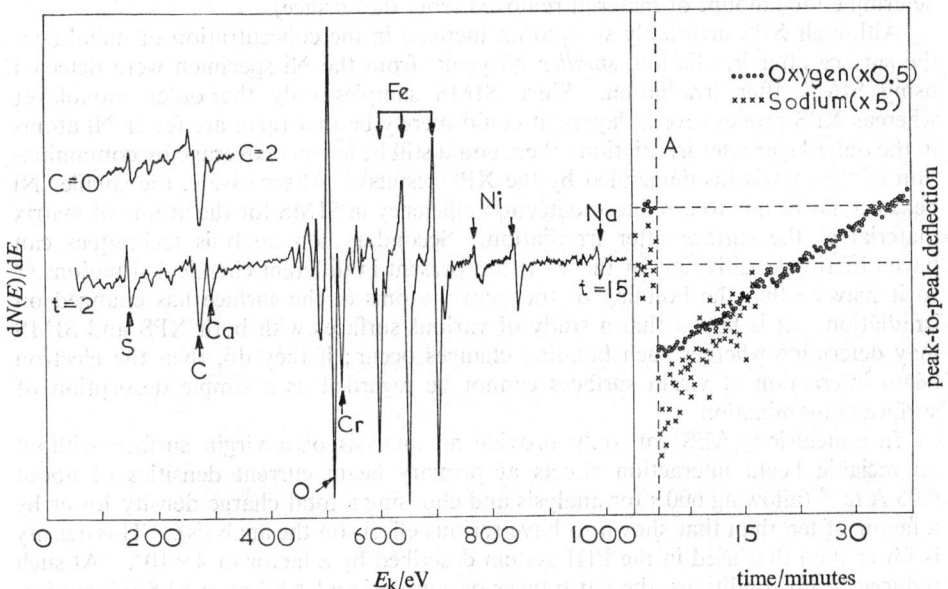

FIG. 9.—The Auger spectrum from a thick oxide film on steel. The time t (in min) take to record the spectrum is shown on the figure. On the right-hand-side, the peak-to-peak size of the major O and Na features have been plotted as a function of time. The line A indicates when the specimen was moved. Beam conditions: 15 µA at 4 keV.

Finally, it must be emphasised that the changes visible in the XPS analyses were the result of very small electron irradiation doses compared with those used during conventional AES analysis. There are many other electron beam interaction effects which require larger doses of electron irradiation, but they are not always easy to observe with XPS due to the long rastering time involved. The analysis of a thick oxide film on a corroded steel sample serves to illustrate this point. Fig. 9 shows the Auger spectrum from the oxide, which had been ion bombarded, with the time spent in the electron beam indicated. Calcium and sulphur clearly segregated to the surface during the recording of part of the spectrum, whilst changes in the oxygen and sodium peak sizes were also gradual. The latter peaks were monitored with a multiplex control (PHI model 20–055), which enables the peak-to-peak size of any Auger feature to be plotted as a function of time. The return to the equilibrium values after moving the specimen is shown on the right-hand side of Fig. 9. No changes in the Ca, S, Na or O levels could be observed from the XPS spectra before and after by rastering over 3×3 mm for 2400 s irradiation, however; this is in agreement with the calculation that any change visible in XPS following such an irradiation dose would require less than a second to occur in AES.

DISCUSSION

The differences between XPS and AES analyses of virgin surfaces reported here have been shown to be due to the interaction of the electron beam with the surface contamination. From the changes in the SIMS and XPS spectra it appears to be likely that water is one of the main desorbates, whilst there may be cracking as well as desorption of hydrocarbons. (Experiments have been planned to confirm the composition of the desorbed material with a mass spectrometer and to attempt to determine the amount of material removed from the surface).

Although XPS invariably showed an increase in the concentration of metal near the surface after irradiation, *smaller* Ni peaks from the Ni specimen were detected using SIMS after irradiation. Since SIMS samples only the outer monolayer whereas XPS probes several layers, it could merely be that there are fewer Ni atoms in the outer layer after irradiation; there could still be less net coverage by contamination of the matrix (as demanded by the XPS results). Alternatively, the smaller Ni peaks could be due to a reduced detection efficiency in SIMS for the atoms of matrix material at the surface after irradiation. Secondary ion analysis techniques can have differing sensitivities for the same ion present in different chemical situations,[23] so it may be that the bonding of the matrix atoms at the surface has changed on irradiation. It is hoped that a study of various surfaces with both XPS and SIMS may determine whether such bonding changes occur; if they do, then the electron beam interaction at virgin surfaces cannot be regarded as a simple desorption of surface contamination.

In conclusion, AES can only provide an analysis of a virgin surface without appreciable beam interaction effects at primary beam current densities of about 0.03 A m^{-2} (allowing 600 s for analysis and choosing a total charge density lower by a factor of ten than that shown to have serious effects on the analysis). This density is lower than that used in the PHI system described by a factor of 4×10^5. At such reduced beam conditions, the advantages of conventional AES over XPS of speed of analysis and spatial resolution largely disappear.

[1] J. P. Coad, H. E. Bishop and J. C. Rivèire, *Surface Sci.*, 1970, **21**, 253.
[2] B. A. Joyce, *Surface Sci.*, 1973, **35**, 1.
[3] R. E. Kirby and D. Lichtmann, *Surface Sci.*, 1974, **41**, 447.
[4] R. E. Kirby and J. W. Dieball, *Surface Sci.*, 1974, **41**, 467.
[5] H. H. Madden and G. Ertl, *Surface Sci.*, 1973, **35**, 211.
[6] J. M. Martinez and J. B. Hudson, *J. Vac. Sci. Tech.*, 1973, **10**, 35.
[7] A. Barrie and C. R. Brundle, *J. Electron Spectr.*, 1974, **5**, 321.
[8] P. W. Palmberg and T. N. Rhodin, *J. Phys. Chem. Solids*, 1968, **29**, 1917.
[9] T. E. Gallon, I. G. Higginbotham, M. Prutton and H. Tokutaka, *Surface Sci.*, 1970, **21**, 224.
[10] S. Thomas, *J. Appl. Phys.*, 1974, **45**, 161.
[11] W. C. Johnson and L. A. Heldt, *J. Electrochem. Soc.*, 1974, **121**, 34.
[12] C. T. H. Stoddart and E. D. Hondros, *Trans. Brit. Ceram. Soc.*, 1973, **73**(2), 61.
[13] J. P. Coad, A.E.R.E. Harwell Report no. R7944, 1975.
[14] S. V. Pepper, *Rev. Sci. Instr.*, 1973, **44**, 826.
[15] B. Goldstein, *Surface Sci.*, 1973, **39**, 261.
[16] G. W. B. Ashwell, C. J. Todd and R. Heckingbottom, *J. Phys. E.*, 1973, **6**, 435.
[17] J. C. Rivière, *Contemporary Phys.*, 1973, **14**, 513.
[18] H. E. Bishop, J. P. Coad and J. C. Rivière, *J. Electron Spectr.*, 1972/73, **1**, 389.
[19] P. W. Palmberg, *J. Electron Spectr.*, 1974, **5**, 691.
[20] J. P. Coad and J. G. Cunningham, *J. Electron. Spectr.*, 1974, **3**, 435.
[21] J. B. Lumsden and R. M. Staehle, *Scripta Metallurgica*, 1972, **6**, 1205.
[22] G. H. Barnes, A. W. Aldag and R. C. Jerner, *J. Electrochem. Soc.*, 1972, **119**, 684.
[23] H. W. Werner and H. A. M. de Grefte, *Surface Sci.*, 1973, **35**, 458.

$$RT \ln\left[\frac{X_2(1-X_b)}{X_b(1-X_2)}\right] + 2\Omega(X_b - lX_2 - mX_1 - mX_3) = 0, \quad (2.6b)$$

$$RT \ln\left[\frac{X_3(1-X_b)}{X_b(1-X_3)}\right] + 2\Omega(X_b - lX_3 - mX_2 - mX_4) = 0, \quad (2.6c)$$

$$RT \ln\left[\frac{X_4(1-X_b)}{X_b(1-X_4)}\right] + 2\Omega(X_b - lX_4 - mX_3 - mX_b) = 0; \quad (2.6d)$$

here X_1 refers to the atom fraction of component A in the first layer, X_2 is the atom fraction of the same component in the second layer, etc. All other symbols are defined earlier.

It should be noted that eqn (2.6a) becomes the same as eqn (2.5) when the second, third, and fourth layer compositions are set equal to X_b, that is, these equations reduce to the monolayer model as they should. This process is completely general and, in fact, similar expressions can be derived involving any number of layers of variable composition obtaining one equation for each layer. Solving eqn (2.6) gives the atom fractions of both components in each atomic layer of the solid. In this way, we can determine the depths profile or as we will refer to it, the equilibrium depth distribution (of composition). These calculations have been performed for a (111) face using the surface free energy values given in the literature for Pb, In, Ag and Au [8, 10] and the results of these are plotted in fig. 1, 2, and 3.

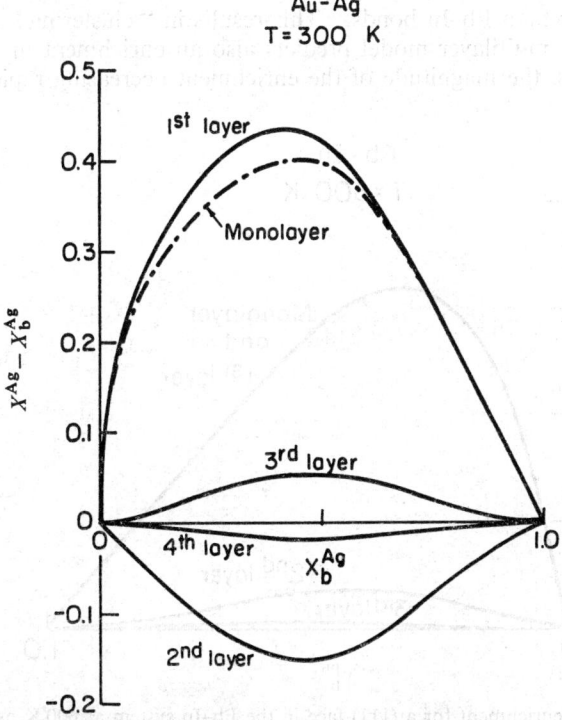

FIG. 1.—Surface enrichment for a (111) face in the Au–Ag system at 300 K as predicted by the monolayer regular and by the 4-layer regular solution models. The enrichment is plotted as a function of the bulk composition. X^{Ag} refers to the atom fraction of Ag in the appropriate surface. In the 4-layer model, the enrichment in each layer is shown.

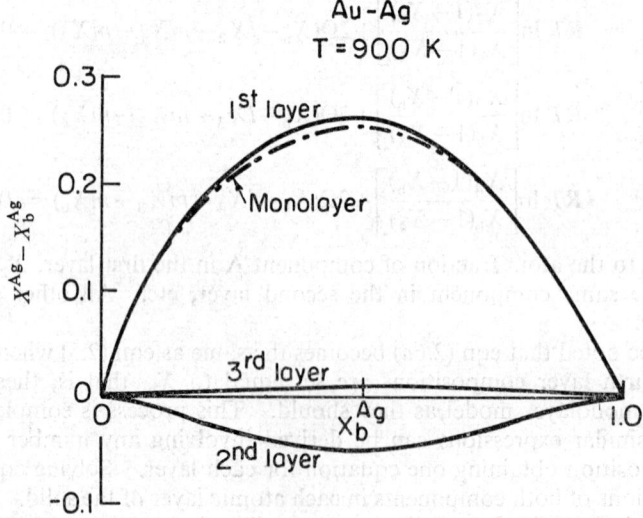

Fig. 2.—Same as fig. 1 except at 900 K.

Due to the lower surface free energy of Pb, the monolayer model and the multilayer model sensibly predict a considerable enrichment of Pb in the top layer. For Pb–In the regular solution parameter is positive which means that Pb–Pb and In–In bonds are stronger than Pb–In bonds. This results in " clustering " of Pb near the surface. Thus the multilayer model predicts also an enrichment in Pb in the 2nd, 3rd, and 4th layers, the magnitude of the enrichment decreasing rapidly toward the bulk.

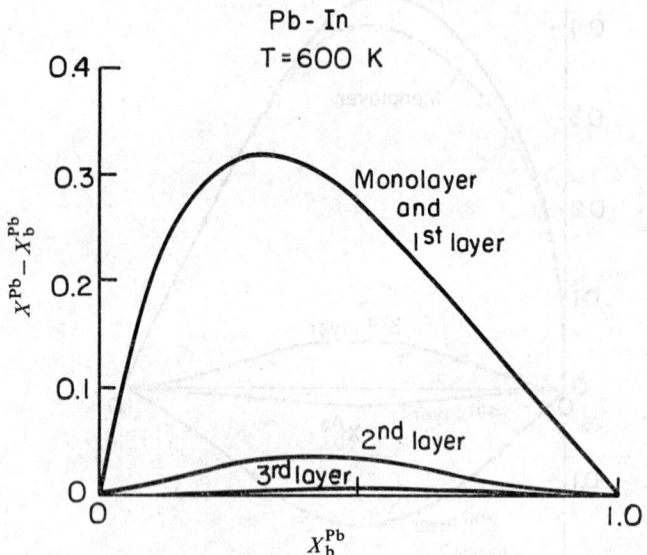

Fig. 3.—The surface enrichment for a (111) face in the Pb–In system at 600 K as predicted by the monolayer regular and the 4-layer regular solution models. The enrichment is plotted against the alloy bulk composition. X^{Pb} refers to the atom fraction of Pb in the appropriate surface. In the 4-layer model, the enrichment in each layer is shown. The surface composition predicted by the monolayer model is very similar to that predicted for the first layer of the 4-layer model.

The calculations give somewhat different results for the Ag–Au system. In this case both models predict an enrichment in Ag in the top layer as before. However, since for Ag–Au the regular solution parameter is negative (Ag–Ag and Au–Au bonds are not as strong as Ag–Au bonds), layering is obtained from the 4-layer model. The first layer is enriched in Ag at the expense of the second layer which is enriched in Au. The third layer then is enriched in Ag. The magnitude of this enrichment and depletion decreases rapidly toward the bulk. Thus, these calculations indicate that due to the large attractive interaction between Au and Ag, the regular solution model may not apply accurately and the Ag enrichment at the surface might well be accompanied by Au enrichment in the second layer. It should be noted that the surface composition is also strongly temperature dependent. With increasing temperature, the surface excess concentration diminishes rapidly.

(3) ANALYSIS OF THE AUGER PEAK INTENSITIES TO OBTAIN SURFACE COMPOSITION AND DEPTH DISTRIBUTION

AES is one of the surface sensitive techniques that can be applied to monitor the surface composition of multicomponent systems.[3] The depth below the surface that is probed by low energy electrons (50–600 eV), that is, the useful energy range of Auger electrons, is of similar magnitude as the expected depth of surface enrichment. Therefore, the intensities of the Auger spectral peaks should contain information on the depth distribution. Careful measurements of intensity ratios, i.e., ratios of the intensities of an Auger peak at one energy divided by the intensity of the Auger peak at another energy, should yield this information. It should be pointed out that the data collected in this manner requires little in the way of absolute calibration and some effects due to the depth distribution can be observed with no calibration necessary whatsoever.

To predict the intensity ratio, the assumption will be made that Auger transitions are excited uniformly with depth. For pure solids this should be a good approximation since the much higher energy incident electron beam is considered to have a much larger attentuation depth than the observed Auger electron. This approximation is also aided by the fact that the average electron energy loss in a solid per collision is small, about 15 eV, so that the incident electron of 2000 eV energy may undergo several collisions but still excite Auger transitions with considerable efficiency. Therefore, if P_1 and P_2 are the probabilities of seeing an atom at depth d_1 and d_2 respectively, then

$$\frac{P_1}{P_2} = \frac{\exp(-d_1/\lambda_E)}{\exp(-d_2/\lambda_E)}, \qquad (3.1)$$

where λ_E is the attentuation depth of the observed electrons of energy E defined in terms of a Beer's law type attenuation. Consequently, for a pure solid the Auger peak intensity I_E at energy E can be written as

$$I_E = \int_0^\infty \kappa(E, \ldots) \exp(-z/\lambda_E)\, dz, \qquad (3.2)$$

where $\kappa(E, \ldots)$ is a complicated function involving properties of the solid, the electron scattering within it, and all experimental parameters. This expression serves to break the Auger intensity into contributions from various depths z and to sum them. For a pure solid, using the assumption given above, integration of eqn (3.2) yields

$$I_E = \kappa(E, \ldots)\lambda_E \qquad (3.3)$$

In order to measure Auger peak intensities as a function of depth for an alloy, two further assumptions will be made. The first is that the presence of neighbouring atoms does not affect the Auger yields. That is, there are no matrix effects. Therefore, the Auger intensity arising from a particular depth will depend only on the number of emitters.

The second assumption is that the escape depths of an electron does not depend upon the medium, but only upon the energy. This was shown experimentally to be approximately correct and many "universal" curves of escape depth versus energy have been published.[9] These assumptions lead to the equation

$$I_E = \kappa(E,\ldots) \int_0^\infty X(z) \exp(-z/\lambda_E)\, dz \qquad (3.4)$$

in which $X(z)$ is the atom fraction of the emitting species at the depth z (the depth distribution).

For a pure solid exhibiting two Auger peaks at energies E and E' the ratio $R^\circ_{E,E'}$ becomes

$$R^\circ_{E,E'} \equiv \frac{I^\circ_E}{I^\circ_{E'}} = \frac{\kappa(E,\ldots)\lambda_E}{\kappa(E',\ldots)\lambda_{E'}} \qquad (3.5)$$

and this ratio is easily measured. The superscript zero is used to denote intensities and ratios from pure metals. For an alloy the corresponding ratios are

$$R_{E,E'} = \frac{I_E}{I_{E'}} = \frac{\kappa(E,\ldots) \int_0^\infty X(z) \exp(-z/\lambda_E)\, dz}{\kappa(E',\ldots) \int_0^\infty X(z) \exp(-z/\lambda_{E'})\, dz} \qquad (3.6)$$

can also be easily measured. Therefore,

$$\frac{R_{E,E'}}{R^\circ_{E,E'}} = \frac{\lambda_{E'}}{\lambda_E} \frac{\int_0^\infty X(z) \exp(-z/\lambda_E)\, dz}{\int_0^\infty X(z) \exp(-z/\lambda_{E'})\, dz} \qquad (3.7)$$

depends only upon the depth distribution and escape depths which in many cases are known or may be estimated. The right-hand side of (3.7) can therefore be calculated for various theoretical depth distributions. It should be noted that if there is no depth distribution, then $X(z) = X_b$ and it follows from eqn (3.7) that $R_{E,E'} = R^\circ_{E,E'}$. Therefore, a change in the ratios of two Auger peaks arising from the same component in an alloy would be indicative of surface segregation of some sort. This same procedure can also be used for comparing Auger peak intensities of the two alloy components such as the ratio of intensity of an Au Auger peak with that of an Ag Auger peak.

In order to reveal surface segregation, if present, or compare the experimentally detected intensity ratios with those predicted by the various models, the Auger peak intensity data that is obtained for the different alloy compositions can be plotted various ways.

(a) One may plot the intensity ratios of two Auger peaks, one for each component of the alloy, $R_{E,E'}(=I_A(E)_{\text{alloy}}/I_B(E')_{\text{alloy}})$ divided by the intensity ratios of the same two Auger peaks of the pure metals, $R^\circ_{E,E'}(=I_A(E)_{\text{pure}}/I_B(E')_{\text{pure}})$, that is, $R_{E,E'}/R^\circ_{E,E'}$ as a function of the bulk atom fraction ratio X^A_b/X^B_b. Plotting the data in this way, values for an alloy having no surface segregation would fall on a line with slope equal to one. If one component is accumulated at the surface and there is a depth distribution of composition near the surface that is different from the bulk composition, the experimental data will lie above or below this "bulk ratio" line.

(b) Another way of displaying the experimental Auger peak intensity data to reveal surface segregation is by plotting the ratio of intensities of two Auger peaks of the same component in the same alloy, $(I_A(E)_{\text{alloy}}/I_A(E')_{\text{alloy}})$ as a function of the bulk atom fraction. If the surface composition changes the same way as the bulk composition does, this ratio would be constant. Surface segregation would be indicated by the systematic variation of this intensity ratio with alloy composition in a nonlinear manner.

(c) A third method to identify surface segregation is by the summation of the intensity ratios

$$\frac{I_A(E)_{\text{alloy}}}{I_A(E)_{\text{pure}}} + \frac{I_B(E')_{\text{alloy}}}{I_B(E')_{\text{pure}}}.$$

Since the two Auger peaks are at different electron energies E and E', they sample the composition over different depths in the alloy. Thus, if the sample is homogeneous, in the absence of surface segregation, these intensity ratios will reflect precisely the bulk composition and their sums should be unity. However, if there is surface segregation than the intensity ratios will not reflect the bulk ratios and their sum may be greater or less than unity.

In addition to these types of data analysis the presence of temperature dependence of the Auger intensity ratios is an indication of changes in the surface composition.

RESULTS AND DISCUSSION

We have studied the Pb–In and Ag–Au systems by AES in some detail. The Pb–In system was studied in the liquid state to assure equilibration of the bulk and the surface phases. The Ag–Au alloy samples had to be heated to 300°C for over 30 min or to above this temperature for shorter times, after suitable cleaning of the surface of impurities (carbon, sulphur and chlorine) by ion sputtering, before equilibration of the surface phase and the bulk phase was achieved. The details of the AES experiments for both of these systems are described elsewhere.[10, 11] In fig. 4 the Pb–In Auger peak intensity ratios are plotted as a function of the bulk atom fraction ratio on a log–log graph according to the first method of data analysis that was described above. All of the experimental points fall below the bulk ratio line indicating surface segregation of Pb as predicted by the regular solution models. In addition, the surface segregation decreases with increasing temperature as shown by the data points in Fig. 4, as predicted by the regular solution models for this system.

The same plot of normalized Au–Ag Auger peak intensity ratios as a function of their atom fraction ratios are shown in fig. 5 and 6 for two different Auger peaks of gold. The solid lines indicate the trend as predicted for the various thermodynamic models. The 4-layer and the monolayer regular solution models give very similar predicted values and the experimental data appears to fit the regular solution model. Unfortunately, the temperature required for achieving surface-bulk equilibration was too high (300°C) to allow a reliable study of the temperature dependence of the surface composition as was carried out for the Pb–In system.

To demonstrate our second method of analysis, i.e., plotting the ratio of two Auger peak intensities of the same component in the same alloy versus bulk composition, the values predicted by regular solution theory for such a ratio are given in fig. 7. Thus, the presence of surface segregation in a binary alloy should show up as a deviation in ratios of this type. The third method of analysis listed above is demonstrated in fig. 8. This figure illustrates that the sum of normalized intensities from both components would not sum to unity for a system obeying the regular solution model. By normalized intensity is meant the intensity obtained from an alloy divided by the intensity from a pure reference.

Detailed studies of the Ni–Au [12] and Cu–Al [13] systems by Auger electron spectroscopy clearly demonstrate the segregation of one of the alloy constituents, gold and aluminium, respectively, to the topmost surface layer. These systems obey the regular solution model of surface composition. There are several contradictory reports on the surface composition in the Cu–Ni system. According to the regular solution models, enrichment of the surface in copper is expected. Copper enrichment was indeed reported by Sachtler et al.,[14-16] Helms, Yee and Spicer [17] and Burton et al.[18] Takasu and Shimizu [19] found copper enrichment at the surface of nickel-rich alloys while copper-rich alloys had excess nickel at the surface. Ertl and Kuppers [20] and Quinto et al.[21] found the surface composition the same as the bulk. It appears that sample preparation must have a controlling influence on the equilibration of the two components, copper and nickel, in this system. It is possible that the contradictory results are due to the phase segregation reported by Sachtler,[14] that would not permit the application of the regular solution model to this binary alloy. As we have discussed above, the low-energy Auger peaks are much more sensitive probes of surface segregation than high-energy Auger peaks because the surface layer makes a larger contribution to their intensity. Some of the contradictory results in studies of the Cu–Ni system may be attributed to the use of different energy peaks for analysis of surface composition.

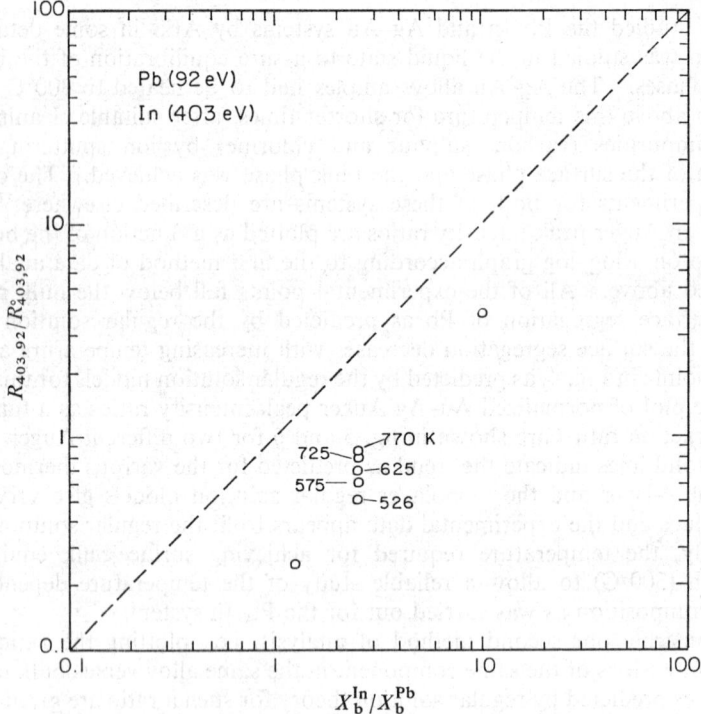

FIG. 4.—Ratios of the In(403 eV) to the Pb(92 eV) intensities. The ratios are all divided by this ratio obtained from pure Pb and In. The multiple points for one alloy demonstrate the temperature dependence of this ratio. The dotted line gives the values expected for a surface with the same composition as that in the bulk.

There are many experimental parameters that may make studies of surface phase diagrams of alloys difficult. Adsorption of gases from the ambient or segregation of impurities by diffusion from the bulk to the surface can markedly change the surface composition. If any of the impurities form stronger bonds with one component as compared to the other, the strongly bound component will be pulled to the surface by

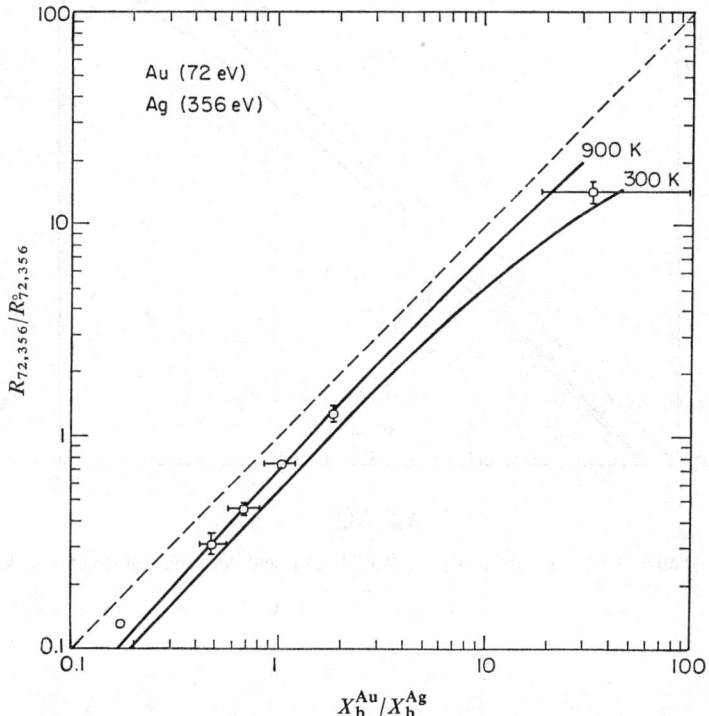

FIG. 5.—Intensity ratio of the Au(72 eV) to the Ag(356 eV) peak. The ratios are all divided by the same ratio for pure Ag and Au and plotted as a function of the bulk composition ratios. The solid lines are predicted for a 4-layer regular solution model. The monolayer regular solution model gives essentially identical values. The dotted lines gives the values expected for a surface with the same composition as that in the bulk.

the impurity segregated there. On removal of the impurity, the surface composition may change again indicating the re-equilibration of the pure surface phase with that of the bulk. For the small crystallites present in the alloy thin-films the surface composition can be influenced by the particle size. In the limit of small particle size the surface composition must approach the bulk composition since most of the atoms must then reside on the surface. As we have pointed out above, a large exothermic heat of mixing would indicate the tendency for layering or ordering near the surface that would disallow the use of the regular solution model. In our analysis of the Ag–Au system we have neglected to include the back-scattering correction. This term arises when one atom (e.g., silver) is placed in the matrix of a higher atomic weight material (e.g., gold) which is a better electron scatterer instead of its own crystal lattice. The Auger peak intensities would be modified, if this correction is made, in such a way as to reduce the magnitude of the silver surface segregation but would not eliminate it. Since the regular solution parameter for the Au–Au system

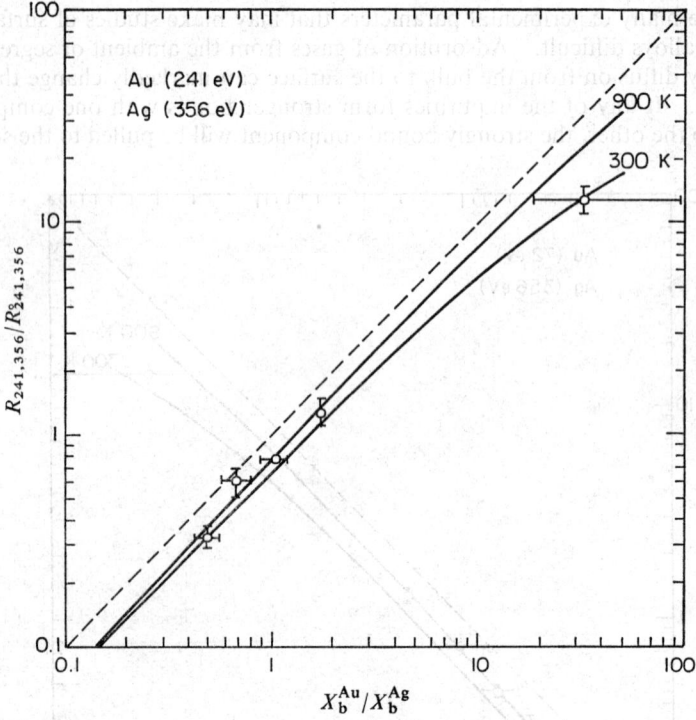

Fig. 6.—Same as fig. 5 except using the Au(241 eV) peak and the Ag(356 eV) peak.

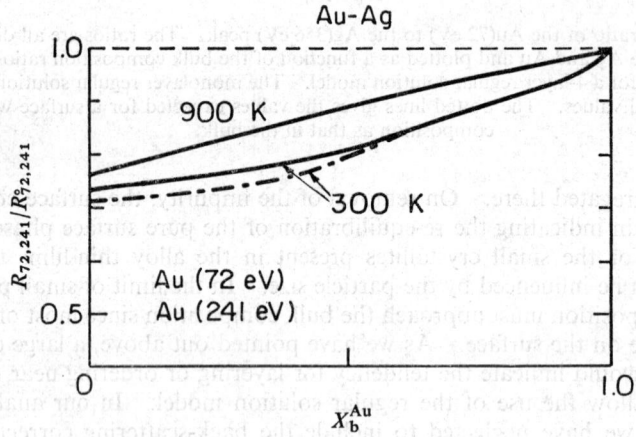

Fig. 7.—The intensity ratio of the Au(72 eV) as predicted by the monolayer and the 4-layer regular solution models. The ratios are divided by the ratio obtained from pure Au. The dotted line is the value expected when the surface composition is identical to that in the bulk.

— from 4-layer model
- - - from monolayer model

is large and negative, indicating the possibility of ordering, it is not too surprising that the agreement between the experimental data and the regular solution theory is less satisfactory than for the more " ideal " (smaller Ω) Pb–In system.

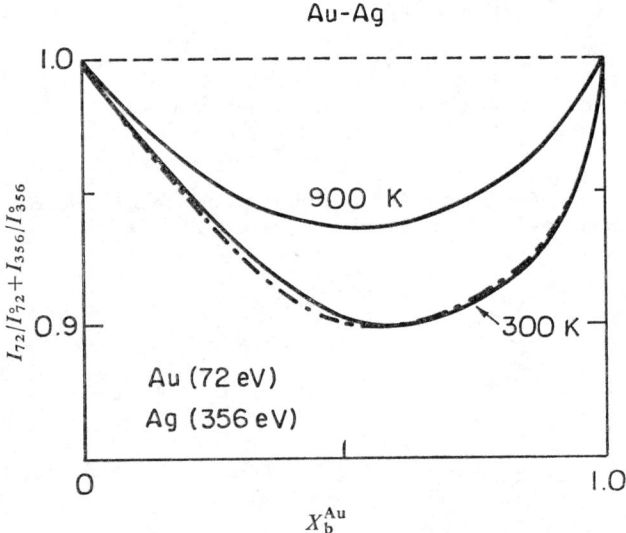

FIG. 8.—The sum of the normalized intensity ratios as predicted from the monolayer and the 4-layer regular solution models for the Ag–Au system. At 900 K the monolayer and 4-layer models give essentially identical results. The dotted line gives the value expected if the surface composition is the same as that in the bulk.

— from 4-layer model
- - - from monolayer model

It would, of course, be of great importance to study the surface composition of alloy systems with complex phase diagrams where ordering and compound formation occurs. Although there have been attempts to describe the surface composition of these complex alloy systems,[14, 22] experimental data have been scarce. The surface composition of a wide variety of complex alloy systems must be studied before realistic thermodynamic models of their behaviour can be developed.

It appears that for homogeneous binary systems with relatively small regular solution parameters, the surface phase diagram can be described adequately with a regular solution model of the monolayer type. Thus, one may use the monolayer regular solution model to predict the surface composition of homogeneous binary alloys.

This work was supported by the U.S. Energy Research and Development Administration.

[1] G. A. Somorjai, *Principles of Surface Chemistry* (Prentice-Hall, Englewood Cliffs, 1972).
[2] R. Defay et al., *Surface Tension and Adsorption* (John Wiley & Sons, New York, 1966).
[3] F. J. Szalkowski and G. A. Somorjai, *Adv. High Temp. Chem.*, 1971, **4**, 137.
[4] F. Meyer and J. J. Vrakking, *Surface Sci.*, 1972, **33**, 271.
[5] R. A. Swalin, *Thermodynamics of Solids* (John Wiley & Sons, New York, 2nd ed., 1972).
[6] F. L. Williams, *Surface Sci.*, 1974, **45**, 377.
[7] R. Hultgren et al., *Selected Values of Thermodynamic Properties of Metals and Alloys* (John Wiley & Sons, New York, 1963).

[8] S. H. Overbury and G. A. Somorjai, *J. Chem. Rev.*, 1975, to be published.
[9] C. J. Powell, *Surface Sci.*, 1974, **44,** 29.
[10] S. Berglund and G. A. Somorjai, *J. Chem. Phys.*, 1973, **59,** 5537.
[11] S. H. Overbury and G. A. Somorjai, to be published.
[12] F. L. Williams and M. Boudart, *J. Catalysis*, 1973, **30,** 438.
[13] J. Ferrante, *Acta Met.*, 1971, **19,** 743.
[14] W. M. H. Sachtler and R. Jongepier, *J. Catalysis*, 1965, **4,** 665.
[15] W. M. H. Sachtler and G. J. H. Dorgelo, *J. Catalysis*, 1965, **4,** 654.
[16] W. M. H. Sachtler, *J. Vac. Sci. Tech.*, 1971, **9,** 828.
[17] C. R. Helms, K. Y. Yu and W. E. Spicer, 1975, to be published.
[18] J. J. Burton and E. Hyman, *J. Catalysis*, 1975, in press.
[19] Y. Takasu and H. Shimizu, *J. Catalysis*, 1973, **29,** 479.
[20] G. Ertl and J. Kuppers, *J. Vac. Sci. Tech.*, 1971, **9,** 829.
[21] D. T. Quinto, V. S. Sundaram and W. D. Robertson, *Surface Sci.*, 1971, **28,** 504.
[22] R. A. Van Santen and W. M. H. Sachtler, *J. Catalysis*, 1974, **33,** 202.

Chemical Shifts of Auger Lines, and the Auger Parameter

By C. D. Wagner*

Shell Development Company, P.O. Box 481,
Houston, Texas, 77001, U.S.A.

Received 28th February, 1975

Earlier observations on the magnitudes of chemical shifts of photoelectron and core-type Auger lines, and the concept of the role of polarizability, have been combined to produce a unified concept of the role of polarization in determining line position. With core-type Auger lines, polarization effects are more important in determining chemical shifts than are changes in electron density on the atom in the ground state. A parameter, termed the Auger parameter, is proposed as a specific property of a chemical and physical state. It is accurately determinable to ±0.1 eV, and with most elements to which it is applicable the range among compounds is several eV. Differences in the Auger parameter are attributable solely to changes in the polarizability of the solid compounds.

INTRODUCTION

One of the significant kinds of information furnished by ESCA leads to the identification of chemical states. The principal spectral feature utilized is the exact position of spectral lines. It has long been known that the position of the line on the kinetic energy or binding energy scale is a function of the electron density at the atom. Recently it has been recognized that another factor of similar importance is extra-atomic relaxation energy. This is of particular significance with Auger lines, which for that reason frequently exhibit far larger chemical shifts than photoelectron lines.[1] This fact leads here to the concept of the Auger parameter: the difference in kinetic energy between an Auger line and a photoelectron line. Since it is a difference in position of two lines, static charge corrections cancel. This quantity is unique for each compound[2] and can be measured more accurately than absolute line positions (kinetic energy) which must be corrected for static charge. Changes in the Auger parameter are due principally to changes in extra-atomic relaxation or polarization energy, as shown in the development below. The following incorporates ideas that originated with our discovery of abnormally-large chemical shifts in core-type Auger lines.[1] (Auger lines with final vacancies in core levels, not valence levels). At that time we explained the effect as due to polarization energy. At about the same time Shirley and co-workers[3-6] were exploring theoretically the magnitude of intra-atomic and extra-atomic relaxation energy. (The latter is another term for polarization energy.) The following is a brief discussion of the role of polarization energy in determining the chemical shifts in photoelectron lines, Auger lines, and the Auger parameter.

The notation used in the discussion is as follows:

E = state total energy
E_B = binding energy
E^V = kinetic energy referenced to vacuum level
E^F = kinetic energy referenced to Fermi level
ε = ground state energy by rigorous non-relativistic Hartree-Fock calculation (for this discussion it is assumed positive).
R = relaxation energy, atomic or intra-atomic

R^{ea} = relaxation energy or polarization energy, extra-atomic
α = kinetic energy of an intense Auger line minus that of an intense photoelectron line, the Auger parameter
PE = photoelectron

subscripts

a = isolated atom
m = isolated molecule
e = conductive element in the solid state
c = insulating compound in the solid state

PHOTOELECTRON BINDING ENERGIES

The photoelectric process is

$$h\nu + Z \rightarrow Z^+ + e^-, \qquad (1)$$

where the energy of the photon is partitioned between the ionization energy of the atom and the kinetic energy of the photoelectron (the energy of the recoiling atom is usually negligible). The energy of Z^+ above the ground state is by definition the electron binding energy and is determined by

$$E_B(Z^+) = h\nu - E(e^-) \qquad (2)$$

Much work has gone into calculating E_B. Koopman's theorem requires that it is the energy of the electron shell in the ground state of the atom, and this can be determined by the Hartree–Fock calculation. Small corrections for relativistic effects and configuration interaction, ordinarily less than one eV, we can ignore. An effect that cannot be ignored is the atomic relaxation or rearrangement energy. The Koopman theorem is termed a "frozen orbital approximation." It ignores the fact that the electrons in the remaining atomic shells relax as the photoelectron leaves, providing stabilization energy for the ion. This relaxation energy has been calculated by Gelius [7] and others; its magnitude in eV turns out to be 75-85 % of the square root of the final energy of the ion, also expressed in eV.

We take eqn (3), then, as a close approximation of the binding energy of an isolated atom, referred to the vacuum level, i.e., the energy required to remove the photoelectron completely from the atomic system:

$$E_B^V(PE_a) = \varepsilon - R, \qquad (3)$$

where ε is the energy of the electron shell in the ground state and R is the intra-atomic relaxation or rearrangement energy. Here we have treated ε and R as positive numerical quantities, ignoring sign convention as they appear in calculations.

The same atom in a gaseous molecule can experience two new effects of significance in the determination of binding energy: (1) bonding to neighbouring atoms will change the atomic energy levels in the ground state by an amount $\Delta\varepsilon$ and (2), electrons from neighbouring atoms can readjust during the ionization in such a way as to reduce further the total energy of the ion. There is no reason to assume that the intra-atomic relaxation energy R changes significantly. Thus

$$E_B^V(PE_m) = \varepsilon_a + \Delta\varepsilon_m - R - R_m^{ea}. \qquad (4)$$

The quantity R^{ea} is the extra-atomic relaxation energy or polarization energy involving electrons from neighbouring atoms. Then the chemical shift between atom and molecule is

$$\Delta E^V(PE_{a \rightarrow m}) = \Delta\varepsilon_m - R_m^e. \qquad (5)$$

Formerly correlations of E_B with electron charge on the atom (an effect of $\Delta\varepsilon$ alone) were attempted without consideration of R^{ea}. Davis and Shirley [3] have shown that the correlation is improved by consideration of R^{ea}.

We now consider the chemical shift in moving the isolated atom to the elemental conducting solid state. The equations are analogous to (4) and (5):

$$\Delta E_B^V(PE_{a \to e}) = \Delta\varepsilon_e - R_e^{ea}, \quad (6)$$

where $\Delta\varepsilon_e$ and R_e^{ea} now represent the appropriate quantities in the new environment, and we again expect R to be close to that for the isolated atom. Regarding $\Delta\varepsilon_e$ for bonding to like atoms in the conducting solid state, Ley et al.[4] believe it is of minor importance compared to R_e^{ea}. They believe it could have either sign, and is unlikely to have a value of more than 1-2 eV.

Experimental values for ΔE_B^V are available from papers of Kowalczyk et al.[6] and Ley et al.,[4] as shown in table 1. The experimental values for the solid are referenced to the Fermi level, and appear in the first column as E_B^F. In order to convert these data to a vacuum reference level, the work function of the solid must be added. When referenced to the same level the ΔE_B is seen to be negative in the solid relative to the gas in all cases.

TABLE 1.—PHOTOELECTRONS—GAS AND SOLID PHASE (data in eV)

	$E_B^F(s)$	$\phi(s)$	$E_B^V(s)$	$E_B^V(s)$	ΔE_B
Ne 1s	863.2[d]	4.5	867.7	870.2[a]	2.5
Ar 2p₃	241.5[d]	4.5	246.0	248.5[a]	2.5
Kr 3p₃	207.9[d]	4.5	212.4	214.4[a]	2.0
Xe 3d₅	670.0[d]	4.5	674.5	676.4[a]	1.9
Li 1s	54.8[b]	2.4	57.2	64.9[b]	7.7
Na 1s	1071.7[b]	2.3	1074.0	1079.1[b]	5.1
Cu 3d	3.0	4.5	7.5	10.44[c]	2.9
Zn 3d	10.2	3.7	13.9	17.3[c]	3.4
Cd 4d₅	10.5	3.9	14.4	17.6[c]	3.2
Pb 4f₇	136.5	4.0	140.5	144.0[c]	3.5
Bi 4f₇	157.0	4.3	161.3	164.9[c]	3.6

[a] ref. (17); [b] ref. (6); [c] ref. (4); [d] ion-implanted in Fe.

In a conductor the polarization energy R_e^{ea} should be given [8] by the equation

$$R_e^{ea} = \frac{e^2}{2r}\left(1 - \frac{1}{k}\right), \quad (7)$$

where e is the charge of the ion, r is the radius of minimum electron screening distance, and k is the dielectric constant.

The implanted noble gases with no chemical bonding should show no $\Delta\varepsilon_e$, and ΔE_B should then be due to the polarization energy. With the noble gases the chemical shift is seen to decrease as the atoms become larger. Values in the table are similar to, but slightly smaller than, the values of Citrin et al.[9] Since the screening conduction electrons can only occur to a dimension dictated by the size of the atom, it is expected that R_e^{ea} will be smallest for the largest atom. Atoms whose valence electrons are part of the conduction band would be expected to have a smaller effective radius. The values of ΔE_B for the metals are considerably larger, especially with the smaller metal atoms, Na and Li. A small portion of ΔE_B for the metals may be derived from $\Delta\varepsilon_e^4$, but most of the effect must be due to screening by conduction electrons.

AUGER KINETIC ENERGIES

The kinetic energy of an Auger electron is given accurately by the difference between the initial and final state. In the following we limit our consideration to core-type Auger lines, where the double-hole final state vacancies are in core levels. We follow the treatment of Shirley[5] and arrive at the expression for a KLL Auger kinetic energy:

$$E^V(\text{Auger}_a) = E_B^V(K^+) - 2E_B^V(L^+) + R^s - F(x) \tag{8}$$

where R^s is his intra-atomic "static" relaxation energy for the L shell and $F(x)$ is an electron–electron interaction term necessary to include because of the double electron vacancy in the final state. In terms of orbital energies and intra-atomic relaxation energies we have

$$E^V(\text{Auger}_a) = \varepsilon(K) - R(K^+) - 2\varepsilon(L) + 2R(L^+) - F(x) + R^s. \tag{9}$$

An atom in a molecule in the gas phase will have chemical shifts in ground-state orbital energies and extra-atomic relaxation effects. From (5) and (9),

$$\Delta E^V(\text{Auger}_{a \to m}) = \Delta\varepsilon_m(K) - 2\Delta\varepsilon_m(L) + R_m^{ea}(L^+L^+) - R_m^{ea}(K^+). \tag{10}$$

The chemical shifts in different shells are experimentally closely the same (identical theoretically in the charged sphere model) and the equation with that approximation simplifies to

$$\Delta E^V(\text{Auger}_{a \to m}) = -\Delta\varepsilon_m(L) + R_m^{ea}(L^+L^+) - R_m^{ea}(K^+). \tag{11}$$

When comparing an Auger process in an isolated atom with that in a conductive elemental solid phase, the equations have the same form. In this case, however, the chemical shifts in orbital energies in the ground state are minimal, as mentioned above. Virtually the entire chemical shift is due to the extra-atomic relaxation energy terms.

From eqn (7), to the extent that r is the same for a one-hole and two-hole state, the relaxation or polarization energy for (L^+L^+) and (K^+) will differ by a factor of four. On this basis, the chemical shift between isolated atom and conducting solid is given by

$$\Delta E^V(\text{Auger}_{a \to c}) = -\Delta\varepsilon_c(L) + \tfrac{3}{4}R_c^{ea}(L^+L^+) = -\Delta\varepsilon_c(L) + 3R_c^{ea}(K^+). \tag{12}$$

This chemical shift in electron kinetic energy may be directly compared to that of the photoelectron line, from eqn (6). There are experimental data bearing on this point, as shown in table 2. Auger lines used as reference are the most intense and

TABLE 2.—AUGER ELECTRONS—GAS AND SOLID PHASE (data in eV)

element	Auger line	$E^F(s)$	$\phi(s)$	$E^V(s)$	$E^V(g)$	ΔE(Auger)	$\dfrac{\Delta E(\text{Auger})}{\Delta E(\text{PE})}$
Ne	$KL_{23}L_{23}(^1D_2)$	818.2[d]	4.5	813.7	804.8[a]	8.9	3.6
Ar	$L_3M_{23}M_{23}(^1D_2{}^3P)$	217.1[d]	4.5	212.6	203.5[a]	9.1	3.6
Xe	$M_4N_{45}N_{45}(^1D_2{}^1G_4)$	545.0[d]	4.5	540.5	532.7[a]	7.8	4.1
Na	$KL_{23}L_{23}(^1D_2)$	994.2[b]	2.3	991.9	977.2[b]	14.7	2.9
Zn	$L_3M_{45}M_{45}(^1D^3P^1G)$	992.2	3.7	988.5	973.3[c]	15.2	4.5
Cd	$M_4N_{45}N_{45}(^1D_2{}^1G_4)$	383.9	3.9	380.0	367.9[e]	12.1	3.8

[a] ref. (17); [b] ref. (6); [c] ref. (18); [d] ion-implanted in Fe; [e] ref. (19); [f] from ΔE in this table and ΔE_B in table 1.

sharp ones in the Auger groups. In order to put all kinetic energies on a vacuum reference level basis, the work function of the solid must be subtracted from the Fermi-referenced Auger energies. Examination of table 2 discloses chemical shifts of the order of three to four times those for photoelectron lines in the same systems. The chemical shift is therefore as great or even greater than that predicted by (12) and (6) if orbital chemical shifts are ignored.

THE AUGER PARAMETER

Early ideas about the magnitude of Auger chemical shifts held that they should be similar to photoelectron chemical shifts and in the same direction. The reasoning was based entirely on changes in ground-state orbital energy, without consideration of polarization or extra-atomic relaxation. However, an early tabulation of differences in kinetic energies between Auger lines and photoelectron lines in sodium salts disclosed substantial differences in chemical shift that correlated with the polarizability of the anion.[2]

The concept of a parameter, defined by the difference between two line energies from the same element in the same sample, is attractive because static charge correction is unnecessary. Work function corrections are also unnecessary, and vacuum level data can be compared to Fermi level data directly. Thus, the concept of the Auger parameter seems useful; it is a parameter characteristic of a chemical and physical state and easily measurable to ± 0.1 eV. When the chemical shift in the Auger line is greatly different from that of the photoelectron line, the range in the Auger parameter α in eV can be almost as great as the range in the Auger line energy.

For the isolated atom, the parameter α is given by the following, based on eqn (2), (3) and (10):

$$\alpha_a = 2\varepsilon(K) - 2\varepsilon(L) - 2R(K^+) + 2R(L^+) - h\nu - F(x) + R_s. \tag{13}$$

For the chemical shift in α on going to the atom in a molecule, from eqn (2), (5) and (11),

$$\Delta\alpha_{a\to m} = R_m^{ea}(L^+L^+) - 2R_m^{ea}(K^+), \tag{14}$$

and the chemical shift in α is then entirely an effect of extra-atomic relaxation.*

In an analogous way, we have the similar expression for the chemical shift on going from the isolated atom to the atom in a conducting solid:

$$\Delta\alpha_{a\to e} = R_e^{ea}(L^+L^+) - 2R_e^{ea}(K^+), \tag{15}$$

or, assuming $R_e^{ea}(L^+L^+) = 4R_e^{ea}(K^+)$,

$$\Delta\alpha_{a\to e} = \tfrac{1}{2}R_e^{ea}(L^+L^+) = 2R^{ea}(K^+). \tag{16}$$

A similar procedure leads to a simple expression for the chemical shift between a conducting element and an insulating compound (any work function corrections cancel),

$$\Delta\alpha_{e\to c} = \tfrac{1}{2}[R_c^{ea}(L^+L^+) - R_e^{ea}(L^+L^+)] = 2[R_c^{ea}(K^+) - R_e^{ea}(K^+)], \tag{17}$$

so that the chemical shift in α will always be negative, reflecting the difference in polarization energy of the media. Comparisons among insulating compounds will reflect these differences more finely. Data on sodium in table 3 show a total spread in α of 9.6 eV between isolated sodium atoms and sodium metal. The sodium salts occupy a 3.4 eV range precisely in the centre between the extremes. At the end nearest Na(g) are the least polarizable anions, and at the end nearest Na(s) are the most polarizable ones.

Some empirical trends are worth noting in that table. All fluoroanions show low

---

* This conclusion was reached in a somewhat similar way in a paper by S. P. Kowalczyk, L. Lay, F. R. McFeely, R. A. Pollak and D. A. Shirley, *Phys. Rev.*, 1974, **B9**, 38.

TABLE 3.—BINDING ENERGIES, AUGER ENERGIES, AND THE AUGER PARAMETER
(Al $K\alpha$ radiation; data in eV)

	binding energy E_B(PE)	kinetic energy E(PE)	Auger energy $E(A)^a$	Auger parameter $E(A)-E$(PE)	$-\Delta\alpha$	ΔE_B(PE)	$-\Delta E$(A)
		1s	$KL_{23}L_{23}$				
Ne$(s)^b$	863.2	623.4	818.2	194.8			
Ne$(g)^c$	865.7	620.9^c	809.3^c	188.4	6.4	2.5^c	8.9^c
Na$(s)^d$	1071.7	414.9	994.2	579.3			
NaI	1071.4	415.2	991.4	576.2	3.1	−0.3	2.8
NaBiO$_3$	1070.6	416.0	991.8	575.8	3.5	−1.1	2.4
NaBr	1071.5	415.1	990.8	575.7	3.6	−0.2	3.4
Na$_2$CrO$_4$	1070.8	415.8	991.4	575.6	3.7	−0.9	2.8
Na$_2$PdCl$_4$	1071.6	415.0	990.4	575.4	3.9	−0.1	3.8
Na$_2$MoO$_4$	1070.7	415.9	991.2	575.3	4.0	−1.0	3.0
NaCl	1071.4	415.2	990.5	575.3	4.0	−0.3	3.7
NaSCN	1071.1	415.5	990.7	575.2	4.1	−0.6	3.5
Na$_2$SeO$_3$	1070.6	416.0	991.2	575.2	4.1	−1.1	2.9
Na$_2$S$_2$O$_4$	1071.0	415.6	990.8	575.2	4.1	−0.7	3.4
Na$_2$S$_2$O$_3$	1071.4	415.2	990.3	575.1	4.2	−0.3	3.9
Na$_2$SO$_3$	1071.1	415.5	990.6	575.1	4.2	−0.6	3.6
Na$_2$WO$_4$	1071.8	414.8	989.8	575.0	4.3	0.1	4.4
Na$_2$TeO$_4$	1070.9	415.7	990.7	575.0	4.3	−0.8	3.5
NaAsO$_2$	1070.7	415.9	990.9	575.0	4.3	−1.0	3.3
NaNO$_2$	1071.4	415.2	990.0	574.8	4.5	−0.3	4.2
Na$_2$IrCl$_6$	1071.7	414.9	989.4	574.5	4.8	0.0	4.8
NaPO$_3$	1071.5	415.1	989.5	574.4	4.9	−0.2	4.7
NaNO$_3$	1071.2	415.4	989.8	574.4	4.9	−0.5	4.4
Na$_2$SO$_4$	1071.0	415.6	990.0	574.4	4.9	−0.7	4.2
NaOAc	1070.9	415.7	990.1	574.4	4.9	−0.8	4.1
NaAlSiO$_4^e$	1071.3	415.3	989.5	574.2	5.1	−0.4	4.7
Na$_3$AlF$_6$	1071.6	415.0	988.2	573.6	6.1	−0.1	6.0
Na$_2$ZrF$_6$	1071.3	415.3	988.9	573.6	5.7	−0.4	5.3
Na$_2$TiF$_6$	1071.4	415.2	988.7	573.5	5.8	−0.3	5.5
NaF	1071.0	415.6	988.8	573.2	6.1	−0.7	5.4
NaBF$_4$	1072.5	414.1	987.3	573.2	6.1	+0.8	6.9
Na$_2$GeF$_6$	1071.5	415.1	988.3	573.2	6.1	−0.2	5.9
Na$_2$SiF$_6$	1071.5	415.1	987.9	572.8	6.5	−0.2	6.3
Na$(g)^d$	1076.8	409.8	979.5	569.7	9.6	5.1	14.7
		$2p_{\frac{3}{2}}$	$KL_{23}L_{23}$				
Mg	49.8	1436.8	1185.8	749.0^f			
Mg(ox)^{g,h}	51.2	1435.4	1179.5	744.1	4.9	1.4	6.3
MgF$_2$	50.7	1435.9	1178.4	742.5	6.5	0.9	7.4
Al	72.8	1413.8	1393.1	979.3^f			
Al(ox)^{g,h}	75.4	1411.2	1386.4	975.2	4.1	2.6	6.7
Si^i	99.4	1387.2	1615.7	228.5			
Si(ox)^{g,h}	103.3	1383.3	1608.7	225.4	3.1	3.9	7.0
WS$_2^j$	162.8	1323.8	2115.0	791.2			
Na$_2$S$_2$O$_3^k$	162.3	1324.3	2112.7	788.4	2.8	−0.5	2.3
			2108.0				

TABLE 3.—continued

	binding energy $E_B(PE)$	kinetic energy $E(PE)$	Auger energy $E(A)^a$	Auger parameter $E(A)-E(PE)$	$-\Delta\alpha$	$\Delta E_B(PE)$	$-\Delta E(A)$
	$2p_{\frac{3}{2}}$		$L_3M_{23}M_{23}$				
Ar$(s)^b$	241.5	1245.1	217.1	972.0^f			
Ar$(g)^c$	244.0	1242.6^c	208.0^c	965.4	6.6	2.5^c	9.1^c
KI	292.6	1194.0	251.0	57.0^f			
KBr	292.5	1194.1	250.6	56.5			
KCl	292.5	1194.1	250.8	56.6			
KSbF$_6$	293.5	1193.1	248.8	55.7			
KNO$_3$	292.7	1193.9	249.5	55.6			
KF	292.3	1194.3	249.8	55.5			
CaCl$_2$	348.1	1138.5	292.1	153.6^f			
CaCO$_3$	346.5	1140.1	292.2	152.1			
CaF$_2$	347.7	1138.9	289.1	150.2			
	$2p_{\frac{3}{2}}$		$L_3M_{45}M_{45}$				
Cu	932.4	554.2	918.8	364.6	0		
Cu$_2$S	932.3	554.3	917.6	363.3	1.3	−0.1	1.2
Cu$_2$O	932.2	554.4	916.9	362.5	2.1	−0.2	1.9
CuCN	932.9	553.7	914.7	361.0	3.6	0.5	4.1^p
CuCl	932.4	554.2	915.2	361.0	3.6	0	3.6
Zn	1021.7	464.9	992.2	527.3			
ZnBr$_2$	1023.2	463.4	987.5	524.1	3.2	1.5	4.7
Zn(ox)	1022.0	464.6	988.0	523.4	3.9	0.3	4.2
ZnF$_2$	1022.6	464.0	986.9	522.9	4.4	0.9	5.3
Zn(acac)$_2$	1021.2	465.4	987.9	522.5	4.8	−0.5	4.3
Zn$(g)^l$	1025.1	461.5	977.0	515.5	11.8	3.4	15.2
Ga^i	1116.8	369.8	1068.3	698.5			
Ga(ox)^g	1117.3	369.3	1062.8	693.5	5.0	0.5	5.5
Ge^i	1217.4	269.2	1145.1	875.9			
Ge(ox)^g	1221.1	265.5	1137.2	871.7	4.2	3.7	7.9
GeO$_2$	1220.4	266.2	1137.9	871.7	4.2	3.0	7.2
Na$_2$GeF$_6$	1221.1	265.5	1135.9	870.3	5.6	3.6	9.2
As^i	1323.0	163.6	1224.8	1061.2			
As$_2$O$_3$	1327.1	159.5	1218.0	1058.5	2.7	4.1	6.8
NaAsO$_2$	1325.4	161.2	1219.6	1058.4	2.8	2.4	5.2
KAsF$_6$	1330.1	156.5	1214.2	1057.7	3.5	7.1	10.6
Na$_2$HAsO$_4$	1326.6	160.0	1217.3	1057.3	3.9	3.6	7.5
	$3p_{\frac{3}{2}}$						
Se^i	161.5	1325.1	1306.9	981.8			
Na$_2$SeO$_3$	164.1	1322.5	1301.4	978.9	2.9	2.6	5.5

TABLE 3.—continued

	binding energy E_B(PE)	kinetic energy E(PE)	Auger energy $E(A)^a$	Auger parameter $E(A)-E$(PE)	$-\Delta\alpha$	ΔE_B(PE)	$-\Delta E(A)$
	$3d_{\frac{5}{2}}$		$M_4N_{45}N_{45}{}^a$				
Ag	368.0	1118.6	358.4	239.8^f	0		
AgOOCCF$_3$	368.6	1118.0	355.3	237.3	2.5	0.6	3.1
Ag$_2$SO$_4$	368.1	1118.5	354.4	235.9	3.9	0.1	4.0^p
Cd	404.7	1081.9	383.9	302.0^f			
CdO	404.9	1081.7	380.2	298.5	3.5	0.2	3.7
CdF$_2$	405.6	1081.0	379.0	298.0	4.0	0.9	4.9
Cd$(g)^o$			371.8				12.1
In^i	444.1	1042.5	410.3	367.8^f			
In(ox)	445.3	1041.3	406.7	365.4	2.4	1.2	3.6
InF$_3$	446.0	1040.6	403.9	363.3	4.5	1.9	6.4
Sn	484.8	1001.8	437.6	435.8^f			
SnS	485.4	1001.2	435.9	434.7	1.1	0.6	1.7
Sn(ox)	486.3	1000.3	432.9	432.6	3.2	1.5	4.7
Na$_2$SnO$_3$	486.5	1000.1	431.9	431.8	4.0	1.7	5.7
NaSnF$_3$	487.2	999.4	431.0	431.6	4.2	2.4	6.6
Sb^i	528.0	958.6	464.4	505.8			
Sb$_2$S$_3$	529.3	957.3	462.3	505.0	0.8	1.3	2.1
Sb$_2$S$_5$	529.1	957.5	462.4	504.9	0.9	1.1	2.0
Sb$_2$O$_3$	529.8	956.8	459.9	503.1	2.7	1.8	4.5
KSbF$_6$	532.7	955.9	454.6	500.7	5.1	4.7	9.8
Te^m	573.2	913.4	491.7	578.3^f	0		
Te(ox)^{m,h}	576.9	909.7	486.5	576.8	1.5	3.7	5.2
Na$_2$TeO$_4$	576.6	910.0	485.7	575.7	2.6	3.4	6.0
Xe$(s)^b$	670.0	816.6	545.0	728.4^f	0		
Xe$(g)^e$	671.9	810.2^c	537.2^c	722.5	5.9	1.9^c	7.8^c
	$4f_{\frac{7}{2}}$		$M_5N_{67}N_{67}$				
WS$_2{}^j$	33.0	1453.6	1728.0	274.4			
Na$_2$WO$_4{}^k$	36.1	1450.5	1722.2	271.7	2.7	3.1	5.8
Pt	71.0	1415.6	2040.7	625.1			
K$_2$PtCl$_4{}^n$	73.2	1413.4	2035.4	622.0	3.1	2.2	5.3

^{a} The most intense Auger line in the group was used in all cases except that $M_4N_{45}N_{45}$ was used instead of the $M_5N_{45}N_{45}$ peak.
^{b} Ion-implanted in iron.
^{c} Ref. (17). The work function for iron was added to all vacuum level kinetic energies to make the values comparable with the solid phase. Fermi-level referenced data.
^{d} Ref. (6). The work function for sodium was added to the vacuum level kinetic energies to make data comparable with the solid phase data.
^{e} Sieve 4A zeolite. ^{f} Add -1000 or -2000.
^{g} Oxidized in air at a temperature sufficient to generate metal oxide peaks more intense than the element lines.
^{h} Element peaks still measurable. ^{i} Vapour-deposited *in vacuo*.

polarizability. The halogens appear in the appropriate order, although the spread between I⁻ and Cl⁻ is only 0.9 eV, compared to 2.1 eV between Cl⁻ and F⁻. Pentavalent NO_3^-, PO_3^- and BiO_3^- become more polarizable as Z increases. Sulphate and tellurate are consistent with this. A complex oxy-anion involving metals, CrO_4^{2-}, MoO_4^{2-}, and WO_4^{2-}, exhibit the opposite behaviour. The increase in polarizability going from SO_4^{2-} or NO_3^- to SO_3^{2-}, $S_2O_3^{2-}$, and NO_2^- fits in this series.

Pearson [10] has designated as " soft " those anions *in solution* that exhibit the most polarizable characteristics. The ones that are least are designated " hard ". Such well-known hard anions as F⁻, OAc⁻, NO_3^-, and SO_4^{2-} are at the low α end of the table, while soft anions such as I⁻, SCN⁻, and $S_2O_3^{2-}$ are near the other end. The correlation is good, even though the anions are in a crystal lattice without solvation. It should be considered also that we are thus far ignoring the effect of the important variable r in the crystalline lattices.

DATA ON THE ELEMENTS

Experimental data have been accumulated on all the elements that give sharp core-type Auger lines. These are assembled in table 3. Data are clearly consistent with theory, for the Auger parameter has large range. The concept is useful for Ne, Na, Mg ($KL_{23}L_{23}$ lines), Ar, K, Ca ($L_3M_{23}M_{23}$), Cu(Cu⁺), Zn, Ga, Ge, As, Se($L_3M_{45}M_{45}$), Ag, Cd, In, Sn, Sb, Te, and Xe($M_4N_{45}N_{45}$ lines). We lack data on elemental K and Ca, but there seems to be considerable dispersion among the salts. The latter fact does not seem to be true with several Cs and Ba compounds tested, so we reserve judgement on them. Data on iodine compounds seem not to be useful. We have attempted to obtain data on the *NNN* and *NOO* series for the heavier elements, but the intensities of the lines are very low. Berthou and Jørgensen [11] have made observations on the energies of these Auger and Coster-Kronig lines but no data on chemical shifts are available.

A serious limitation is, of course, the energy of the X-radiation. If higher photons or electrons were used, each series could be extended. This was tried with a chromium-plated anode, emitting 5415 eV radiation. Auger lines with energies up to 2300 eV could then be recorded with our instrument. Sample data are shown for Al, Si, and S(*KLL*) and W and Pt($M_5N_{67}N_{67}$). With such a technique we should be able to extend the *KLL* series to include Al, Si, P, S and Cl; the $L_3M_{45}M_{45}$ series to include Br, Kr, Rb, Sr, Y, Zr, Nb and Mo, and the $M_5N_{67}N_{67}$ series to include all the heavy metals. The latter series has much more intense lines than the *NOO* or $N_5N_{67}N_{67}$ series. It is not expected that the Auger parameter for the first series of transition metals or the rare earths will ever be very useful because multiplet splitting and

---

[j] S(*LMM*) line used as reference connecting Cr X-ray spectrum to that of Al $K\alpha$. With the latter the C 1s line = 284.6 eV was used as the charge reference.

[k] Na(*KLL*) and O(*KLL*) lines used as references to connect Cr spectrum to Al spectrum. The C 1s line at 284.6 was used as the absolute reference. The line at 2108.0 is from the central sulphur of $S_2O_3^{2-}$.

[l] Ref. (18). Work function added to vacuum level kinetic energies. Photoelectron data were only for the 3d line (ref. (4)). The value for $2p_{\frac{1}{2}}$ inserted was calculated assuming ΔE_B is the same for the 3d and $2p_{\frac{3}{2}}$ lines.

[m] Te powder used with oxide coat—elemental and oxide lines both observable.

[n] K(*LMM*) and Cl(*LMM*) Auger lines used as references to connect Cr X-ray spectrum to the Al X-ray spectrum. The C 1s line was used as the absolute reference.

[o] Ref. (19). Work function added to the vacuum level kinetic energy.

[p] Ref. (11). Berthou and Jørgensen find a chemical shift of 4.05 eV for CuCN and 4.1 eV for AgCN.

shake-up processes in the paramagnetic compounds prevent sharp Auger and photo-electron lines in the spectra. The elements Li through F do not have core-type Auger processes. Analysis for all other elements should be benefited by this concept.

EXPERIMENTAL

Data were obtained with a Varian IEE-15 spectrometer, equipped with a high-intensity Mg or Al anode. Energy data obtained with the Mg anode were corrected by 233.0 eV to make them comparable with those obtained with the Al anode. Compounds used were the purest obtainable commercially, ground in a nitrogen atmosphere, mounted on double-sided Scotch tape, and transferred to the instrument in the absence of air. Mg, Cu, Zn, Ag, Cd and Sn were used in the form of machined cylinders, cleaned and then abraded with fine silicon carbide paper in a nitrogen atmosphere. Ne, Ar and Xe were implanted in abraded-iron in an ion discharge. Ga, Ge, As, Se, In and Sb were vapour-deposited on stainless steel. Tellurium was available as a metal powder with an oxide layer.

The instrument was calibrated to give elemental binding energies: $Au\,4f_{\frac{7}{2}} = 83.8$, $Ag\,3d_{\frac{5}{2}} = 368.0$, $Pt\,4f_{\frac{7}{2}} = 71.1$, $Cu\,2p_{\frac{3}{2}} = 932.2$ eV. Measured peaks in general had intensities 5000-50 000 counts above background and were determinable in position to ± 0.05 eV.

Binding energy data on conductors were taken directly. Data on non-conductors were corrected for static charge by using adventitious carbon, $C\,1s = 284.6$ eV, the value it attains on inert metals in our system. The maximum charge correction required was 5.7 eV for magnesium fluoride. Most corrections were less than 3 eV.

Comparisons of absolute Auger energies may be made with data by others previously published, particularly those of Castle and Epler,[12] Schön,[13] Yin et al.,[14] Kowalczyk et al.,[15] and Aksela et al.,[16] on elements Cu-Se and oxides [12] Cu_2O, ZnO, Ga_2O_3 and GeO_2. The maximum deviation between the data in table 3 and those cited is 1.7 eV. When data for any line cited by others are averaged, the maximum deviation is 1.0 eV.

[1] C. D. Wagner and P. Biloen, *Surface Sci.*, 1973, **35**, 82.
[2] C. D. Wagner, *Anal. Chem.*, 1972, **44**, 967.
[3] D. W. Davis and D. A. Shirley, *Chem. Phys. Letters*, 1972, **15**, 185.
[4] L. Ley, S. P. Kowalczyk, F. R. McFeely, R. A. Pollak and D. A. Shirley, *Phys. Rev. B*, 1973, **8**, 2392.
[5] D. A. Shirley, *Phys. Rev. A*, 1973, **7**, 1520.
[6] S. P. Kowalczyk, R. A. Pollak, F. R. McFeely, L. Ley and D. A. Shirley, *Phys. Rev. B*, 1973, **8**, 3583.
[7] U. Gelius, *Phys. Scripta*, 1974, **9**, 133.
[8] N. F. Mott and R. W. Gurney, *Electronic Processes in Ionic Crystals* (Clarendon, Oxford, 1948).
[9] P. H. Citrin, D. R. Hamann and S. Hüfner, *Bull. Amer. Phys. Soc.*, 1973, **1**, 55.
[10] R. G. Pearson, *J. Chem. Ed.*, 1968, **45**, 581.
[11] H. Berthou and C. K. Jørgensen, *J. Electron. Spectr.*, 1974, **5**, 935.
[12] J. E. Castle and D. Epler, *Proc. Roy. Soc. A*, 1974, **339**, 49.
[13] G. Schön, *J. Electron. Spectr.* 1972, **1**, 377; 1973, **2**, 75.
[14] L. Yin, T. Tsang, I. Adler and E. Yellin, *J. Appl. Phys.*, 1972, **43**, 3464.
[15] S. P. Kowalczyk, R. A. Pollak, F. R. McNeely, L. Ley and D. A. Shirley, *Phys. Rev. B*, 1973, **8**, 2387.
[16] S. Aksela, M. Pessa and M. Karras, *Z. Physik*, 1970, **237**, 381.
[17] K. Siegbahn, C. Nordling, G. Johansson, J. Hedman, P. F. Heden, K. Hamrin, U. Gelius, T. Bergmark, L. O. Werme, R. Manne and Y. Baer, *ESCA Applied to Free Molecules* (North-Holland Publishing Company, Amsterdam, 1969).
[18] S. Aksela and H. Aksela, *Phys. Letters A*, 1974, **48**, 19.
[19] H. Aksela and S. Aksela, *J. Phys. B, Atom Mol. Phys.*, 1974, **7**, 1262.

GENERAL DISCUSSION

Dr. C. M. Quinn (*Birmingham*) said: As discussants are no doubt aware by this time, Lloyd, Richardson and I at Birmingham University have recently begun a study of angular dependence in p.e. of surfaces. I am very impressed by Willis's paper since it seems to go a considerable distance along the road to our understanding electron emission processes at surfaces. I would like to ask him one question and make one remark about the interpretation of off-normal angular resolved emission data.

My remark concerns the influence of momentum broadening normal to the surface whenever an evanescent final state occurs in the photoelectric process. Under such conditions the surface photoelectric process applies and the degree of normal-momentum broadening reflects the localization in the normal direction into the crystal of the incoming scattering function. Thus it seems to me that a characteristic of surface photoelectric emission should be an "angular broadening" in which spectral features due to the surface effect occur over a range of angles in angular-resolved spectra, a range reflecting the normal-momentum broadening. For copper single crystals, our results to date suggest that we may be observing this type of effect in certain instances.

My question concerns the problem of refraction. For a copper (001) surface and analyser tracking from (001) to (100) directions, one expects band-structure in the $\Gamma W K W \Gamma$ section of the Brillouin-zone to be observable. We expected for identical photon angles that we should be able to monitor the ΓW band-structure at $\sim 26°$ and then again at $\sim 63°$ off-normal. However, the spectra for the analyser settings are different and this is also true over a range on either side of these positions. Can this be due to a large refraction effect?

Dr. W. F. Egelhoff, Prof. J. W. Linnett and **Dr. D. L. Perry** (*Cambridge*) (*communicated*): The results of some studies we have made of the angular dependence of the photoelectron spectra of a number of different adsorbates on W(100) may be of interest in the context of the remarks made by Quinn. The spectrum of a clean W(100) surface is, as would be expected, strongly dependent, in both peak position and peak intensity, on both photon energy and on angle.[1] This may be understood, in part, in terms of the effects described by Quinn. On the other hand, the difference spectra for adsorbed monolayers, that is the changes in the photoelectron spectrum induced by adsorption, show markedly less sensitivity both to photon energy and to angle. In general, except for the case of hydrogen adsorbed on W(100),[2] we have observed no angular dependence of peak position for adsorbate spectra, only changes in the absolute and relative intensities of the different spectral features. As outlined in our paper, for cases of relatively weak molecular adsorption, e.g., α-CO and the molecular complexes of HCHO, CH_3OH and NH_3, there is no photon and angular dependence, the intensity of the adsorbate induced photoelectron emission varying with that of the clean surface. For more strongly held adsorbates on W(100), such as β-CO, β-N_2 and carbon, which exhibit very broad spectral features, angular changes result in changes of peak shape due to relative changes in the intensity of

[1] W. F. Egelhoff, J. W. Linnett and D. L. Perry, unpublished results.
[2] W. F. Egelhoff and D. L. Perry, *Phys. Rev. Letters*, 1975, **34**, 93.

the different components contributing to the total peak.[1] Our method of studying the angular sensitivity of a photoelectron spectrum is not ideal (the angle of incidence and angle of photoelectron detection cannot be varied independently, but are fixed at 90° to each other, see fig. 1 of paper) and does not permit a full angular analysis. However, since this would be expected to increase the complexity of observed angular dependent effects, we conclude from the lack of any angular dependent energy shifts that band structure effects, as observed for photoemission from crystalline solids, are not as important for photoemission from adsorbed monolayers. This would suggest that, although the essential differences between gas and solid phase photoemission processes should be recognised, direct comparison of gas phase with adsorbed phase spectra, as has been carried out for a number of adsorbed systems, should not necessarily lead to significant error. Consideration has to be given, of course, to differences in both initial state and final state screening effects and to the definition of a common reference level.

Dr. D. P. Woodruff (*Warwick*) (*communicated*): As we have heard quite a lot about the angular resolved photoemission measurements of Quinn and his co-workers, some simple calculations on the refraction effect in these experiments seems worthwhile. The kinetic energy difference between an electron inside a crystal and outside in the vacuum is commonly referred to as the inner potential: it is approximately equal to the sum of the work function and the Fermi energy, although it is actually a function of the electron energy. However, its value is reasonably well-established from its use as an adjustable parameter in multiple scattering LEED calculations. For copper the value is close to 15 eV in the energy range of interest.[2] Using this value it is easy to see why Quinn and co-workers do not observe identical spectra at, or indeed near to, emission directions of 26.5° and 63.5° to the surface normal of an (001) surface in the $\langle 100 \rangle$ azimuth. The maximum energy of photoelectrons outside the crystal resulting from HeI photons is approximately 16 eV and such electrons emitted at 26.5° to the surface normal *inside* the crystal will emerge at $\sim 39°$ to the surface normal. Further down the spectrum emitted in the same internal directions, electrons of kinetic energy 5 eV outside the crystal will emerge at $\sim 63°$ to the surface normal. Moreover, all electrons emitted at 63.5° to the surface normal inside the crystal will be totally internally reflected at the surface when the photon source is the HeI line. The minimum photon energy compatible with the observations of *any* electrons emitted at 63.5° to the surface normal inside the crystal is ~ 65 eV (and then only electrons emitted from near the Fermi level will emerge and at grazing angle). The absence of the symmetry is, therefore, scarcely surprising not only because any spectrum at fixed angle outside corresponds to a range of emission angles inside the crystal, but also because electrons travelling in the second ΓW direction inside the crystal cannot escape into the vacuum.

It is perhaps worth commenting, in addition, that for electrons emitted from the surface region, the band structure approach is really rather inappropriate. Band structure calculations are normally made using schemes which make full use of the 3-dimensional periodicity of the bulk, whereas no such periodicity exists in the surface region due to the presence of the surface. It is, therefore, unlikely that, even in the absence of a refraction effect, identical spectra would be observed in the two ΓW directions at different angles to the surface normal which are symmetrically equivalent in the bulk.

[1] W. F. Egelhoff, J. W. Linnett and D. L. Perry, *Faraday Disc. Chem. Soc.*, 1974, **58**, 35.
[2] e.g., J. B. Pendry, *J. Phys. C* (Solid State Phys.), 1971, **4**, 2514; G. Capart, *Surface Sci.*, 1971, **26**, 429.

GENERAL DISCUSSION

Dr. R. A. Armstrong (*NRC, Ottawa*) said: In addition to the method described by Willis, band gaps may also be detected by measuring the fraction $R(E)$ of low energy electrons (0-20 eV) normally incident which are elastically reflected.[1-3]

In particular $1-R(E)$ for $W(110)$[3] matches very well the low energy part of the $N(E)$ curve given by Willis in fig. 3. Adsorption of oxygen followed by heating has been found to produce new structure in the band gap region [4] similar to that described by Willis.

Dr. U. Landman (*Rochester*) said: I would like to point out, in connection with Andersson's paper, that analysis of inelastic low energy electron diffraction (ILEED) data though complicated, has been used for the study of the dispersion relation and damping of surface plasmons. In the analysis a proper description of both the elastic and inelastic vertices is required due to the two-step nature of the phenomena. Detailed analysis of ILEED data from Al(111) and Al(100) yielded results [5, 6] which are in general agreement with current theoretical models.

Prof. E. A. Stern (*University of Washington*) (*communicated*): In Andersson's presentation the question is raised about the explanation of the collective energy loss suffered by electrons scattered from absorbed alkali metal layers. In a private discussion with Dr. Andersson I suggested the following explanation which agrees with his experimental results.

Andersson points out that his eqn (6) and (7) for thin alkali layers disagree with the experimental results. This is not surprising since eqn (6) and (7) are for a thin layer surrounded on both sides by free space, not the experimental situation. It is straightforward to derive the expression for the experimental configuration and we will now do so.

The wave number k_m of the surface plasmon of radial frequency ω most strongly coupled by an electron of velocity v is [7]

$$k_m^{-1} \approx (\omega/v)^{-1} \approx 6.8 \,\text{Å}. \tag{1}$$

The numerical value is evaluated for an electron of bombarding energy 25 eV and an energy loss of $\hbar\omega = 3$ eV. The value of k_m^{-1} is the distance over which the electromagnetic (EM) fields of the plasmon decay in space normal to the boundary. This distance is somewhat larger than interatomic dimensions. A conduction electron of the medium travels a distance about three times k_m^{-1} per time cycle of the plasmon vibration frequency. The first result means that quantum effects can be neglected in calculating the response of the medium to the plasmon fields. The second result means that non-local effects are not overwhelming but should be taken into account, i.e., $D(\mathbf{r})$ depends not only on $E(\mathbf{r})$ but also on values of $E(\mathbf{r})$ in the vicinity of $\mathbf{r}$. However, since we would be satisfied with a semiquantitative theory, we will simplify the problem significantly and maintain the qualitative features by using the local relationship between D and E. Only the frequency dependence of the dielectric constant need be considered, i.e., $\varepsilon(\omega)$.

Such a problem can be solved in general by Maxwell's equations.[7] Since

---

[1] E. G. McRae, *Surface Sci.*, 1974, **42**, 427.
[2] S. Andersson and J. B. Pendry, *J. Phys. C.*, 1973, **6**, 601.
[3] R. A. Armstrong, *Surface Sci.*, 1975, **47**, 666.
[4] R. A. Armstrong, unpublished.
[5] C. B. Duke and U. Landman, *Phys. Rev. B*, 1973, **8**, 505.
[6] C. B. Duke, U. Landman, L. Pietronero and J. O. Porteus, abstracts of the 34th Annual Conference on Physical Electronics, Bell Labs. Murray Hill, N. J., (February 1974).
[7] E. A. Stern and R. A. Ferrell, *Phys. Rev.*, 1960, **120**, 130.

$k_m \ll \omega/c$, the retardation of the EM fields can be neglected and Poisson's equation is sufficient.

Referring to fig. 1 we look for solutions of Poisson's equation corresponding to a wave travelling along the boundary in the x-direction. The z-dependence normal to

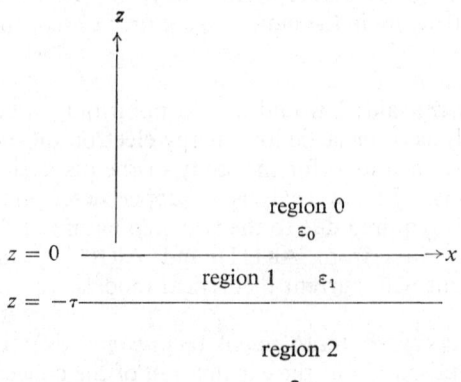

FIG. 1.—The geometric configuration for which the surface plasmon dispersion relation is derived. Region 1 is a thin film of thickness τ sandwiched between two semi infinite regions 0 and 2. The frequency dependent dielectric constants for the corresponding regions are ε_0, ε_1 and ε_2.

the surface is determined from Poisson's equation. In the 0-region, $z > 0$, which has the dielectric constant ε_0, (which we will set equal to 1 in this case for free space, but will not do so as yet to obtain a more general formula) the solution we desire is of the form

$$\phi_0 = A\, e^{ikx}\, e^{-kz}. \qquad (2)$$

In region 1, the alkali metal layer, $0 > z > -\tau$, the desired solution is of the form

$$\phi_1 = e^{ikx}(Be^{-kz} + Ce^{kz}). \qquad (3)$$

In region 2, the substrate, $z < -\tau$, the desired solution of Poisson's equation is

$$\phi_2 = e^{ikx}(De^{kz}). \qquad (4)$$

The forms in (2)-(4) are all solutions of Poisson's equation

$$\nabla^2 \phi = 0 \qquad (5)$$

in their respective regions. On the boundaries between the regions eqn (5) is no longer valid because of surface charges. We handle this in the usual manner by imposing the boundary conditions that both the tangential E and the normal components of D be continuous. These boundary conditions lead to the relationship

$$\frac{[1-\varepsilon_0(\omega)/\varepsilon_1(\omega)]}{[1+\varepsilon_0(\omega)/\varepsilon_1(\omega)]} = \frac{[1+\varepsilon_2(\omega)/\varepsilon_1(\omega)]}{[1-\varepsilon_2(\omega)/\varepsilon_1(\omega)]} e^{2k\tau}. \qquad (6)$$

This expression goes to the correct limits as $\tau \to \infty$ corresponding to bulk surface plasmons (if they can exist) at the two surfaces bordering on regions 0 and 2. When $k\tau \lesssim 1$ there is appreciable coupling between the two surfaces modifying the frequency of the excitation. When $\varepsilon_2 = \varepsilon_0 = 1$, $\varepsilon_1 = 1 - \omega_p^2/\omega^2$ and $k\tau \ll 1$, corresponding to a thin alkali metal film surrounded by free space on both sides, (6) leads to Andersson's eqn (6), as it should.

Because of our approximations, the ε's in (6) are equal to the optical dielectric constant. In the experiments of Andersson $\varepsilon_0 = 1$, $\varepsilon_1(\omega) = 1 - \omega_p^2/\omega^2$, where ω_p is given by eqn (7) of Andersson's paper, and $\varepsilon_2(\omega)$ is the optical dielectric constant of Ni. Making these substitutions in (6) we find

$$1-\frac{\omega_p^2}{\omega^2} = -\frac{(\varepsilon_2+1)}{2\tanh k\tau} \pm \frac{1}{2}\sqrt{\left(\frac{\varepsilon_2+1}{\tanh k\tau}\right)^2 - 4\varepsilon_2}. \qquad (7)$$

Since at the experimental surface plasmon frequency ε_2 is negative and $\omega < \omega_p$, eqn (7) has only one solution corresponding to the *negative* square root. When the substrate is a dielectric instead of a metal, so that $\varepsilon_2 > 0$, two modes exist. However, when the substrate is metallic only one mode exists.

For a monolayer or less coverage, $\tau = 3.71$ Å according to Andersson's table 1 and thus $k_m\tau \simeq \frac{1}{2}$ and $\tanh k\tau \approx \frac{1}{2}$. Eqn (7) reduces to

$$\omega \approx \omega_p/2 \qquad (8)$$

for the only solution with the negative square root, and at $\hbar\omega \approx 3$ eV where the real part of $\varepsilon_2 \approx -4.4$ for Ni. The energy loss corresponding to this frequency is about 3 eV, in rough agreement with the experimental results.

The mode corresponding to the negative square root in (7) has the correct positive dispersion with k in agreement with the experimental measurements.

In summary, by taking account of the Ni substrate the collective excitation noted by Andersson can be explained as the excitation of surface plasmons.

Dr. R. W. Joyner (*Bradford*) said: Coad, Gettings and Rivière rightly emphasize electron-beam-induced desorption, which is likely to be important for weakly held adlayers. Two other effects of electron beams interactions with surfaces are also worth discussing. Even at LEED energies (< 100 eV) the electron beam can modify the structure of the adsorbed layer, as is observed [1,2] for CO on copper (100) at 80 K and also for oxygen on Cu (100) at low temperatures. This phenomenon may be more general than is recognized.

We have also been interested in the ability of electron beams to "induce" adsorption of carbon at metal surfaces. If a tungsten surface, in a low pressure of carbon monoxide (e.g., 5×10^{-6} Pa) is exposed to the electron beam of an Auger spectrometer the carbon Auger peak is observed to grow slowly with time, eventually reaching a size many times larger than that observed for the chemisorbed CO monolayer. It seems likely that the electron beam dissociates weakly held (α)-carbon monoxide. The resulting W—C bond is very strong while the oxygen fragment desorbs from the surface.

We have used several approaches to quantify the extent of carbon build-up after saturation exposures to electron beams. Simplest and most successful has been to determine Auger spectra in the energy distribution, $N(E)$ mode and to normalize carbon peak areas to the area of the tungsten multiplet peak at ~ 180 eV. We have then used as a calibration the known coverage of the saturated carbon monoxide monolayer which is $\sim 6.0 \times 10^{14}$ molecules cm^{-2}. The saturation coverage thus obtained is $3.6 \pm 0.7 \times 10^{15}$ atoms cm^{-2}, which is very close to the density of carbon atoms in the closely packed, basal (0001) plane of graphite (3.83×10^{15} atoms cm^{-2}). It seems, therefore, clear that the effect of the beam is to generate a tightly packed graphite like monolayer at the tungsten surface. It seems probable that a similar graphite like layer is formed on platinum [5] and also on other metals.

[1] J. C. Tracy, *J. Chem. Phys.*, 1972, **56**, 2 748.
[2] M. A. Chesters and J. Pritchard, *Surface Sci.*, 1971, **28**, 460.
[3] M. J. Braithwaite, R. W. Joyner and M. W. Roberts, paper at this Discussion.
[4] R. W. Joyner, J. Rickman and M. W. Roberts to be published.
[5] R. M. Lambert and C. M. Comrie, *Surface Sci.*, 1973, **38**, 197.

Dr. C. D. Wagner (*Houston*) said: The rates of chemical changes brought about in AES and XPS by means of radiation-chemical effects, i.e., effects of charged particles, can be compared by some calculations based on simple assumptions.

We assume these effects from XPS arise by passage through the sample of photoelectrons and Auger electrons emitted by the sample atoms, that all of the energy of these electrons is spent in the sample, and that the distribution of this energy with depth is the same as the photon flux at each depth. We assume similarly with AES that the energy distribution versus depth is the same as the flux of unscattered beam electrons versus depth. This latter assumption is more in question, but the actual analysis is more complex and it is not believed such an analysis would lead to a very different value for energy adsorbed at the surface.

Let us assume a sample of unit density, and take three cases: (1) An XPS system, with a 10 kV 100 mA electron beam, an aluminum anode, a screening window transmitting 0.3 of the AlK photons, and the sample 3 cm from the anode. We assume a photon mean free path of 10^{-3} cm in the sample. (2) An AES system, with 3 kV 10 μA electron beam, 30 μm diameter at the sample. The mean free path for inelastic scattering of the electrons in the sample is assumed to be 50 Å. (3) An AES system the same as (2) but with a rastered beam, 3 $\times$ 3 mm. (These AES systems are similar to those used by Coad *et al*.).

We calculate first the particle or photon flux on the sample, then the average energy absorbed per second per gram of sample, assuming it is all absorbed in the average path length (mean free path) of the photon or particle. This, multiplied by e, and converted to rad s^{-1}, gives the dose rate at the surface. Finally, on the assumption that 10 eV is required to produce a radiation chemical change in a molecule, we can calculate the fraction of surface molecules converted per second.

The calculation is straightforward once one has data for the conversion efficiency of 10 kV electrons into Al X-ray photons. Dolby [1] has determined this experimentally to be 2.3×10^{-4} photons per electron per steradian. With this and the value 6.24×10^{19} eV per gram megarad, one can readily calculate the data in table 1.

It is clear that even with a rastered AES beam one would have to use about a nanoampere of current in order to approximate the rate of radiation-chemical damage at the surface in XPS.

TABLE 1.—COMPARISON OF XPS AND AES IN RADIATION EFFECTS AT SAMPLE SURFACES

	XPS	AES	scanned AES
electrons or photons/cm^2 [a]	4.8×10^{12}	8.8×10^{12}	6.8×10^{12}
eV/g s [a]	5.3×10^{19}	3.9×10^{29}	3.0×10^{23}
Mrad/s	0.31	2.3×10^9	1760
fraction molecules changed/s	3.2×10^{-6} M [b]	2.4×10^4 M	1.8×10^{-2} M

(*a*) Average energy absorbed, i.e., total energy divided by weight of layer to depth of mean free path.
(*b*) M = molecular weight.

Dr. J. S. Johannessen (*Stanford*) said: My comments on the paper on chemical shifts of Auger lines [2] are meant to underline the importance which Strausser, Spicer and I believe such chemical shifts will have as a means of interpretation of the chemical structure of compounds and of the chemical structure and morphology of interfaces, as seen by AES. High energy resolution AES makes it possible to distinguish between chemical states of elements in different chemical environments. This is

[1] R. M. Dolby, *Brit. J. Appl. Phys.*, 1960, **11**, 65.
[2] C. D. Wagner, paper at this Discussion.

clearly demonstrated by the comprehensive study presented by Wagner,[1] and exemplified by the two cases I will discuss.

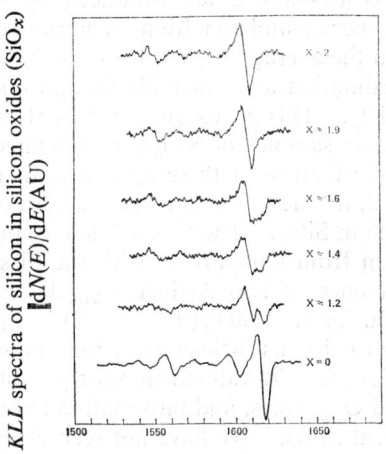

FIG. 1.—*KLL* spectra of silicon oxides SiO$_x$. The samples are thermally grown dioxide ($x = 2$), Si(100) wafer ($x = 0$), and evaporated SiO powder in diluted O$_2$ background ($x = 1.9, 1.6, 1.4, 1.2$). The vertical scale is the same for all spectra except the lowest ($x = 0$) which has been reduced by a factor of 1.4.

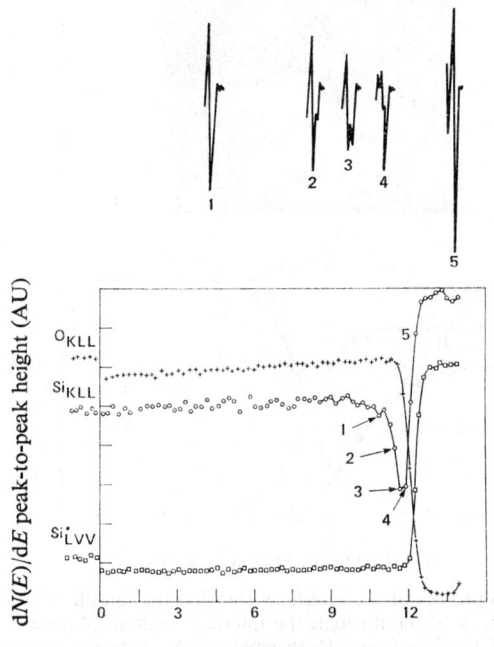

FIG. 2.—Chemical depth profiles through 1000 Å of thermally grown SiO$_2$ on Si(100). Above the depth profiles is shown a series of Si$_{KL_{2,3}L_{2,3}}$ Auger lines from the SiO$_2$–Si interface region. Sputtering rate 77 Å/min.

[1] C. D. Wagner, paper at this Discussion.

308   GENERAL DISCUSSION

We have used AES in a study of silicon oxides of stoichiometry SiO_x, where $0 < x \leqslant 2$.[1] Silicon dioxide samples are grown on Si(100) wafers by oxidation in dry O_2, and unsaturated oxides $x < 2$ are produced by vapour deposition. A detailed description of the experimental conditions is given elsewhere.[2] We find that silicon atoms in SiO_x are on the average present in two distinct chemical states,[1] one associated with Si–Si$_4$ coordination as in metallic Si, and one state associated with Si–O$_4$ coordination as in SiO_2. This is demonstrated by the Si_{KLL} spectra shown in fig. 1. Here, spectra from six samples of SiO_x are compared, ranging from $x = 2$ to $x = 0$. The characteristic feature of these spectra is that, as the stoichiometry factor x becomes less than 2, a " new " Auger line emerges on the high energy side of the 1611 eV $Si_{KL_{2,3}L_{2,3}}$ line in SiO_2. The " new " line is the 1618 eV $Si_{KL_{2,3}L_{2,3}}$ of elemental Si, as can be seen from comparison with the Si spectrum shown at the bottom of fig. 1. The presence of two distinct Si_{KLL} lines in unsaturated silicon oxides leads us to the conclusion that SiO_x ($0 < x < 2$) is a phase-separated system where the minor component exists as inclusions in the major component.[1] Fig. 1 illustrates the usefulness of AES. Overall stoichiometry is obtained from the peak-to-peak heights of Si_{KLL} and O_{KLL} lines, and information about the chemical state of Si and O is given by chemical shifts. We have not seen chemical shifts in the O_{KLL} Auger line in any of our samples.

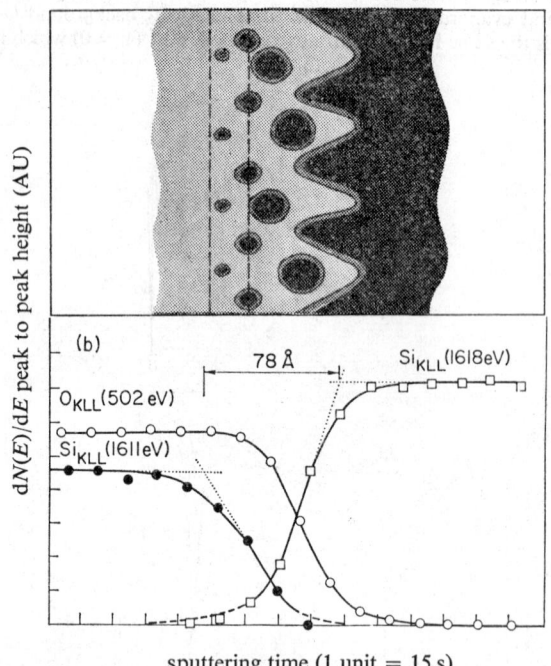

FIG. 3.—Details of the depth profiles at the interface, showing the fall-off of the O_{KLL} line and the change from 1611 eV to 1618 eV Si_{KLL} through the interface region. Above is shown a tentative model of the morphology of the interface. Dark areas are Si, grey areas are connective regions between Si and SiO_2, shown as light grey. Vertical broken lines indicates electron escape depth at 1600 eV.

[1] J. S. Johannessen, W. E. Spicer and Y. E. Strausser, *Appl. Phys. Letters*, 1975, **27**, 452.
[2] Y. E. Strausser and J. S. Johannessen, ARPA/NBS IV, Surface Analysis For Silicon Devices, April 23–24 (1975).

Chemical shifts of Auger lines have proved very useful in analysis of interfaces as they appear in chemical depth profiles of layered structures.[1] A typical depth profile is shown in fig. 2. Here a 1000 Å thick oxide thermally grown on Si(100) is sputtered through, and peak-to-peak heights of Si_{KLL}, O_{KLL}, and Si_{LVV} lines are plotted as function of sputtering time. Knowing the rate of sputtering, 77 Å/min in this case, the time scale can be converted to distance or depth into the sample. The bulk of the oxide layer shown in fig. 2 is homogeneous SiO_2. However, at the interface the stoichiometry appears to be SiO_x where $x < 2$. This is demonstrated by the dip in the Si_{KLL} profile, caused by the presence of silicon atoms in two chemical states. The $Si_{KL_{2,3}L_{2,3}}$ lines plotted above the depth profiles in fig. 2 show how the 1611 eV line decays and the 1618 eV line grows as the interface is sputtered through. A closer examination of the interface region, fig. 3, shows that the Si atoms change their chemical state in an orderly fashion over a spatial extent of 80 Å. The interface in fig. 3b appears to be wider than the " real " interface due to escape depth and ion beam effects. When these effects are taken into account we arrive at a " real " interface width on the order of 30 Å.[1] Based on our experimental data and on transport properties of MOS-channel devices we have proposed a model of the chemical structure and the morphology of the SiO_2–Si interface.[1] A tentative sketch of our model is shown above in fig. 3. The vertical broken lines indicate the escape depth of 1600 eV electrons. On the time scale one unit corresponds to 19 Å. The dark areas are silicon, the grey areas are connective regions between silicon and silicon dioxide shown light grey in fig. 3. We suggest that the excess positive charge associated with the SiO_2–Si interface in MOS-devices is confined to bonding defects in the connective regions at the interface.

The chemical shift of 7 eV we have observed in unsaturated silicon oxides and at SiO_2–Si interfaces is in excellent agreement with that of oxidized Si surfaces reported by Wagner.[2] We hope that the two examples we have discussed will underline the importance of chemical shifts in AES analysis.

Dr. P. R. Norton (*AECL*) said: Cracking of CO adsorbed on a clean, annealed Pt (111) surface has been observed during XPS experiments. The elemental carbon peak (C 1s at 284.6 eV) increased substantially relative to the CO C 1s peak (286.4 eV) over a one-hour period with the X-ray source operating at 200 W. A much smaller increase in a 531.4 peak relative to the 532.5 eV CO O 1s peak was also noted. Oxygen presumably desorbed during the cracking process. No dissociation was detected in the absence of an X-ray beam. These results are qualitatively similar to those of Lambert and Comrie [3] and it is possible that the cracking mechanism might be the same; i.e. that the slow secondary electrons cause the observed cracking rather than the primary photoexcitation process.

Prof. M. W. Roberts (*Bradford*) said: I am struck by the observation that sodium is present in significant quantities on surfaces studied by SIMS (e.g., see fig. 5). We have recently observed [4] the presence of calcium on Au surfaces, furthermore there is strong evidence that this calcium is responsible for the development of simple surface structures on initially polycrystalline gold. We are unsure of the source of the calcium and one possibility is the filament of the ion-bombardment gun. Has

[1] J. S. Johannessen, W. E. Spicer and Y. E. Strausser, to be published.
[2] C. D. Wagner, paper at this Discussion.
[3] R. M. Lambert and C. M. Comrie, *Surface Sci.*, 1973, **38**, 197.
[4] S. A. Isa, R. W. Joyner and M. W. Roberts, *J.C.S. Faraday I*, 1976, **72**, in press.

Coad any comment on the likely source of the alkali metal impurities in his experiments?

Dr. J. P. Coad (*Harwell*) said: In reply to Roberts' comments, there is some doubt about the source of the sodium in the SIMS spectra. SIMS is very sensitive to sodium and doubtless the surface contamination will give a contribution to the observed peak. However it has been suggested that some sodium may also come from the ion gun—I would not discount this possibility and we have not attached any great significance here to the sodium level. I would also like to clarify one point in the paper; the XPS kinetic energy scales are not compensated for work function nor corrected for possible charging effects so the energies of the oxygen peaks, for example, are not comparable with other work.

A number of contributors have described electron beam effects; there are many others referred to in the Introduction to the paper, including one to work on silicon similar to that described by Johannessen (ref. (10)). The carbon level on a surface under electron irradiation may either increase or decrease depending on the conditions; in the examples in this paper the carbon decreased, but in other cases (particularly if the vacuum is poor as in most electron microscopes) the carbon may increase. It should be pointed out that some beam effects also occur in XPS, and a number of examples were given by Wallbank, Johnson and Main (*J. Electron. Spectr.*, 1974, **4**, 263. However, as can be seen from Wagner's comparison between AES and XPS, the extent of the beam effects in XPS is several orders of magnitude less than in AES.

Prof. S. C. Fain (*Seattle*) said: Overbury and I have compared Auger data from his Ag–Au alloys [1] with data taken by McDavid in my laboratory, most of which was published last year.[2] The two sets of Auger data seem to agree when some allowance is made for backscattering differences due to different primary energies (4 keV [1] as compared with 2 and 3 keV [2,3]). Our data imply much less segregation than do fig. 1 and 2 of Somorjai and Overbury or similar predictions published by van Santen.[4]

TABLE 1.—NORMALIZED AUGER ELECTRON INTENSITIES MEASURED AT ROOM TEMPERATURE FOR THREE GOLD PEAKS OF DIFFERENT ENERGIES

gold Auger peak eV	primary beam energy eV	Ag–Au before silver overlayer	Ag–Au with ⩽ 6Å silver overlayer	Ag–Au after 1 h 300°C aneal	Cu–Au before overlayer	Cu–Au with ⩽ 8Å gold overlayer	Cu–Au after 1.5 h 300°C anneal
72	2 000	0.73	0.16	0.69	0.63	0.90	0.64
239	2 000	0.72	0.30	0.68	0.51	0.82	0.54
2024	3 000	0.73	0.55	0.67	0.39	0.63	0.42

Under very carefully controlled experimental conditions, Auger data on alloys can be obtained with sufficient precision to determine the depth profile in the first few layers without assuming a particular theory for segregation. In our work on Cu–Au alloys McDavid and I were able to show that surface enrichment with Au occurs and is confined essentially to the top layer for the large range of composition

---

[1] S. H. Overbury and G. A. Somorjai, *Surface Sci.*, in press.
[2] S. C. Fain, Jr. and J. M. McDavid, *Phys. Rev. B*, 1974, **9**, 5099.
[3] J. M. McDavid, *Ph.D. Thesis* (University of Washington, 1975).
[4] R. A. van Santen, *J. Catalysis*, 1974, **34**, 13.

studied.[1] For the Cu–Au alloy system there is a large size difference (12%) of the pure metals which is not taken into account in regular solution theory.

Two questions which naturally arise in Auger electron spectroscopy of alloy surfaces are: (1) Are the published escape depths reasonable? and (2) Are the surface layers of samples in thermodynamic equilibrium? We investigated these questions for our Ag–Au and Cu–Au samples by evaporating at room temperature very thin overlayers on alloys. Results are given in table 1 for the normalized Auger electron intensities measured at room temperature for three gold peaks of different energies. The results before the overlayers give examples of data used to conclude that there is essentially no segregation at the surfaces of our Ag–Au samples, but considerable gold segregation for our Cu–Au samples. The overlayer thicknesses were estimated by the evaporation times; there was no significant diffusion into the samples at room temperature. Measurements of peak intensities taken during heating to 300°C indicated that most of the overlayer diffusion took place before the sample reached 300°C. The heater had about a 20 min time constant.[1]

The answers to the two questions as provided by the overlayer data are: (1) Analysis of the overlayer intensities indicates that estimates of 4, 8, and 30 Å escape depths for the three energies in table 1 are reasonable. (2) The changes in Auger intensities during annealing the overlayer indicate that our thin film samples are representative of some equilibrium temperature slightly above room temperature, but definitely less than 300°C. This latter point means that our Ag–Au data should be more comparable with Somorjai and Overbury's fig. 1 (300 K) than fig. 2 (900 K). Some improvements to Williams' 4-layer regular solution theory which may be important for alloys such as Ag–Au are an entropy calculation that allows for short range order such as that of Sundaram and Wynblatt [2] and an inclusion of modifications to the A–A and B–B bonds due to charge transfer in the alloys.[3]

Prof. G. A. Somorjai (*Berkeley*) said: The work of Fain and McDavid has demonstrated the utility of an intensity analysis similar to the one presented here. Their work on Cu–Au, as well as their overlayer studies of Ag on Ag–Au, clearly demonstrates that Auger electron spectroscopy can be utilized to yield quantitative depth profile for an alloy exhibiting surface segregation. However, our data on Ag–Au demonstrate the need to consider backscattering effects in the analysis. Improvements in our understanding of the Auger intensity analysis will lead to more accurately determined depth profiles for various equilibrated alloy surfaces. Such information may then be used to improve our understanding of surface thermodynamics and allow us to predict surface compositions of binary solutions.

Dr. T. A. Carlson (*Oak Ridge*) said: The Auger parameter, as defined by Wagner, is a valuable parameter with which to correlate chemical bonding. Unlike the study of binding energy shifts alone for nonconducting solids, where differences in work functions, charging and the exact definition of binding energy are uncertain between the different materials, the measurement of the differences between Auger and photoelectron energies is made on the same solid. Dress and I have made a number of studies similar to Wagner's for Al, Si and sulphur compounds using silver L_α X-rays. I offer one word of caution. The Auger process involves electrons from two different shells of different principal quantum numbers. When comparison with Auger spectra is made, it is assumed that chemical shifts in photoelectron spectra are

[1] J. M. McDavid and S. C. Fain, Jr., *Surface Sci.*, 1975, **52**, 161.
[2] V. S. Sundaram and P. Wynblatt, *Surface Sci.*, in press.
[3] S. C. Fain, Jr. and J. M. McDavid, *Phys. Rev. B*, 1974, **9**, 5099.

independent of the location of the core vacancy. This has generally proven to be a good assumption, but if the interpretation of the Auger parameter is to be treated with complete confidence, more studies on comparative chemical shifts of the different core electrons of the same atom are needed.

Mr. K. Y. Yu (*Stanford*) said: We have made surface composition measurements on the Cu–Ni alloy by using two techniques (AES and UPS). After proper annealing at elevated temperature (650°C), the surface layers of the alloy are considerably enriched in Cu, in agreement with the predictions of the regular solution model. Fig. 1 shows the UPS results on the 10% Cu/90% Ni (bulk composition) sample.

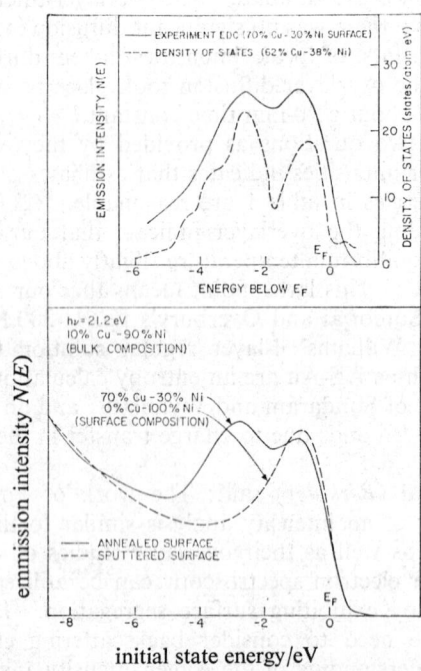

FIG. 1.—Photoemission spectra at $h\nu = 21.2$ eV of the annealed surface (solid line) and sputtered surface (dashed line) of the Cu–Ni alloy (bulk composition 10% Cu/90% Ni). The surface compositions indicated on the figure are obtained from independent Auger measurement using the $M_1M_{4,5}M_{4,5}$ transitions. The insert of the figure shows a comparison of the UPS spectra with a calculated density of state using the coherent potential approximation (CPA) theory.

The "surface" compositions listed on the figure are deduced from Auger measurements using the $M_1M_{4,5}M_{4,5}$ transition at ~100 eV.[1] This transition is more surface sensitive than the LMN series of transitions (~900 eV) commonly used by other workers. On the annealed surface, the Auger measurements give a surface composition of 70% Cu and 30% Ni, (compare with 10% Cu in the bulk). The Cu rich surface layers can be removed by argon sputtering, yielding a surface that is effectively 100% Ni. In fig. 1 the UPS data (at $h\nu = 21.2$ eV) show respectively the Ni d-peak (0.5 eV below E_F) and the Cu d-peak (2.5 eV below E_F) having the same strength on the annealed surface (solid line), and the complete removal of the Cu d-peak by sputtering (dashed line). The AES and UPS measurement are thus self-consistent. The monitoring of surface composition by UPS is only applicable to

[1] C. R. Helms and K. Y. Yu, *J. Vac. Sci. Tech.*, 1975, **12**, 276.

alloys systems having *d*-bands separated in energy (e.g., Cu/Ni, Ag/Pd, etc.). This method does have the advantage of varying the escape length (hence the surface sensitivity) by changing the exciting photon energy, which is not always possible with AES.

The measurements of surface composition of the Cu/Ni alloy system have been controversial in the past. Most of the previous results report the surface composition to be the same as the bulk composition. We attribute the discrepancy between our results and past results to (*a*) our use of the more surface sensitive $M_1 M_{4,5} M_{4,5}$ transition, which allows us to investigate change in compositions of the first couple of layers and (*b*) the extremely sluggish diffusion process of Cu or Ni in the alloy. A short anneal cycle at temperatures lower than 650°C may not be sufficient to bring about a truly equilibrium surface composition, with the result that the surface composition resembles more closely the bulk composition.

Dr. C. R. Brundle (*IBM*) said: Following on the remarks of Yu, I would like to propose that the capabilities of UPS for studying order-disorder transitions in alloys be examined. Several years ago while attempting to measure UPS escape depths through condensed layers of Hg on Au (at 80 K), I noticed some interesting band structure effects. The Au HeI, HeII, and XPS spectra are well known. The 21.2 eV spectrum has a lot of fine structure in the *d*-bands caused by the nature of the final-states and the role of *k*-conservation in reaching them. The 40.8 eV spectrum is much more like the XPS spectrum, and both are more representative of the initial D.O.S. because the final states are sufficiently high in energy to be essentially continuum-like. Condensation of Hg on Au merely attenuates the Au *d*-bands, as expected, with the sharp Hg *d*-bands growing at about 9 eV below E_F. On warming to 300 K, most of the Hg desorbs, but a chemisorbed layer remains. Over a period of time, the Hg *d*-band decreases further; and the Au *d*-band structure in the HeI spectrum loses its fine-structure and ends up looking like the HeII spectrum. This is presumably a consequence of the disorder occurring due to lattice penetration by Hg which removes the significance of *k*-conservation. By heating the sample and then re-cooling to 300 K, we can remove all trace of the Hg *d*-band in the spectrum (either because of desorption or penetration below the sampling depth), but the Au part of the HeI spectrum changes only slowly with time back toward the original shape, presumably indicating a slow re-ordering to crystallinity. It seems that UPS might have a role to play here both in studying the electronic changes occurring and in following the kinetics of order-disorder processes.

Dr. J. S. Johannessen (*Stanford*) said: It is shown by Coad and co-workers [1] that AES can only provide an analysis of a virgin surface without appreciable beam interaction effects at primary current densities of about 3×10^{-6} A/cm^2 (at 3 keV) or at a total exposure of $\sim 10^{16}$ electrons/cm^2. We agree with the general conclusion of their analysis. However, we feel that the limitations of AES, set by primary beam interaction effects, are strongly dependent on the chemical composition of the sample under investigation, and that the primary beam parameters may be optimized in each experimental situation. As an illustrative example we will discuss electron stimulated desorption (ESD) of oxygen from silicon oxides of stoichiometry SiO$_x$, where $x \leqslant 2$. It is well known that SiO$_2$ decomposes under prolonged electron exposure.[2,3]

[1] J. P. Coad, M. Gettings and J. C. Rivière, paper at this Discussion.
[2] S. Thomas, *J. Appl. Phys.*, 1974, **45**, 161.
[3] Y. E. Strausser and J. S. Johannessen, ARPA/NBA Workshop 1V, Surface Analysis For Silicon Devices, April 23–24.

The most pronounced beam interaction effect is dissociation and subsequent desorption of oxygen.[1] This results in time dependent changes in the Auger spectra, as shown in fig. 1. In the upper part of fig. 1 we show how oxygen is depleted from the surface of SiO_2 and SiO_x, where $x = 1.4$. Notice that the initial rate is the same for the two samples, but that more oxygen is desorbed from the unsaturated oxide ($x = 1.4$), at long exposure times. This is a general feature. The smaller the stoichiometry factor x is, the more volatile the sample appears to be under electron

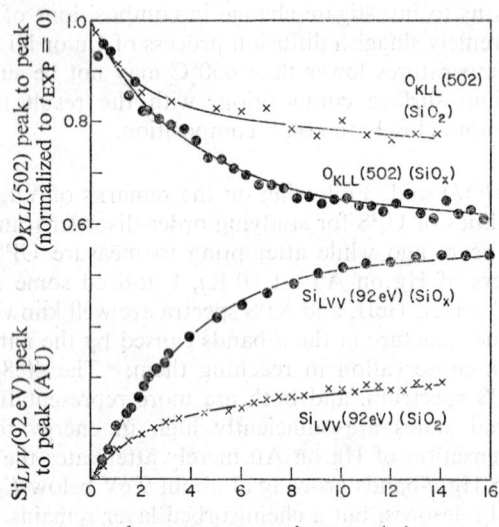

FIG. 1.—The effect of ESD on the O_{KLL} and Si_{LVV}(92 eV) lines from SiO_2 and $SiO_x(x = 1.4)$ shown as function of exposure time for a beam flux larger than 2×10^{17} electrons/cm² at 2 keV.

exposure, provided the sample has been exposed to normal atmosphere (e.g., "as-received" virgin sample). Depletion of oxygen from the surface of the samples is accompanied by a growth of the "free" silicon LVV line at 92 eV, as shown in the lower part of fig. 1. In the unexposed samples ($T_{EXP} = 0$) there is no "free" silicon, even though we know that there is excess "free" silicon in unsaturated oxides.[2] The reason for this is, of course, that the samples have been exposed to atmosphere, and hence the samples are oxidized to a depth sufficient to screen off their bulk stoichiometry. Fig. 1 is a typical example of excess exposure at a primary flux larger than 2×10^{17} electrons/cm² s at 2 keV. We have investigated the effect of primary beam flux and energy on the interaction with silicon oxide surfaces in terms of a desorption cross section of oxygen. The desorption cross section is related to the primary electron flux Φ (electrons/cm² s) and defined by

$$\sigma = -\frac{1}{\Phi}\left(\frac{1}{H}\frac{d}{dt}H\right)_{t=0} \tag{1}$$

where H is the O_{KLL} peak-to-peak height and t is the exposure time. Both Φ and H are experimentally obtainable quantities, and Φ is varied by scanning the beam over a known sample area. In fig. 2 we show how the cross section of eqn (1) depends on

[1] Y. E. Strausser and J. S. Johannessen, ARPA/NBA Workshop 1V, Surface Analysis for Silicon Devices, April 23-24.
[2] J. S. Johannessen, W. E. Spicer and Y. E. Strausser, *Appl. Phys. Letters*, 1975, **27**, 452.

the electron flux at 2 keV. Below 2×10^{17} electrons/cm² s the cross section appears to be a well defined quantity. At larger values of Φ the cross section increases rapidly, probably due to thermal enhancement of the desorption process. The

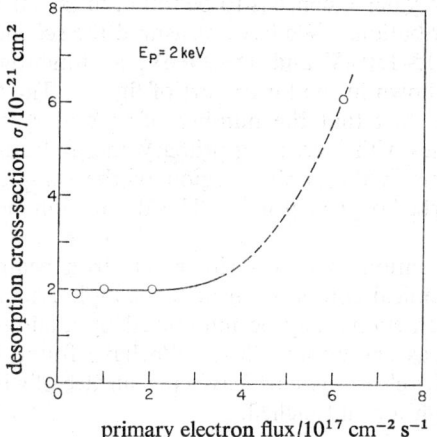

Fig. 2.—The total desorption cross section of oxygen at 2 keV shown as function of primary flux for a silicon oxide.

magnitude of the total desorption cross section we observe, 2×10^{-21} cm² for oxygen, is in reasonable agreement with oxygen desorption from metal oxides such as SrO

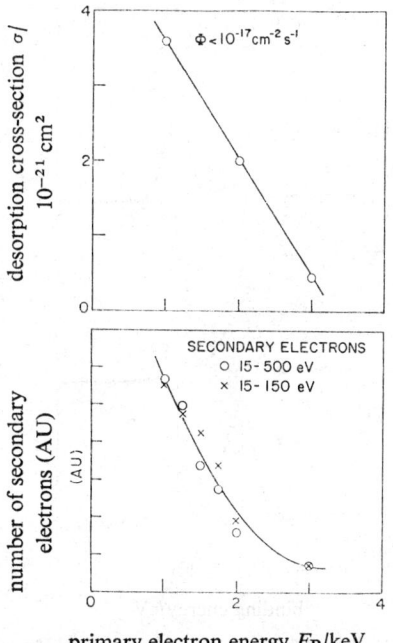

Fig. 3.—The upper part shows the dependence of the cross section on primary electron energy. The lower part shows how the secondary electron density depends on primary electron energy. The secondary electron energies are given in the figure.

and MgO [1] where dissociation is a part of the process. The desorption cross section is also a function of the primary beam energy as shown in the upper part of fig. 3. The linear relation is probably fortuitous. However, the important thing to notice is that σ decreases with increasing energy to a small but finite value at higher energies. The reason for this energy dependence is to be found in the energy dependence of the secondary electron distribution. We have measured the secondary electron yield for energies in the range 15–150 eV and 15–500 eV, as function of primary electron energy. The result is shown in the lower part of fig. 3. The total current was kept constant at 5 μA. The fact that the number of secondary electrons in the given energy intervals decreases with increasing primary energy leads to a smaller number of ionized oxygen atoms in the surface region of the sample, and hence to a yet smaller number of desorbed oxygen atoms. This desorption mechanism is elaborated on in the literature.[1]

In conclusion we mention that the primary electron beam interactions in AES analysis depend the chemical composition of the sample under investigation and its pre history. Beam interactions may be minimized by a suitable choice of primary beam parameters such as energy and flux. We have found that at 2 keV a total exposure of less than 10^{19} electrons/cm^2 leaves the surface of silicon oxides practically speaking virgin, although not untouched.

Prof. C. S. Fadley (*Hawaii*) said: In discussing this paper, Johannessen has presented some very interesting data concerning the application of Auger spectroscopy to silicon oxidation studies, and has also suggested the possibility of a certain degree of silicon island formation during the growth of very thick oxide layers.

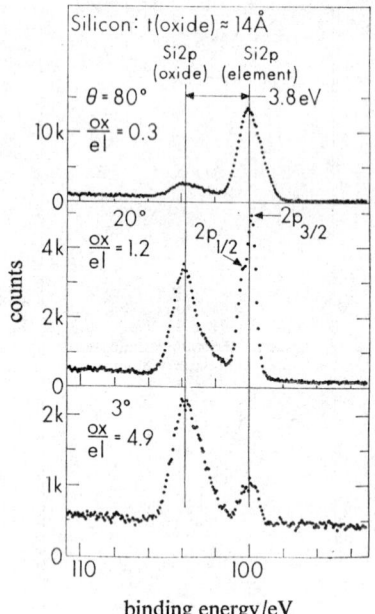

FIG. 1.—Si2p XPS spectra obtained at different electron exit angles from a silicon specimen with a 14 Å thick thermally-grown oxide layer. $\theta = 90°$ corresponds to exit perpendicular to the surface.

[1] T. E. Madey and J. T. Yates Jr., *J. Vac. Sci. Tech.*, 1971, **8**, 525.

Hill, Baird, Grunthaner, Wagner and I [1] have been studying the oxidation of silicon by means of angular-dependent XPS. Oxide layers of up to 88 Å have been thermally grown on highly polished silicon substrates; layer thicknesses were measured by ellipsometry. Core- and valence-level spectra were then obtained at various angles of electron emission with a specially modified Hewlett Packard spectrometer, and pronounced enhancements of surface-layer features were noted for grazing angles of escape (low θ values). Fig. 1 shows Si2p spectra obtained for a specimen with a 14 Å thick oxide layer at exit angles of $\theta = 80°$, 30°, and 3°. There is a complete inversion of the oxide-element relative intensities over this angle range, with the chemically-shifted Si2p (oxide) peak being very weak at 80° and the Si2p (element) peak being correspondingly weak at 3°. The net change in relative intensity between 80° and 3° is approximately a factor of 15. Another example of the marked increase in surface sensitivity achieved at low angles is provided by fig. 2, where Si2p spectra are presented for a silicon surface cleaned by an HF etch immediately before being loaded into the spectrometer via an inert atmosphere. At $\theta = 49°$, the oxide peak is barely visible, but at 5° it has an intensity roughly one quarter that of the element peak; the net surface enhancement here is a factor of 20.

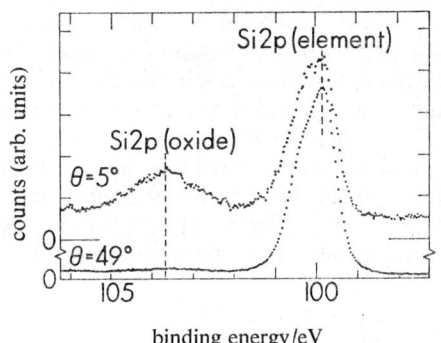

FIG. 2.—Si2p XPS spectra obtained at two electron exit angles from a silicon specimen cleaned by a hydrofluoric acid etch immediately before placement in the spectrometer.

Of further interest in fig. 1 is the fact that the oxide peak shows marked asymmetry toward the low-binding-energy side, an asymmetry which increases upon going to lower angles. Specimen charging in the oxide layer was ruled out as a cause of this asymmetry by applying a flux of low energy electrons from a Hewlett Packard flood gun to the surface; no changes in shape or significant shifts were observed over a broad range of currents and energies. Another source of this asymmetry could be the presence of oxidized silicon atoms in chemical states associated with lower binding energies than that for SiO_2. If this is true, then the enhancement of the asymmetry at low angles further indicates that these lower-binding-energy oxide states are present in the outer regions of the oxide layer. A more detailed account of this work is in preparation.[1]

It is thus of interest to watch for these lower-binding-energy oxides in any study of silicon oxidation, and it might further be fruitful to investigate both high resolution Auger and XPS spectra. In fact, C. D. Wagner's values of the Auger parameters in Si and SiO_2 suggest that the $KL_{2,3}L_{2,3}$ Si Auger peaks for this system will be spread over a fairly broad kinetic energy range; the Si2p chemical shift of 3.8 eV added to the change in Auger parameter of 3.1 eV predicts a range of 6.9 eV for these Auger peaks.

[1] J. M. Hill, R. J. Baird, F. Grunthaner, L. F. Wagner and C. S. Fadley, to be published.

Prof. R. N. O'Brien (*Victoria*) said: My question is directed to the theoretically oriented among us and concerns the possibility of using electron spectroscopy techniques to examine charge storage in electrets.

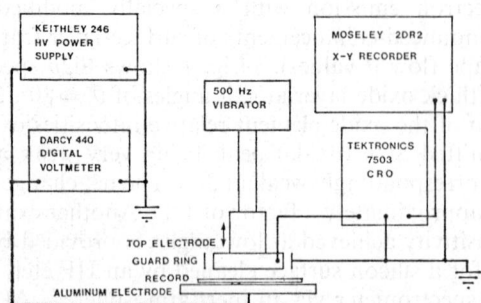

FIG. 1.—Block diagram of vibrating capacitor.[2]

The most common modern technique for making electrets [1] is to inject electrons into a good insulator (10^{16}–10^{22} ohms cm^{-1}) under a field of 10-20 kV cm^{-1} or somewhere near the electric break down strength of reasonably dry air. The amount of charge injected is usually measured by a vibrating capacitor of the type reported by Reedijk and Pearlman.[2] The assumption is made that the charge is all manifested on the surface and the charge density is found from $\sigma_A = \varepsilon K V_B/d$ where ε is the permittivity of free space, V_B is the bucking voltage supplied to smooth the alternating current trace on an oscilloscope (in the experimental set-up shown in block diagram, fig. 1) and d is the thickness of the electret all in cgs units. Charge densities of 1 electron per 100 CF_2 surface groups have been calculated from the expression for Teflon electrets.

The trap depths of trapping sites are in the range of 0.4 to 1.5 eV. These are currently measured by heating the electret at a constant rate and measuring the current flow with time (temperature) to get a Thermally Stimulated Current Emission spectrum.[3] The exponential part of the leading edge of glow peaks are analyzed by Arrhenius plot for trap depths. The general technique is not very satisfactory and the field would benefit from a second, hopefully more reliable technique. It would seem that one of the electron spectroscopy techniques should serve and I ask the opinion of the theorists on this.

[1] F. Gutman, *Rev. Mod. Phys.*, 1948, **20**, 457; L. M. Baxt and M. M. Perlman, *Electrets and Related Electrostatic Charge Storage Phenomena* (The Electrochemical Society, Inc., New York, 1968).
[2] C. W. Reedijk and M. M. Perlman, *J. Electrochem. Soc.*, 1968, **115**, 49.
[3] M. M. Perlman and S. Unger, in *Electrets, Charge Storgae and Transport in Dielectrics*, ed. by M. M. Perlman (The Electrochemical Society, Inc., New York, 1973).

AUTHOR INDEX*

Andersson, S., **255**.
Armstrong, R. A., 303.
Barber, M., 173.
Braithwaite, M. J., 89.
Briggs, D., **81**, 153, 158, 159.
Brodén, G., **112**, 167.
Brundle, C. R., 45, 46, **51**, 139, 141, 145, 149, 152, 159, 162, 313.
Carley, A. F., **51**.
Carlson, T. A., 30, 45, 196, 311.
Clark, D. T., **183**, 197, 198.
Clarke, T. A., **119**.
Coad, J. P. **269**, 310.
Collins, D. M., 166, 168.
Dilks, A., **183**.
Egelhoff, W. F., **127**, 301.
Evans, S., **102**, 148, 149, 151, 161, 197, 200.
Fabian, D. J., **37**.
Fain, S. C., 310.
Fadley, C. S., **18**, 47, 48, 49, 143, 155, 316.
Gay, I. D., **119**.
Gettings, M., **269**.
Hillier, I. H., 45, 50, **173**, 197, 200.
Johannessen, J. S., 306, 313.
Jona, F., **210**.
Jostell, U., **255**.
Joyner, R. W., 49, **89**, 157, 169, 172, 305.
Kowalczyk, S. P., **7**.
Landman, U., 147, 162, 167, **230**, 239, 242, 303.
Lang, W. C., **37**.
Lavery, R., **173**.
Law, B., **119**.
Ley, L., **7**.

Linnett, J. W., **127**, 301.
Martin, R. L., **7**, 45.
Mason, R., **119**, 142, 169, 170.
McFeely, F. R., **7**.
Norris, P. R., **37**.
Norton, P. R., **71**, 147, 148, 150, 309.
O'Brien, R. N., 318.
Overbury, S. H., **279**.
Padalia, B. D., **37**.
Parry, D. E., **102**.
Peeling, J., **183**.
Perry, D. L., **127**, 172, 301.
Pianetta, P., 153, 170.
Quinn, C. M., 139, **201**, 301.
Rhodin, T. N., **112**.
Richardson, N. V., 165, **201**.
Rivière, J. C., **269**.
Roberts, M. W., **89**, 144, 157, 158, 163, 167, 309.
Shirley, D. A., **7**, 46.
Somorjai, G. A., 146, **279**, 311.
Stern, E. A., 244, 303.
Stone, F. S., 161.
Tapping, R. L., **71**, 147.
Thomas, H. R., **183**.
Thomas, J. M., **102**, 162.
Tricker, M. J., 162.
Wagner, C. D., **291**, 306.
Watson, L. M., **37**.
Willis, R. F., **245**.
Wood, M. H., **173**.
Woodruff, D. P., 47, **218**, 240, 241, 302.
Yu, K. Y., 44, 137, 165, 312.

* References in heavy type indicate papers submitted for discussion.

GENERAL DISCUSSIONS OF THE FARADAY SOCIETY

Date	Subject	Volume
1907	Osmotic Pressure	Trans. 3
1907	Hydrates in Solution	3
1910	The Constitution of Water	6
1911	High Temperature Work	7
1912	Magnetic Properties of Alloys	8
1913	Colloids and their Viscosity	9
1913	The Corrosion of Iron and Steel	9
1913	The Passivity of Metals	9
1914	Optical Rotary Power	10
1914	The Hardening of Metals	10
1915	The Transformation of Pure Iron	11
1916	Methods and Appliances for the Attainment of High Temperatures in a Laboratory	12
1916	Refractory Materials	12
1917	Training and Work of the Chemical Engineer	13
1917	Osmotic Pressure	13
1917	Pyrometers and Pyrometry	13
1918	The Setting of Cements and Plasters	14
1918	Electrical Furnaces	14
1918	Co-ordination of Scientific Publication	14
1918	The Occlusion of Gases by Metals	14
1919	The Present Position of the Theory of Ionization	15
1919	The Examination of Materials by X-Rays	15
1920	The Microscope: Its Design, Construction and Applications	16
1920	Basic Slags: Their Production and Utilization in Agriculture	16
1920	Physics and Chemistry of Colloids	16
1920	Electrodeposition and Electroplating	16
1921	Capillarity	17
1921	The Failure of Metals under Internal and Prolonged Stress	17
1921	Physico-Chemical Problems Relating to the Soil	17
1921	Catalysis with special reference to Newer Theories of Chemical Action	17
1922	Some Properties of Powders with special reference to Grading by Elutriation	18
1922	The Generation and Utilization of Cold	18
1923	Alloys Resistant to Corrosion	19
1923	The Physical Chemistry of the Photographic Process	19
1923	The Electronic Theory of Valency	19
1923	Electrode Reactions and Equilibria	19
1923	Atmospheric Corrosion. First Report	19
1924	Investigation on Oppau Ammonium Sulphate-Nitrate	20
1924	Fluxes and Slags in Metal Melting and Working	20
1924	Physical and Physico-Chemical Problems relating to Textile Fibres	20
1924	The Physical Chemistry of Igneous Rock Formation	20
1924	Base Exchange in Soils	20
1925	The Physical Chemistry of Steel-Making Processes	21
1925	Photochemical Reactions in Liquids and Gases	21
1926	Explosive Reactions in Gaseous Media	22
1926	Physical Phenomena at Interfaces, with special reference to Molecular Orientation	22
1927	Atmospheric Corrosion. Second Report	23
1927	The Theory of Strong Electrolytes	23
1927	Cohesion and Related Problems	24

GENERAL DISCUSSIONS OF THE FARADAY SOCIETY

Date	Subject	Volume
1928	Homogeneous Catalysis	24
1929	Crystal Structure and Chemical Constitution	25
1929	Atmospheric Corrosion of Metals. Third Report	25
1929	Molecular Spectra and Molecular Structure	25
1930	Optical Rotatory Power	26
1930	Colloid Science Applied to Biology	26
1931	Photochemical Processes	27
1932	The Adsorption of Gases by Solids	28
1932	The Colloid Aspects of Textile Materials	29
1933	Liquid Crystals and Anisotropic Melts	29
1933	Free Radicals	30
1934	Dipole Moments	30
1934	Colloidal Electrolytes	31
1935	The Structure of Metallic Coatings, Films and Surfaces	31
1935	The Phenomena of Polymerization and Condensation	32
1936	Disperse Systems in Gases: Dust, Smoke and Fog	32
1936	Structure and Molecular Forces in (*a*) Pure Liquids, and (*b*) Solutions	33
1937	The Properties and Functions of Membranes, Natural and Artificial	33
1937	Reaction Kinetics	34
1938	Chemical Reactions Involving Solids	34
1938	Luminescence	35
1939	Hydrocarbon Chemistry	35
1939	The Electrical Double Layer (owing to the outbreak of war the meeting was abandoned, but the papers were printed in the *Transactions*)	35
1940	The Hydrogen Bond	36
1941	The Oil-Water Interface	37
1941	The Mechanism and Chemical Kinetics of Organic Reactions in Liquid Systems	37
1942	The Structure and Reactions of Rubber	38
1943	Modes of Drug Action	39
1944	Molecular Weight and Molecular Weight Distribution in High Polymers. (Joint Meeting with the Plastics Group, Society of Chemical Industry)	40
1945	The Application of Infra-red Spectra to Chemical Problems	41
1945	Oxidation	42
1946	Dielectrics	42 A
1946	Swelling and Shrinking	42 B
1947	Electrode Processes	Disc. 1
1947	The Labile Molecule	2
1947	Surface Chemistry. (Jointly with the Société de Chimie Physique at Bordeaux.) Published by Butterworths Scientific Publications, Ltd.	
1947	Colloidal Electrolytes and Solutions	Trans. 43
1948	The Interaction of Water and Porous Materials	Disc. 3
1948	The Physical Chemistry of Process Metallurgy	4
1949	Crystal Growth	5
1949	Lipo-Proteins	6
1949	Chromatographic Analysis	7
1950	Heterogeneous Catalysis	8
1950	Physico-chemical Properties and Behaviour of Nuclear Acids	Trans. 46
1950	Spectroscopy and Molecular Structure and Optical Methods of Investigating Cell Structure	Disc. 9
1950	Electrical Double Layer	Trans. 47
1951	Hydrocarbons	Disc. 10
1951	The Size and Shape Factor in Colloidal Systems	11
1952	Radiation Chemistry	12
1952	The Physical Chemistry of Proteins	13
1952	The Reactivity of Free Radicals	14
1953	The Equilibrium Properties of Solutions of Non-Electrolytes	15
1953	The Physical Chemistry of Dyeing and Tanning	16
1954	The Study of Fast Reactions	17
1954	Coagulation and Flocculation	18

GENERAL DISCUSSIONS OF THE FARADAY SOCIETY

Date	Subject	Volume
1955	Microwave and Radio-Frequency Spectroscopy	19
1955	Physical Chemistry of Enzymes	20
1956	Membrane Phenomena	21
1956	Physical Chemistry of Processes at High Pressures	22
1957	Molecular Mechanism of Rate Processes in Solids	23
1958	Interactions in Ionic Solutions	24
1957	Configurations and Interactions of Macromolecules and Liquid Crystals	25
1958	Ions of the Transition Elements	26
1959	Energy Transfer with special reference to Biological Systems	27
1959	Crystal Imperfections and the Chemical Reactivity of Solids	28
1960	Oxidation-Reduction Reactions in Ionizing Solvents	29
1960	The Physical Chemistry of Aerosols	30
1961	Radiation Effects in Inorganic Solids	31
1961	The Structure and Properties of Ionic Melts	32
1962	Inelastic Collisions of Atoms and Simple Molecules	33
1962	High Resolution Nuclear Magnetic Resonance	34
1963	The Structure of Electronically-Excited Species in the Gas-Phase	35
1963	Fundamental Processes in Radiation Chemistry	36
1964	Chemical Reactions in the Atmosphere	37
1964	Dislocations in Solids	38
1965	The Kinetics of Proton Transfer Processes	39
1965	Intermolecular Forces	40
1966	The Role of the Adsorbed State in Heterogeneous Catalysis	41
1966	Colloid Stability in Aqueous and Non-Aqueous Media	42
1967	The Structure and Properties of Liquids	43
1967	Molecular Dynamics of the Chemical Reactions of Gases	44
1968	Electrode Reactions of Organic Compounds	45
1968	Homogeneous Catalysis with Special Reference to Hydrogenation and Oxidation	46
1969	Bonding in Metallo-Organic Compounds	47
1969	Motions in Molecular Crystals	48
1970	Polymer Solutions	49
1970	The Vitreous State	50
1971	Electrical Conduction in Organic Solids	51
1971	Surface Chemistry of Oxides	52
1972	Reactions of Small Molecules in Excited States	53
1972	The Photoelectron Spectroscopy of Molecules	54
1973	Molecular Beam Scattering	55
1973	Intermediates in Electrochemical Reactions	56
1974	Gels and Gelling Processes	57
1974	Photo-effects in Adsorbed Species	58
1975	Physical Adsorption in Condensed Phases	59
1975	Electron Spectroscopy of Solids and Surfaces	60

For current availability of Discussion volumes, see back cover.

QD
1
F32
#60